Empirical Research on Semiotics and Visual Rhetoric

Marcel Danesi
University of Toronto, Canada

A volume in the Advances in Multimedia and
Interactive Technologies (AMIT) Book Series

Published in the United States of America by
IGI Global
Information Science Reference (an imprint of IGI Global)
701 E. Chocolate Avenue
Hershey PA, USA 17033
Tel: 717-533-8845
Fax: 717-533-8661
E-mail: cust@igi-global.com
Web site: http://www.igi-global.com

Library of Congress Cataloging-in-Publication Data

Names: Danesi, Marcel, 1946- editor.
Title: Empirical research on semiotics and visual rhetoric / Marcel Danesi.
Description: Hershey, PA : Information Science Reference, [2018]
Identifiers: LCCN 2017049343| ISBN 9781522556220 (hardcover) | ISBN
   9781522556237 (ebook)
Subjects: LCSH: Visual communication--Research. |
   Rhetoric--Philosophy--Research. | Semiotics--Research.
Classification: LCC P93.5 .E565 2018 | DDC 302.2/22--dc23 LC record available at https://lccn.loc.gov/2017049343

This book is published in the IGI Global book series Advances in Multimedia and Interactive Technologies (AMIT) (ISSN: 2327-929X; eISSN: 2327-9303)

British Cataloguing in Publication Data
A Cataloguing in Publication record for this book is available from the British Library.

All work contributed to this book is new, previously-unpublished material. The views expressed in this book are those of the authors, but not necessarily of the publisher.

For electronic access to this publication, please contact: eresources@igi-global.com.

Advances in Multimedia and Interactive Technologies (AMIT) Book Series

Joel J.P.C. Rodrigues

National Institute of Telecommunications (Inatel), Brazil &
Instituto de Telecomunicações, University of Beira Interior,
Portugal

ISSN:2327-929X
EISSN:2327-9303

MISSION

Traditional forms of media communications are continuously being challenged. The emergence of user-friendly web-based applications such as social media and Web 2.0 has expanded into everyday society, providing an interactive structure to media content such as images, audio, video, and text.

The **Advances in Multimedia and Interactive Technologies (AMIT) Book Series** investigates the relationship between multimedia technology and the usability of web applications. This series aims to highlight evolving research on interactive communication systems, tools, applications, and techniques to provide researchers, practitioners, and students of information technology, communication science, media studies, and many more with a comprehensive examination of these multimedia technology trends.

COVERAGE

- Digital Images
- Digital Communications
- Web Technologies
- Internet Technologies
- Digital Games
- Gaming Media
- Mobile Learning
- Multimedia Technology
- Multimedia Streaming
- Multimedia Services

IGI Global is currently accepting manuscripts for publication within this series. To submit a proposal for a volume in this series, please contact our Acquisition Editors at Acquisitions@igi-global.com or visit: http://www.igi-global.com/publish/.

Titles in this Series

For a list of additional titles in this series, please visit: www.igi-global.com/book-series

Real-Time Face Detection, Recognition, and Tracking System in LabVIEW™ Emerging Research and Opportunities
Manimehala Nadarajan (Universiti Malaysia Sabah, Malaysia) Muralindran Mariappan (Universiti Malaysia Sabah, Malaysia) and Rosalyn R. Porle (Universiti Malaysia Sabah, Malaysia)
Information Science Reference • copyright 2018 • 140pp • H/C (ISBN: 9781522535034) • US $155.00 (our price)

Multimedia Retrieval Systems in Distributed Environments Emerging Research and Opportunities
S.G. Shaila (National Institute of Technology, India) and A. Vadivel (National Institute of Technology, India)
Information Science Reference • copyright 2018 • 140pp • H/C (ISBN: 9781522537281) • US $165.00 (our price)

Handbook of Research on Advanced Concepts in Real-Time Image and Video Processing
Md. Imtiyaz Anwar (National Institute of Technology, Jalandhar, India) Arun Khosla (National Institute of Technology, Jalandhar, India) and Rajiv Kapoor (Delhi Technological University, India)
Information Science Reference • copyright 2018 • 504pp • H/C (ISBN: 9781522528487) • US $265.00 (our price)

Transforming Gaming and Computer Simulation Technologies across Industries
Brock Dubbels (McMaster University, Canada)
Information Science Reference • copyright 2017 • 297pp • H/C (ISBN: 9781522518174) • US $210.00 (our price)

Feature Detectors and Motion Detection in Video Processing
Nilanjan Dey (Techno India College of Technology, Kolkata, India) Amira Ashour (Tanta University, Egypt) and Prasenjit Kr. Patra (Bengal College of Engineering and Technology, India)
Information Science Reference • copyright 2017 • 328pp • H/C (ISBN: 9781522510253) • US $200.00 (our price)

Mobile Application Development, Usability, and Security
Sougata Mukherjea (IBM, India)
Information Science Reference • copyright 2017 • 320pp • H/C (ISBN: 9781522509455) • US $180.00 (our price)

Applied Video Processing in Surveillance and Monitoring Systems
Nilanjan Dey (Techno India College of Technology, Kolkata, India) Amira Ashour (Tanta University, Egypt) and Suvojit Acharjee (National Institute of Technology Agartala, India)
Information Science Reference • copyright 2017 • 321pp • H/C (ISBN: 9781522510222) • US $215.00 (our price)

701 East Chocolate Avenue, Hershey, PA 17033, USA
Tel: 717-533-8845 x100 • Fax: 717-533-8661
E-Mail: cust@igi-global.com • www.igi-global.com

Table of Contents

Detailed Table of Contents

Chapter 1

Facebook and other social media sites have been used by young Arabs for many purposes such as exchanging ideas and information, reporting breaking news, posting special events, launching political campaigns, announcing family gatherings, and sending seasons' greetings. Another emerging type of timeline posts is creative writing in English. Some Arab Facebook users post lines of verse, short anecdotes or points of view, express emotions, personal experiences, and/or inspirational stories or sayings written in literary style. A sample of Facebook creative writing pages/clubs and creative timeline posts was collected and analyzed to find out the forms and themes of creative writing texts. A sample of Facebook Arab creative writers was also surveyed to find out the reasons for their creative writing activities in English. This chapter describes the data collection and analysis procedures and reports results quantitatively and qualitatively. Implications for developing creative writing skills in foreign/second language learners using Facebook and other social media are given.

Chapter 2

The primary focus of the studies on adult beauty pageants involves their creation, negotiation, and implication vis-à-vis national and/or political identity within the pageant industry; or else they examine national pageants which are primarily scholarship-based. The present chapter is an attempt to understand the many psychological costs that come from practicing beauty within the realm of pageantry, and the rationale behind entering into an expensive venture for which there is little to gain, but much to lose emotionally. It will address two main questions: What are the physical and emotional (or metaphysical) costs of entering into and, later, winning, a beauty pageant? Who enters into beauty pageants and why? The objective is to examine the incentive to publicly parade oneself against dozens of other women, at the risk of simply being dismissed at the hands of quasi-objective opinion.

Chapter 3

Terry Marks-Tarlow, Insight Center, USA & Italian Universita Niccolo Cusano London, UK

Myth is a universal conveyor of culture whose stories capture the human heart and whose embodied set of guidelines serve to conduct everyday life. When Freud added the Oedipus myth to his theory of psychosexual development, his method of psychoanalysis subsequently launched worldwide. Whereas Freud viewed the myth of Oedipus quite literally as a prohibition against infanticide, patricide, and incest, this chapter views the myth more metaphorically to examine how the riddle of the Sphinx informs self-referential thinking as a collective stage of human consciousness. Two contemporary theoretical lenses are adopted: 1) interpersonal neurobiology, which proposes that mind, brain, and body develop from relational origins, and 2) second-order cybernetics, which examines how observers become entangled in their very processes under observation. From within these perspectives, the Sphinx's riddle appears as a paradox of self-reference whose solution requires humankind to leap from concrete to metaphorical thinking. Only upon retaining recursive loops in consciousness can humans attain full self-reflection as a beacon towards full actualization.

Chapter 4

Caterina Clivio, Columbia University, USA
Marcel Danesi, University of Toronto, Canada

When looked at cumulatively, it can be said that American pragmatist philosopher Charles Sanders Peirce strove to understand cognition via his sign theory and especially his notion of existential graphs. Peirce put forth ideas for a discipline that would incorporate notions of psychology and semiotics into a unified ontological and epistemological theory of mind. The connecting link was his system of diagrammatic logic, called "existential graphs." For Peirce a graph was more powerful than language as a means of understanding because it showed how its parts resembled relations among the parts of cognitive acts. Existential graphs show that cognition cannot be extracted from a linear or hierarchical succession of structures, but the very process of thinking itself in actu. In fact, Peirce called his graphs "moving pictures of thought" because they allow us to see how are thoughts are unfolding.

Chapter 5

Miriam Tribastone, University of Amsterdam, The Netherlands
Sara Greco, Università della Svizzera Italiana, Switzerland

By presenting the case study of the Charlie Hebdo attack in news discourse, this chapter combines a semantic analysis of the most frequent frame-activating words through text linguistics tools with frame analysis, developed according to the model proposed by Entman in the news making context. The linguistic perspective adopted in this chapter combines the works by Fillmore and Congruity Theory. As shown in the present work, both linguistics and news framing benefit from such integration.

The most fascinating semiotic applications of recent years came not from semioticians but from those who practice semiotics without knowing they do so (what the author calls the Monsieur Jourdain syndrome). Military and surveillance applications, genome sequencing, and the practice of phenotyping are immediate examples. The entire domain of digital computation, now settled in the big data paradigm, provides further proof of this state of affairs. After everything was turned into a matter of gamification, it is now an exercise in data acquisition (as much as possible) and processing at a scale never before imagined. The argument made in this chapter is that semiotic awareness could give to science and technology, in the forefront of human activity today, a sense of direction. Moreover, meaning, which is the subject matter of semiotics, would ground the impressive achievements we are experiencing within a context of checks-and-balances. In the absence of such a critical context, the promising can easily become the menacing. To help avoid digital dystopia, semiotics itself will have to change.

Chapter 7

Computationalism should not be the view that (human) cognition is computation; it should be the view that cognition (simpliciter) is computable. It follows that computationalism can be true even if (human) cognition is not the result of computations in the brain. If semiotic systems are systems that interpret signs, then both humans and computers are semiotic systems. Finally, minds can be considered as virtual machines implemented in certain semiotic systems, primarily the brain, but also AI computers.

Chapter 8

This chapter introduces Peirce's notion of proposition, "Dicisign." It goes through its main characteristics and argues that its strengths have been overlooked. It does not fall prey to some of the problems in the received notion of propositions (their dependence upon language, upon compositionality, upon human intention). This implies that the extension of Peircean Dicisigns is wider in two respects: they comprise 1) propositions not or only partially linguistic, using in addition gesture, picture, diagrams, etc.; 2) non-human propositions in biology studied by biosemiotics.

While people use language to let others know how and what it is that we think, language is also the means by, and also the substrate within which, humans think. This chapter explores the use of language as the basis for cognition, based on both a chosen word's denotative meaning and also its rhetorical (metaphorical) connotative meanings. The artificial dichotomy between language and speech is deconstructed. Peircean semiotics is used to argue that language is indexical in its primary referential functions, including sociolinguistic functions. Three new words, all of which were coined in the twenty-first century, are examined from a sociolinguistic and a semiotic point of view.

Mathematical understanding goes beyond grasping numerical values and problem solving. By incorporating visual representation, students can be able to grasp how math can be understood in terms of geometry, which is essentially a visual device. It is important that students be able to incorporate visual representations alongside numerical values to gain meaning from their own knowledge. However, it is also vital that students understand mathematical terminology, via a dialogical-rhetorical pedagogy that now comes under the rubric of "Math Talk," which in turn is part of a system of teaching known as knowledge building, both of which aim to recapture, in a new way, the Socratic method of dialogical interaction. This chapter explores how knowledge building, as a methodology, can assist in furthering student understanding and how math talk leads to a deeper understanding of mathematical principles.

In 1935, Walter Benjamin introduced the aura as the abstract conceptualization of uniqueness, authenticity, and singularity that encompasses an original art object. With the advent of technological reproducibility, Benjamin posits that the aura of an object deteriorates when the original is reproduced through the manufacture of copies. Employing this concept of the aura, the author outlines the proliferation of plaster casts of sculptures in eighteenth- and nineteenth-century Europe, placing contextual emphasis on the cultural and prestige value of originals and copies. Theories of authenticity in both art history and material culture are used to examine the nature of the aura and to consider how the aura transforms when an original object is lost from the material record. Through an object biography of a fifteenth-century sculpture by Francesco Laurana, the author proposes that the aura does not disappear upon the loss of the original, but is reincarnated in the authentic reproduction.

Chapter 12

Does art tend towards immersion? Positing James Turrell's Roden Crater (2015) as the modern epitome of the landscape art-object, the evolution of the medium is traced through prominent examples its transformations: Titian's Venus and the Organist with Dog (1550), De Loutherbourg's Eidophusikon (1781), and Barker's Panorama (1792). Discussion regarding Roden Crater's predecessors serve to illustrate distinct innovations that greatly influenced its construction of sensory experience, spanning the use of dialogue to the integration of physicality. This chronology is used to demonstrate an overarching tendency of media towards immersion, and to reflect how the development of contemporary culture evolves towards progressively psychological experiences.

Preface

INTRODUCTION

In an age of interdisciplinarity, semiotics and Visual Rhetoric (VR), have acquired particular relevance and salience in the investigation of cognition and culture. Although each one has its own agenda of research and its set of principles, by and large both are used by scholars and researchers in various disciplines, from science to education, as complementary or supplementary tools of critical analysis. The basic method of VR, which can be traced back to semiotician Roland Barthes' 1964 article "The Rhetoric of the Image;" like semiotics generally its aim is to provide general principle of analysis to help scholars unravel the embedded meanings of visual images of all kinds. Given their simple, yet effective methods of analysis, semiotics and VR, have become staples as techniques within various subfields of psychology, anthropology, computer science, and cognitive science generally.

This book brings together some of the more interesting applications and discussions of semiotics and VR of the last few years. Except for a few of the studies, most have been published in IGI Global periodicals. They have all been updated, revised, and elaborated here. They add considerably to the ever-expanding fund of knowledge that involves the interaction between cognition and human behavior. If there is one thematic subtext that unites them all, it is as following: *Meaning structures underlie all kinds of human behaviors, activities, and artifacts.* Indeed, the wide, scope of the topics treated here highlights the versatility and breadth of semiotics and VR today. Each one sheds significant light on some aspect of relevance today, from education to the modeling of the mind with computers.

The title of this volume includes the word "empirical," and this needs some elaboration here, given that it means something vastly different in semiotics and VR than it does in the cognitive sciences generally. When used in, say, psychology, the term "empirical" typically implies that quantitative analysis is involved in the conduct of some experiment or form of inquiry. This designation is certainly the correct one in the physical sciences, which involve some form of quantification of observed phenomena. This same meaning has been adopted by many of the cognitive and social sciences. However, the original meaning of the term referred to the notion that knowledge is derived from sense-experience. Semiotics and VR are also concerned with this same very notion, but do not necessarily resort to statistical techniques to bolster their analyses. So, like the quantitative empirical sciences, semiotics and VR are basically empirical in the sense that they are based on observations utilizing principles of analysis that decode the meaning inherent in them. That meaning of empirical unites the chapter in this volume.

CHAPTER CONTENTS AND OBJECTIVES

The opening chapter by Reima Al-Jarf ("Exploring Discourse and Creativity in Facebook Creative Writing by Non-Native Speakers") deals with one of the most crucial aspects of foreign-language learning—imparting the discourse notions to students that will allow them to truly articulate their ideas meaningfully in the new code. It is in this sense that semiotics and VR (the use of images to stimulate ideas in students) come into play, as implied through Al-Jarf's insightful chapter. She describes a pedagogical experiment based on the use of Facebook and other social media by young Arab students, which she describes as immersion environments in which the students can develop creative ideas to write about a host of themes in English that characterize everyday life, from political campaigns and family gatherings to seasons' greetings. The social media sites provide, in other words, sources of authentic discourse that get the students to become involved with everyday reality through the creative resources of both the language they are learning and their own subjective interpretations of it. From this, Al-Jarf has been able to get them to write verse, anecdotes, personal experiences, and inspirational stories written in literary style. The results are impressive indeed, as can be witnessed by reading the texts of the students themselves. Here we see that by engaging students semiotically, that is n the meaning structures of the new language directly, they seem to be better able to articulate their ideas in that language proficiently, at least when compared to other kinds of learning environments. The subtext in this is, really, that the brain is a semiotic organ that needs nourishment through creativity and meaningfulness. In this way, it is better able to process new information and interpret it correctly and practically. The implications for foreign language teaching are self-evident.

Mariana Bockarova ("Not Just a Pretty Face: The Cost of Performative Beauty and Visual Appeal in Beauty Pageants") deals with one of the most pertinent topics within both semiotics and VR—the allure of beauty pageants. Starting with Barthes, this topic has been perhaps analyzed to death, so to speak. But Bockarova gives it a new, and valuable empirical twist—she interviews those who decide to be a part of such pageants, providing insights into why they do it. Through the words of the contestants, Bockarova develops several key analyses that explain their motivations insightfully—from archetypal fantasies that have always conditioned women's views of their bodies to more economically-based motivations, such as the need to achieve prosperity (which she calls the Cinderella complex). From the interviews and Bockarova's framework we come to understand the many psychological costs that come from practicing beauty, from financial to emotional. In effect, Bockarova truly deconstructs beauty pageantry without any particular ideological axe to grind. This chapter is truly a model of hoe to conduct empirical semiotics in a way that sheds light on human travail, rather than just remain interesting at a theoretical level.

One of the most emblematic myths of human history is the Oedipus legend that is informed by and ensconced in the Riddle of the Sphinx. Treatises have been written on the significance, psychological, and various other implications of the myth and the riddle. Terry Marks-Tarlow ("The Riddle of the Sphinx Revisited: Self-Referential Twists to and Ancient Myth") not only takes a fresh look at the many implications of the riddle and the myth, but takes it one step further to argue that it is more revelatory of human cognition than may at first seem thinkable. She bases her interpretation on the cognitive theory of metaphor, which is itself a key t understanding how unconscious thought evolves and how it brings forth mythical concepts and reformulates them I such a way as to make them reverberate latently in the mind. Ultimately, the riddle and the myth are early treatises in human consciousness, which did not differential between waking and dreaming states. The riddle is actually, in this sense, not a riddle but a paradoxical self-referential statement about the human condition. Overall, this chapter truly penetrates the essence of human consciousness and its mythic-metaphorical origins and evolutionary propensities.

Caterina Clivio and Marcel Danesi ("Existential Graphs and Cognition") take a brief look at one of the most important concepts that unite semiotics and VR—Peirce's notion of Existential Graphs (EGs), or diagrams that allow us to represent ideas in image-schematic form. These are used in logic and mathematics, showing how semiotics is becoming more and more of a metalanguage within these disciplines. This chapter revisits EGs from the viewpoint of image schema theory, arguing that they are themselves models of thought unfolds as we draw our visual representations. In effect, EGs, are indirect theories of cognition.

Miriam Tribastone and Sara Greco ("Framing in News Discourse: The Case of the Charlie Hebdo Attack") take a perceptive look at news reporting following the Charlie Hebdo attach in Paris. Using the concept of framing, which comes from mass communications theory, suggesting that the sum and substance of some newsworthy issue or event is realized by juxtaposing it within a broader filed of meaning. So, the attack was not portrayed by the media as an isolated criminal act but as part of a narrative that involves the perception of specific groups. The authors have collected a substantial sample of news reports—100 news articles, published on January 8, 2015 in English, German, and Italian—from which they derive five frames: *act of war on terrorism, attack on Western freedom, the price to pay for the offence, the result of disastrous integration policies,* and *Islam internal issues*. They analyze the keywords in each frame as occurring subconsciously across languages, so that the same event is portrayed as an unjustifiable terroristic attack against Western society. The frames uncovered by the researchers raise fundamental (and urgent) questions. Above all else, they reveal the power of implicit argumentative and rhetorical discourses.

Mihai Nadin ("Meaning in the Age of Big Data") constitutes a challenge and a paradigm for semiotics—in other words, it diagnoses the problems connected with using the discipline for various self-serving purposes, while at the same time arguing for a more scientific rigorous use of the tools of semiotics in fields such as computer science. Nadin argues cogently that the world of big data and digital technologies is passing semioticians by, who continue to study self-serving artifacts of culture from their particular ideological angles. Rather, semiotics is now, more than ever, an important tool for diagnosing the world we live in and for tapping into innovations in that world that are changing it drastically. By emphasizing the role of meaning in human life, semiotics would provide a lens through which the impressive achievements we are experiencing can be assessed with relevant checks-and-balances. To counteract digital dystopia, semiotics itself will have to change.

William J. Rapaport's in-depth treatment of computationalism ("Syntactic Semantics and the Proper Treatment of Computationalism") brings us into the heart of the true significance of semiotics—as a means to understand the modeling of human cognition through computers. This is a key study for both semioticians and computer scientists, because computationalism cannot be understood as the view that cognition *is* computation, but that it should be understood as a comput*able* phenomenon. So, despite warnings from hyper-sensitive anti-computationlists, computationalism can be true even if cognition is not the result of computations in the brain. Semiotic systems are systems that interpret signs, and both humans and computers are, in this sense, semiotic systems. Human minds can thus be considered as "virtual machines" implemented within certain semiotic systems: brains and AI computers equally.

Frederik Stjernfelt ("Signs Conveying Information: On the Range of Peirce's Notion of Propositions – Dicisigns") revisits one of the least discussed, yet clearly important, sign types put forward by Charles Peirce—Dicisigns. These are signs of actual existence, necessarily involving a rheme to describe the fact which it is interpreted as indicating. While this might seem to have no relevance to theories of mind, as Stjernfelt cogently argues, it is perhaps one of the most important models of how propositions and arguments gain sense in human interaction and, indeed why they exist in the first place. A proposition is a Dicigin, and it is not limited to verbal reasoning, but crosses all modalities, including nonverbal ones. Moreover, Dicisigns, as Stjernfelt so persuasively shows, exist in other species—a truly remarkable insight. Dicisigns vary considerably, but they seem to be part of natural selection, which has had to adapt to Dicisigns, rather than the other way around. We should thus see evolution as having increasingly adapted to the structures of Dicisigns, and not the other way around.

Joel West ("Neologisms: Semiotic Deconstruction of the New Words 'Lizardy,' 'Staycation,' and 'Wannarexia' as Peircean Indexes of Culture") uses the coinage of new colloquialisms to argue for the view that language, thought, and culture are inextricable. Indeed, when a new word comes into usage, it both reflects and changes cultural modalities which, in turn, affect cognitive states. Using Peirce's notion of indexicality, West argues that words point to cultural references all the time, incorporating them into their semantics. Many semioticians and rhetoricians would consider iconicity (resemblance to the referential domain) as the primary mode of word coinage, relegating indexicality (the establishment of relations among words and concepts) to a secondary status. It But Peirce himself saw indexicality as a primary form in verbal creativity in some domains of reference. West explains how this is so through new words that are models of how language and cognition interact in deciphering the world around us.

Stacy Costa ("Math Talk as Discourse Strategy") takes a perceptive look at how language and mathematics are intertwined in the learning process, thus uniting semiotic and education ideas into a model of learning called Math Talk. This is a dialogical-rhetorical pedagogical method that is part of a system of teaching known as Knowledge Building, both of which aim to recapture, in a contemporary way, the Socratic method of dialogical interaction. Math Talk leads to a deeper understanding of mathematical principles and constitutes an important rhetorical solution to the many problems faced by children attempting to grasp mathematical ideas in the classroom. By discussing them dialogically with other students and the teacher, the ideas are more easily acquired.

Victoria Bigliardi (The Reincarnation of the Aura: Challenging Originality With Authentic Plaster Casts of Lost Sculptures") looks at one of the most interesting and influential concepts in art criticism Walter Benjamin's controversial concept of the aura, a state of interpretation that indicates how the the originality of a work is lost in its reproductions. The aura has disappeared because art has become reproducible, no longer having a unique effect in time and space. In effect, according to Benjamin, art cannot be reproduced, because it disappears it is is reproduced. Bigliardi tackles this concept by arguing that reproductions do indeed have their own aura. She uses Laurana's *Portrait of a Woman* as an example of a vanishing original, from which the aura has not disappeared. She uses biographical arguments, positing that the value and prestige of the original are not lost when casting is employed. Instead, Benjamin's aura may be temporarily concealed or displaced.

Finally, Aaron Rambhajan ("Order of Experience: The Evolution of the Landscape Art-Object") looks at the meaning of art from a contemporary view of its contextualization in a meaningful substratum of life. Rambhajan uses James Turrell's *Roden Crater* (2015) as the epitome of landscape art-object, an art form that speaks directly to people today, since it evokes sensory experience. In effect, art is an immersive artifact, saturating us with meaningful psychological experiences via landscapes. Unlike the didactic practices of traditional art that solely encompass the act of *seeing*, Turrell's work encompasses instead the acts of *viewing* and *sensing* spaces. Turrell's art thus serves as a metaphor of contemporary culture, which is immersed in all kinds of media.

CONCLUDING REMARKS

As can be seen the scope and eclecticism of these chapters is truly remarkable. But they are all united by a common subtext—humans produce meaningful artifacts, from art works to theories of mind. Through these, we come to an understanding of who we are. In his 2001 book, *Media Unlimited: How the Torrent of Images and Sounds Overwhelms Out Lives,* Todd Gitlin decries how the modern-day media bombard us with a constant barrage of images that wash over us but which end up influencing how we see the worlds. We truly need to understand these images, signs, and the texts that carry them, since they have deep implications for everything from education to Artificial Intelligence and social evolution. The various chapters in this book shed light on these images and signs and how the disciplines of semiotics and VR can help us decode them in specific ways or else how we can better understand the world in which we live.

In the end, semiotics aims to understand the quest for meaning to life—a quest so deeply rooted in us that it subtly mediates how they experience the world. This quest gains actual expression in the signs, sign systems, texts, practices, and the like found throughout human societies. Paradoxically, these are highly restrictive and creative at the same time. Signs learned in social contexts are highly selective of what is to be known and memorized from the infinite variety of things that are in the world. So, we tend to we let our culture (which is a network of signs) do the thinking for us when we use signs unreflectively. But there is a paradox here—a paradox that lies in the fact that we can constantly change, expand, elaborate, or even discard the habits of thought imprinted in sign systems. We do this by creating new signs (words, symbols, etc.) to encode new knowledge and modify previous knowledge. The study of signs and sign systems is of extreme importance. This volume shows essentially why this is so.

Chapter 1
Exploring Discourse and Creativity in Facebook Creative Writing by Non-Native Speakers

Reima Al-Jarf
King Saud University, Saudi Arabia

ABSTRACT

Facebook and other social media sites have been used by young Arabs for many purposes such as exchanging ideas and information, reporting breaking news, posting special events, launching political campaigns, announcing family gatherings, and sending seasons' greetings. Another emerging type of timeline posts is creative writing in English. Some Arab Facebook users post lines of verse, short anecdotes or points of view, express emotions, personal experiences, and/or inspirational stories or sayings written in literary style. A sample of Facebook creative writing pages/clubs and creative timeline posts was collected and analyzed to find out the forms and themes of creative writing texts. A sample of Facebook Arab creative writers was also surveyed to find out the reasons for their creative writing activities in English. This chapter describes the data collection and analysis procedures and reports results quantitatively and qualitatively. Implications for developing creative writing skills in foreign/second language learners using Facebook and other social media are given.

INTRODUCTION

Writing in a foreign/second language (L2) is a difficult task for many students, because they are inhibited, afraid of making mistakes, have insufficient grammar and vocabulary knowledge, or because they are incapable of generating ideas. To enhance students' writing skills, in general, and creative writing, in particular, researchers and teachers have utilized several instructional strategies and practices, such as using wordless picture books (Henry, 2003), plot scaffolding (O'Day, 2006), collaborative creative writing activities, assignments and projects (Vass, 2002; Feuer, 2011; Bremner, Peirson-Smith and Bhatia, 2014; Arshavskaya, 2015), the integration of cooperative learning and journalizing (Bartscher, Lawler, Ramirez and Schinault, 2001; and Racco, 2010;), learning about photography and using it as

DOI: 10.4018/978-1-5225-5622-0.ch001

inspiration for students' creative writing (Haines, 2015), the cluster method (Sahbaz and Duran, 2011), the integration of creative and critical written responses to literary texts in different genres (Racco, 2010; Wilson, 2011), incorporating journal and/or personal letter writing from the perspective of people that have been marginalized in the students' dominant culture (Stillar, 2013), developing a creative writing instructional program based on speaking activities (Bayat, 2016), using a semiotic analysis theory-based writing activity in which cartoon caricatures are selected as visual texts for analysis (Sarar Kuzu, 2016), using nonfiction mentor texts to assist students in writing their own creative informational texts about animals (Dollins, 2016), inviting students to write poetry across the curriculum (Bintz, 2017), and analyzing song lyrics containing vivid details, using a graphic organizer, and a kinesthetic activity to help students devise similes and metaphors and construct vivid sensory details in their fiction and creative nonfiction writing (Del Nero, 2017).

In addition to the above classroom techniques, several technologies have been integrated in writing instruction. Two decades ago, word processors, e-mail, specially designed software and Powerpoint presentation were utilized to develop L1 and L2 students' writing skills (Casella, 1989; Gammon, 1989; Scott, 1990; Owen, 1995; Keiner, 1996; Hodges, 1999; and Biesenbach-Lucas and Weasenforth, 2001). At a later stage, online journal writing, computer labs, online courses, online discussion boards, special software, wiki projects and school blogs were used to enhance students' writing skills. For example, adolescents in Guzzetti and Gamboa's (2005) study used online journal writing as a literacy practice, and for social connection, identity formation and representation. Online journal writing proved to be effective in developing students' writing skills. Pifarré, Marti and Guijosa (2014) found that the wiki environment helped develop an effective and creative online collaborative learning community among secondary school students. Even seven-year old children who used Kodu code-based software to create imaginary worlds, characters and story lines, and translated their video game-like creations into dynamic short stories excelled in producing sophisticated short stories that were not expected from them (Salcito's (2012). Likewise, 10-16-year old students in India reported that OmmWriter (a minimalist text editing tool) was an effective tool for feeling better (with music), happier and more relaxed, as it helped them avoid distractions that often come with technology and the internet. It also reduced writing distractions, gave a relaxing writing experience by offering three different writing soundtrack options, as well as three different keyboard-typing sounds (Gonçalves, Camposand Garg, 2015). Elementary school students participating in school blogs reported that blogging enthused them, and gave them access to new kinds of writing and new audiences (Barrs and Horrocks, 2014). Furthermore, Sessions, Kang and Womack (2016) found that fifth grade students with iPad applications wrote more cohesive, sequential stories using more sensory details than students who wrote with paper and pencil. iPad applications also affected students' motivation to write, and made the writing process more social and engaging.

Not only were technologies used to enhance the writing skills of English native-speaking students, but they were also used to help ESL college students develop their writing skills. For example, Al-Jarf's (2007) found that use of *Nicenet*, an Online Course Management System, as a supplement to in-class instruction encouraged Saudi freshman students of different proficiency levels to write poems and short stories in English as a foreign language (EFL). In another study, Yunus, Salehi and Chenzi (2012) found that the integration of an online discussion board in ESL writing classrooms helped broaden Malaysian students' knowledge, increased their motivation, and enhanced their confidence in acquiring writing skills in ESL. In Oman, Jayaron and Abidin (2016) found no statistically significant differences in writing performance and linguistic complexity between EFL post-foundation level students involved in a synchronous online discussion forum and students engaged in an asynchronous blog. However, use of the online discussion forum facilitated EFL writing and had a positive effect on the learning process. In Egypt, the WebQuest model proved to be effective in developing memoir writing skills as a creative non-fiction genre by ESL 8th grade students focusing on a specific life, changing event, using an engaging opening, using the first person. Students using the WebQuest were exposed to rich, relevant and elaborate language input which enhanced their ability to write spontaneously and creatively (Al-Sayed, Abdel-Haq, El-Deeb and Ali, 2016). In the USA, findings of a study by Dzekoe (2017) showed that computer-based multimodal composing activities (CBMCAs) helped advanced-low proficiency ESL students discover specific rhetorical and linguistic elements that they used to revise their written drafts. The CBMCAs activities helped them develop language and voice to convey ideas that they were struggling to express using the written mode alone. The students made revisions in the content which correlated with text quality.

Online collaborative writing tasks, in articular, proved to be effective in developing EFL/ESL students' creative writing abilities. For example, L2 college students engaged in four in-class Web-based collaborative writing tasks using Google Docs valued their collaborative in-class writing tasks. The tasks included: considering group members' feelings when changes are made, negotiating politely and respectfully, managing time in a group, utilizing each group member's strengths, negotiating the writing process with each other, and troubleshooting strategies for communication breakdowns (Bikowski and Vithanage, 2016). In another study by Stoddart, Chan and Liu (2016) wiki-based collaborative writing projects provided L2 students an environment that fostered student satisfaction, motivation and learning.

Since 2005, social media platforms, such as Facebook, Twitter, Instagram, Google+, LinkedIn, Tumblr, Pinterest, Behance, Snapchat and others have been popular in online communication among adult Internet users. They have been used for many purposes such as exchanging ideas and information, reporting breaking news, posting and commenting on special events, launching political campaigns and special causes, announcing family gatherings, sending seasons' greetings, posting photos and videos of interest to them and posting comments and viewpoints on social, political, economic, and local and world events. They have also been used by students and instructors as a communication and instructional medium.

A review of the literature has shown a rising interest in the academic applications of social media sites. Numerous studies have investigated issues such as the influence of social media on high school students' social and academic development (Ahn, 2010), their educational use in higher education (Hung and Yuen, 2010), their use in recruiting undergraduate students (Ferguson, 2010), their impact on academic relations at the university (Rambe, 2011), to investigate appropriate professional behavior of faculty and college students on social media (Malesky and Peters, 2012), to teach international business via social media projects (Alon and Herath, 2014) and others.

In the area of foreign/second language learning, in particular, a multitude of Facebook pages are available, which students can join to develop their listening, speaking, reading, general writing skills, and knowledge of grammar and vocabulary. Facebook also has a multitude of creative writing pages for students, amateurs as well as professional creative writers. Even on non-academic Facebook pages, many English and Arabic posts are characterized by their creative expression and artistic style.

A review of the literature has revealed numerous studies that integrated social media in second language teaching and learning such as the use of social networking in intensive English program classrooms from a language socialization perspective (Reinhardt and Zander, 2011); learning English on Facebook (Al-Jarf, 2012); whether Facebook facilitates language acquisition and social access by international students (Lee and Ranta, 2014); integrating ethnic culture Facebook pages in EFL instruction (Al-Jarf, 2014); and the effect of WhatsApp on EFL learners' critique writing skills and their motivation for learning (Awada, 2016). Diploma students in a Malaysian university reported five Facebook activities that improved their English language proficiency: writing posts and comments in English, reading news feeds in English, participating in interest-based Facebook groups, watching movies in English, and communicating with foreign Facebook friends (Shafie, Yaacob and Singh, 2016).

Another group of studies focused on the effects of Facebook on L2 students' writing skill development such as the effect of peer feedback on Facebook on improving revised drafts by undergraduate students enrolled in a fundamental English course (Wichadee, 2013); and the integration of some new forms of online writing tasks on Facebook while 3rd-year EFL student-teachers are working as a community (Abdallah, 2013). Japanese university EFL engaged in free writing sessions on Facebook made more significant gains in writing fluency than students who used paper-and-pencil. However, neither group made significant improvement in lexical richness and grammatical accuracy (Dizon, 2016).

Despite the importance of social networking sites, as a medium for creative writing, there is a dearth of studies that investigated the effects of using social media such as Facebook, Twitter, Penterest, Google+ and WhatsApp on the production of creative texts, and that explore the forms, themes, discoursal features of creative texts posted on adult social media. To fill the gap in this area of research, the present study aims to examine a sample of Facebook creative writing in English by a sample of non-native English-speaking Arab adults. It aims to find out the types of forms, themes and discoursal features of the Facebook creative texts, written by non-native writers of English; describe the Facebook creative writing environment and Facebook activities that initiated creative writing; report the personal and social factors that affect creativity in L2; and provide guidelines for nurturing creativity in EFL/ESL students based on the findings. Specifically, the study aims to answer the following questions: (i) What kind of creative writing forms and themes are posted on Facebook by Arab non-native speakers of English? (ii) What are the characteristics of the Facebook creative discourse by Arab non-native speakers of English? (iii) What are the characteristics of the Facebook creative writing environment in English as a foreign/ second language (EFL/ESL)? (iv) What Facebook activities initiated creative writing by Arab non-native speakers of English? (v) What personal and social factors impact the Facebook creative writing activity by Arab non-native speakers of English?

To answer the above questions, the study will describe how the samples of Facebook creative writing pages/clubs, creative texts and creative writers were selected, how the sample of creative texts was analyzed into theme and types, and how discourse features were identified. It will also survey a sample of Facebook creative writers to find out the personal and social factors that affect their creative endeavors. It is noteworthy to say that the aim of this study is not to conduct an extensive literary critical analysis of the creative texts, rather the aim of the text analysis is to explore the characteristics of the Facebook environment that was conducive to creative writing by Arab non-native speakers of English.

The present study is significant as it starts a new line of research in the study of Facebook discourse and in using social media websites such as Facebook in creative writing, particularly creative text forms and themes, and the factors conducive to creative writing on social media. The study of the effects of social media on the production of creative discourse by Arab non-native English-speakers is especially interesting and results of the analysis will contribute to our understanding (especially writing instructors) of their creativity on Facebook. The study will describe ways in which peer collaboration can resource, stimulate and enhance the production of creative texts. The benefits of teaching writing through social media, especially to non-mainstream students who are usually inhibited by traditional writing instruction in the classroom will be demonstrated as well. Insights can be gained from the results and used to improve the general teaching of creative writing to EFL students in general, thus expanding the peda-gogical repertoire available to writing instructors.

METHODOLOGY

To answer the questions of the present study, a sample of Facebook creative writing pages/clubs, a sample of creative texts selected from those pages/clubs, and a sample of Arab creative writers who are members of those clubs were selected. Each of which is described in detail below.

Sample of Creative Writing Facebook Pages/Clubs

Several search terms were used to search Facebook for English creative writing pages such as: *"creative writers," "creative writing," "creative writing page," "creative writing group," "creative writers' club,"* and *"creative writing community."* More than 450 creative writing and "creative writers" Facebook pages were found, of which 70% received fewer than 50 likes. Those belong to universities, colleges, schools, non-profit and community organizations, centers for creativity, and individuals. Most of them were created by US users and are for creative writers who are native speakers of English. Numerous Facebook creative writing pages created and used by non-native speakers of English in India, Pakistan, Nigeria, the Czech Republic, Egypt, Algeria and Saudi Arabia were found. For purposes of the present study, only creative writing Facebook pages created by Arab writers who are non-native speakers of English in Saudi Arabia, Kuwait, Egypt, and Lebanon were found. The sample selected had to meet several criteria: (a) Pages/clubs that are part of a professional English language teaching and learning pages/clubs and that are part of formal writing courses; (b) Creative writing pages with fewer than 10 likes or members; (c) those created two months before writing this article; and (d) those with members who are native speakers of English were excluded. Thus, the final sample consisted of five Facebook creative writing pages that met the above criteria. The following is a brief description of each:

1. *Riyadh Writing Club (RWC)* was created by two Saudi female creative writers. It joined Facebook on April 6, 2011, and has received 798 likes. The RWC page is complementary to the creators' blog. The aims of the page/club are:

to bring together the talented female writers in Riyadh, Saudi Arabia. Its intention is to activate imagina-tion and enable receiving constructive criticism from like-minded, talented, and creative people.

2. *Kuwait Writing Club (KWC)* was created by a female creative write on March 31, 2013 by a group of Kuwaiti creative writers. It has received 856 likes. It is complementary to the creators' blog. KWC aims to:

… get together and pick a topic to write about for the next meeting. The writing can be anything from a poem to a song or a short story, as long as it stays on the topic chosen by the members. In this way, we

as creative people are forced to activate our imagination and receive constructive criticism from like-minded, talented, and creative people.

3. *Riyadh' Creative Writers and Designers on Facebook (RCWD)* was created by a single male creative writer. It joined Facebook on March 15, 2010 and has received 726 likes. The aim of the page is to *"express thought in words."*
4. *English Inspiration Club in Egypt (EIC)* was created by a group of Egyptians. It joined Facebook on September 13, 2011, and has received 554 likes.
   a. The page aims to spread the quality of Education for English in Egypt…learn English to get more knowledge and better communication with other cultures… (and learn) from each other.
   b. EIC is not limited to creative or inspirational writing, images with inspirational quotes, grammar rules, usage tips, quizzes and vocabulary enrichment exercises are also posted.
5. *Creative Writing (CW)* was created by a retired Lebanese teacher and members are Lebanese students. The page was created on November 2, 2014 and has received 421 likes. It aims to encourage talented students to express themselves.

Sample of Creative Texts

To have a sufficient sample of creative texts, all timeline posts (entries) and related comments (sub-entries) in the five Facebook creative writing pages/clubs were collected. The author browsed through all of the posts in each page and only posts that are creative were selected. Posts with announcements, advertisements, links to articles, videos, compliments, news stories, texts written in a foreign language other than English, viz French, and texts with academic or journalistic style were excluded. Creative texts copied or cited from other authors, sources or websites were also excluded. The sample of creative texts included only original texts written by Arab members of the five clubs. Thus the final corpus of Facebook discourse included a total of 1196 creative texts (posts) which included: 383 creative texts from RWC, 364 creative texts from KWC, 304 creative texts from RCWD, 91 creative texts from EIC in Egypt, and 54 creative texts from CR. Poems constituted 24% of the creative posts, and 76% were prose–short narratives/stories, reflections (spiritual and otherwise), religious supplications, points of view, words of wisdom, inspirational sayings, motivational thoughts and advice, and short narratives (short stories). The creative texts varied in length: From one line to about a page. However, most consisted of few lines. The unit of creative discourse analysis chosen was the single post (entry) regardless of its length and excluding the comments (sub-entries) it received.

The author faced several challenges in selecting the sample of creative texts. First, Facebook pages, page/club members and posts are not permanent in nature. Some page creators choose to delete or deactivate their page/account and some authors choose to delete some of their posts for no obvious reason. Some creative texts that were located by the author disappeared although they were present a month or even a week earlier. What helped keep track of those was that the author saved all of the posts on the pages in PDF format. A second challenge was locating creative texts and sorting out stories by author and topic. Although Facebook has a timeline that displays all of the stories posted each month by all members of a particular club, it does not have an analytics tool or App for sorting out stories by author or topic, and does not give statistics nor lists all posts by a single author as it is the case in online forums. In addition, posts are not archived under tags, as it is the case in blogs, to facilitate the browsing and searching processes. Although hashtags are common on Facebook, they are not used at all in the sample posts. To locate and sort out creative and non-creative posts, the author had to print all of the club pages, browse through the posts one by one, read each post and count and analyze those that are creative. Locating, browsing through all posts, sorting out creative and non-creative texts and calculating the number of creative texts posted by each writer and analyzing the texts manually was time-consuming.

Sample of Creative Writers

Only members of the creative writing clubs who have creative contributions were selected. The sample consisted of total of 234 Facebook creative writers (189 females and 45 males) distributed as follows: 71 creative writers from RWC (62 females and 9 males), 62 creative writers from KWC (41 females and 21 males), 46 creative writers from RCWD (34 females and 12 males), 31 creative writers from CR (all females), and 24 creative writers from EIC (21 female and 3 male). Creative writers from RWC are all Saudi, those of KWC are Kuwaiti, those of the EIC are Egyptian, and those of CR are all Lebanese. Members of RCWD are from different Arab countries. All the subjects are native speakers of Arabic. Most of the subjects are in their twenties. Some are studying English literature, others are studying journalism, business, information technology or social work. Some are professionals: English teachers, journalists, businessmen, information technology specialists, and lawyers. Five subjects are in high school and three are as young as 14 and 15 years old. Most went to an English-medium school at least half of their school years and studied in an English-medium college. Other than posting in the creative writing club page, members of the RWC and KWC have a group blog where they publish their creative work.

Some of the challenges that the author faced while collecting personal data about the subjects were that many did not post information about themselves such as age, major (specialty), education, where they live or nationality on Facebook; and others made their own Facebook page private and thus their personal data could not be accessed. Some did not provide contact details and did not respond to Facebook messages.

QUESTIONNAIRE SURVEYS

All of the Arab creative writers who are members of the 5 Facebook pages/clubs were surveyed and each was sent a copy of the study questionnaire-survey. The questionnaire-survey requested information about their age, major area of study, type of work, where they went to school, whether they went to an English-medium school, how they develop and polish their writing skill in English, why they chose Facebook to post their creative work, in what ways Facebook helped them to write creatively, whether they knew other members prior to joining the Facebook creative writing club, and the role that Facebook club members play in their creative writing process and productivity.

The questions were sent to each member through his/her Facebook account and answers were received through my Facebook Messenger. Seventy percent of the subjects responded to the author's questions probably because they do not know her or because they were busy.

There were more female than male respondents and all of the respondents agreed to participate in the study and answer the survey questions, to have excerpts of their work cited in this article and to have their first names referred to herein. Respondents were assured that their identities and associated responses to the questions will remain confidential and will be used for research purposes only. As a matter of fact, some were happy to be part of a research study and did not mind the citation of their work as it is publicly published on Facebook.

DATA ANALYSIS

Identifying Creative texts, Their Forms, Themes and Features

To help identify creative texts, an operational definition of creative texts was used. Features given by the following definitions were taken into consideration in sorting out creative and non-creative posts. According to Fraser (2006):

Creativity concerns novelty and originality...Creative writers surface original ideas through construct-ing their own creative texts. They generate novel responses and multiple interpretations. Creative texts reveal unique voices that range from the playful to the dramatic in their creative exploration of what it means to be human.

Your Dictionary (2012) defines creative writing as:

Writing that expresses ideas and thoughts in an imaginative way. The writer gets to express feelings and emotions instead of just presenting the facts." The dictionary adds that "the best way to define creative writing is to give a list of things that are and that are not considered creative writing.

Witty and LaBrant (1946) define creative writing as:

...a composition of any type of writing at any time primarily in the service of such needs as: (1) the need for keeping records of significant experience, (2) the need for sharing experience with an interested group, and (3) the need for free individual expression which contributes to mental and physical health.

In a study by Cheung, Tse and Tsang (2003) in which they asked 449 Chinese language teachers to define creativity in writing, the teachers identified imagination, inspiration and original ideas as com-ponents of effective writing.

Based on the above definitions, creative texts that focus on a writer's free self-expression, i.e., those in which the writers express feelings and emotions and reveal unique voices and those characterized by novelty, imagination, inspiration and originality were considered creative and were selected. Themes that express emotions (such as love, nostalgia, disappointment, misery, cruelty, abandonment, alienation and so on), personal experiences, observations, philosophy, interpersonal relations, social issues, and/ or beliefs were considered creative.

As to types of creative writing, i.e. genres, Donovan (2015) identified 14 types of creative writing: journals, diaries, essays, storytelling, poetry, memoir, vignettes, letters, scripts, song lyrics, speeches, journalism, blogging and free writing. In addition, creative texts can be novels, novella, short stories, epics, biography, creative personal essays, flash fiction, playwriting/dramatic writing, screenplays, wise sayings, reflections and so on. Academic, technical and journalistic writing were not considered creative in the present study.

To be considered narrative (novels and fiction), a text should have characters, point of view, plot, setting, dialogue (fiction), style (fiction), theme and motif no matter how short it is. To be considered a poem, a text should be characterized by rhyme schemes, sound elements, figurative language and/or images.

Sophistication level was determined on the basis of types of sentence structures, sentence length and complexity, word choice, style, use of innovative expressions, figurative language and rhythm. Correctness of grammar, spelling, punctuation and capitalization and appropriateness of the text layout to a particular genre were noted.

For validation purposes, two raters who are professors of English literature were asked to sort out a random sample of 30% of the texts from each Facebook page/club into creative and non-creative, identify the forms, themes, and discoursal features of those creative texts. To help the raters identify the types of creative texts, i.e. their genre, it was necessary to give them some background about the aims of the study, the authors of the excerpts and the Facebook creative writing clubs. Providing the raters with the page links where the excerpts are located helped them comprehend and identify their types more accurately than giving them the excerpts in isolation, i.e. out of context. For example, when a rater was given the excerpt below from RWC in isolation, i.e. without providing any information about the author and where it was taken from, she thought it was a *"diary,"* not a *"poem."*

I will write my way home tonight; I will write my way to you. Love, you may ask: But where do you find me? I find you in music. Every sweet song, and every strum of the guitar that comes through my headphones brings you out. I find you in literature. In every word...

The type of genre was misidentified, because of the word *"write."* But when she saw it on Facebook, and she browsed through the RWC page, she was able to identify the genre more accurately. Results of the analyses by the author and the two raters were compared and disagreements were resolved by discussion.

Identifying the Factors Affecting Facebook Creativity

Creative writers' responses to the open-ended questions were sorted out and personal, affective and social reasons that affected their creative writing activities were identified. Affective factors in creative writing include personal attitude, motivation, and emotional state; whereas social factors in creative writing include desire to share feelings and thoughts with readers and members of an online community, and support and feedback given by the Facebook community. Conclusions based on a content analysis of the subjects' posts and comments received were also made.

Quantitative and Qualitative Data Analyses

The total number of creative texts and percentages of creative texts belonging to each category (poem, reflections or short narrative) were calculated. The themes and discoursal features of the creative texts collected are reported and described qualitatively. Responses to the questionnaire results are also reported qualitatively. Where excerpts are cited, the authors' real first names are also cited.

RESULTS AND DISCUSSION

Type of Creative Writing Forms and Themes Posted

Analysis of the creative texts collected from the five Facebook creative writing pages/clubs has shown that creative writers in the present study post poems, short narratives (stories), reflections (spiritual and otherwise), religious supplications, points of view, words of wisdom, inspirational sayings, motivational thoughts and advice written in literary style. They express emotions such as love, nostalgia, disappointment, misery, cruelty, abandonment, alienation, write about personal experiences, observations, philosophy, interpersonal relations, social issues, and/or Islamic beliefs (See examples 1, 4, 9, 10, 12, 15 in the Appendix). The same author may write several types of genres: Poems, reflections and short narratives, as in the case of Ali, Sameera and Zahra (See Examples 1 & 2; 3 & 4; 17 & 18).

Being Muslim, the influence of Islamic culture and Islamic beliefs is very evident in the themes that creative writers in the present study write about, and in how they view the world, other humans, relationships and other life issues expressed, especially in their reflections, advice and motivational sayings as in Examples 1, 2, 4, & 13. Ali from RCWD even states his Islamic faith/stance very clearly in one of his posts:

The post below is written to MYSELF alone...I write ALL things with ME or God as the subjects or advise in general to people...Either I am addressing Him or finding fault in Me or Sharing wisdom I know, heard or experienced and find it to be reality...My life is about me and my Creator ...

Moreover, it was found that in addition to free creative posts by individual club members, of RWC and KWC founders post *"projects"* that constitute themes for club members to write voluntarily about. A total of 11 projects have been posted on the RWC Facebook page: *"Being Human, Coffee, John, New Beginnings, The Paranormal, The World, Turning Point, Glass, Nostalgia, Eve, Elle,"* and 10 projects on the KWC Facebook page: *"Voiceless, Traitor, Box, Terminal, Home, Jay, Socks, Birth, Melancholy, Glass."* Twenty four more projects have been posted in the RWC blogs (*2:47, Ash, Conspiracy, Fear, Free topic, Here, I/We, Letter, Light/Yet, Maha, No, Reply, Rib, Seven Deadly Sins, Skin, Sounds, Speak, Still, Uncategorized, Wanderlust, War, Water, Winter, Woman*), and 29 more projects have been posted in the KWC blogs (*Aftertaste, Blood, Book, But Daddy I love her, But Daddy I love him, Choice, color, Depression, Echo, Higher power, Ink, Ja, Joy, Justice, Lipstick, Maze, Monkey, Mountain, Noah, Nostalgia, Puppet-Blood-Lighter, Revolution, Sciamachy, Secret, Seeds, Smoke, Superhero, Supervillian, Wave*), but those were not included in the samples nor analysis. A total of 53 Facebook creative texts were written under the 11 projects in RWC, and 65 creative texts were posted under the 10 projects in KWC. Examples 8-11 show texts (all poems) posted under the theme *"Coffee"* in RWC and examples 15-16 show texts posted under the theme *"Birth"* in KWC. The content in many of the posts has nothing to do with the general project theme. Examples 8-9 are unrelated, whereas examples 10 and 11 mention *"coffee"* in the content. Some of the posts are poems with verses laid out next to each other, rather than under each other as it customary in poem layout (See examples 8-11) may be because of formatting limitations on Facebook. However, the full texts of the same posts in the blogs have a poem format.

Unlike the other four clubs, members of CW posted paragraphs on themes such as *"honesty, respect, self-confidence, alone with ambition, money cannot buy happiness, the meaning of apology."* They also wrote paragraphs in which they criticized social issues and dysfunctional systems in their society and called for reform as in the following topic: *"employment of disabled persons, Lebanon can be the best, a word on Independence Day, are they receiving their rights, handicapped people are humans too" (See Examples 17 & 18).*

As in Stillar's (2013) study, it seems that the aims of the CW creating writing activities is to raise critical consciousness by having students write journals and/or personal letters from the perspective of individuals that have been marginalized or vilified in the students' dominant culture. Such activities, Stillar indicated, encouraged the students to adopt new perspectives on polarizing topics while being an enjoyable and effective activity that helped them practice writing in English, taking into consideration that CW members chose their own topics. In addition, the types of forms and themes posted by CW members are partially consistent with findings of a study by Khalaf (2014). In her study, Khalaf analyzed the changing views of Lebanese students as reflected in their personal narrative texts that they created during her creative writing workshops over a period of 16years (1997-2012). She found that students' texts focused on three main thematic issues: Idealism (1998-2005), Activism (2005-2008) and, Disillusionment (2008-present). In this study, CW members wrote narrative and expository paragraphs and their themes fall under the *"Idealism"* and *"Activism"* categories only.

Activities That Triggered Creative Writing

Members of the Facebook creative pages/clubs in the present study write freely and voluntarily. What they write about is not initiated by a teacher, class or an assignment, and they are not pressured by deadlines or commitments. However, members of the KWC hold monthly face-to-face meeting for members in Kuwait, where each member is requested to bring 3 samples of his/her writing for discussion. They also hold creative writing seminars and workshop for members such as *"how to write a short story."*

In CR, Lebanese students post mainly short expository paragraphs and few short stories (journals) about abstract themes and human values. Here, the themes are chosen by the members themselves and not imposed by anybody. In one of the CW posts, there was a call for joining an *"English Open Day."*

In RCWD and EIC members are free to post anything. However, some members of EIC begin a story and ask other members to complete it. Some request other club members to design creative logos and comment on them as in Figure 1. The creator of RCWD is also an artist and his creative posts are reflections on some paintings or scenes posted, as Figure 2 shows.

Characteristics of Creative Discourse

Creative texts written by members of RWC, KWC and RCWD are more sophisticated and native-like than those written by EIC and CW members. Members of the three former clubs are of a higher proficiency level in English. They exhibit more verbal originality and verbal flexibility as in Examples 1 to 11 and 15-16 even when very simple language is used as in Examples 1, 2, 3, 4, 5, 16. They used more sophisticated themes and details, innovative expressions *(strappy heals, flung across, perched, haunting),* innovative images and figures of speech *(fire-eyed; desert of your mind; well of wisdom; you wore nothing but your skin)* and almost made no grammatical, spelling or punctuation mistakes. As mentioned earlier, the lines of the poems in RWC are laid out next to each other like prose as in Examples 8 to 11, may be because of the Facebook formatting limitations. However, this does not affect comprehensibility, vividness, and effectiveness of their language and style, and does not affect the rhythm of the poems.

Figure 1. Logo and creative comment from EIC

Figure 2. Poem describing a painting from RCWD

On the other hand, members of the EIC are mainly students learning the basics of the English language. Therefore, their creative texts are written at their own ability level and they are less imaginative and innovative in their themes and style. They made more grammatical and spelling errors as in Zoha, Emi and Mamadou's poems in Examples 12, 13 and 14. Although Zoha's poem is sophisticated in theme, expressions, vocabulary and style, she made mistakes in verb tenses, and spelled the first person singular pronoun *"I"* in lower case (See underlined errors in Example 12). Emi capitalized every single word for no reason, did not know the English equivalents to the Arabic and Islamic words *"hijab"* and *"Gheebah"* and sometimes used meaningless expression such as *"Please Winder And More Moddest Cloths."* Mamadou spelled hurry as *"harry,"* used non-standard forms such as *"gonna," "ur"* and *"FRV"* and capitalized the last line which makes is difficult to tell whether it is a note or part of the poem. Writers like Zoha, Emi and Mamadou do not seem to worry about grammatical, spelling and punctuation mistakes as they write for communication and their aim, as expressed in their other posts, is to practice writing in English.

Similarly, members of CW are students learning English. Their posts are not as creative and imaginative as those of RWC and KWC. Members of CW have not mastered English usage yet. Some structures and expressions that they sometimes used are transferred from Arabic (L1). In some cases, they did not know which words collocate with each other, and in which contexts synonyms and near-synonyms are used. For example, Zahra in Example 17 says: *"put me a zero"* and *"without listening me"* which are transferred from Arabic in structure and meaning. She also made some structural mistakes such as *"a sword who," "taking in consideration"* and *"After the long cherished path of that was full of difficulties, I decided to let the volcano that was in my heart erupt,"* which made the meaning of the sentence unclear, due to adding the preposition "of" in *"path of that."* Zahra also wrote *"my partner"* rather *"my classmate"; "I was innocent and not mistaken"* instead of *"innocent and not guilty";* and *"fire me from class"* instead of *"expel me from class."* However, such weaknesses did not affect the message she tried to convey while narrating the incident she had been through. As in Example 18, CW members sometimes mixed two themes in one short paragraph, wrote general statements, gave few supporting details, and their paragraphs lack cohesion and coherence. However, student-writers, like Zahra, were able to use figurative language in their posts such as *"like the sea waves," "like finding a treasure," "lend you a shoulder to cry," "sword who (that) cried for revenge," "give our planet its rights," "the cage of disabilities"* and *"open the lock to let them fly freely."*

Another feature of creative texts in the present study is that informal and conversational English is more evident than formal English in the subjects' creative writing. This is evident in their use of contracted verb forms, imperative verbs and the second person pronoun *"you."* The reader feels that those writers are talking to him//her. Use of spoken English makes ideas, images and expressions more vivid.

Findings about the sophistication level of the language used by creative writers in the present study are consistent with findings of a study by Elgort (2017) in which she compared the language and discourse characteristics of course blogs and traditional academic assignments produced in English by L1 and advanced L2 graduate student writers enrolled in the same M.A. course. The text types produced by L1 and L2 students differed in their lexical sophistication, syntactic complexity, use of cohesion and agency. Overall, the traditional course assignments were more formal, lexically sophisticated and syntactically complex, while the blog posts contained more semantic and situational redundancy, resulting in higher readability, and communicated a clearer sense of agency. There was a significant difference between the textual artefacts produced by L1 and L2 writers, such as using a more traditional impersonal academic style of the L2 texts. The blog posts were rated lower on the use of language than traditional assignments for the L2, but not L1 writers.

By contrast, the sophistication level of the language used by creative writers in the present study is inconsistent with the assumptions and results of Tin's study (2011) in which he investigated opportunities for creative language use and the emergence of complex language in creative writing tasks with high formal constraints (acrostics) and those with looser formal constraints (similes). Formal constraints lead to complex and creative language use, transforming familiar utterances into unfamiliar ones, shaping and reshaping learners' language syntactically and lexically, paradigmatically, and syntagmatically. Tin concluded that for learners' language to develop in complexity, conditions that require them to access L2 directly to construct new ideas are needed for both L2 forms and meaning to co-evolve. In the present study, the sophistication level of the creative texts is not due to any formal constraints as the three Facebook creative writing clubs are for amateur creative writers and not part of a formal creative writing course. No constraints on style, length, theme, content, time or space are imposed on the writers for posting their work. No constraints are imposed on linguistic accuracy and correctness as even less proficient writers have the courage to post their poems without being inhibited by their linguistic inadequacies. Talent, wide reading, self-motivation and desire for self-expression that the subjects reported, seem to affect creativity.

Factors Affecting Creativity in L2 Writers

In her analysis of the processes of creative writing, Morgan (2006) argued that intuition and analysis, the conscious and unconscious, working together, and the social and personal are all involved in these processes. She concluded that personal as well as social, cultural and disciplinary factors are at play in the development of creative work. In another study, Jonesa and Issroffb (2005) investigated the role of affective factors in three main areas of collaboration: In settings where learners are co-located, in on-line communities and to support and develop socio-emotional skills. Their results stressed affective issues in learning technologies in a collaborative context. Learner attitude, motivation, and emotional state were found to be very important. Furthermore, Ruan (2014) indicated that motivation, self-efficacy, and writing anxiety constituted Chinese EFL students' awareness of personal variables influencing writing ability in EFL.

As in Morgan, Jonesa and Issroffb, and Ruan's studies, results of the questionnaire-survey in the present study revealed that both personal and social factors play a role in the subjects' creativity. Creative writers in the present study tended to be intrinsically motivated and enthusiastic. Area of specialty, educational level, and proficiency level in English do not seem to be factors leading to creative writing. All creative writers reported that they are avid readers, write a lot and have an urge for expressing themselves. Members of the RWC and KWC share a blog where they post their creative work, and several members of RCWD have their own blog as well; Ali has four blogs. Other members, such as Eman, reported that they post poem on the Poetry.com website. Creative writers of lower proficiency level in EIC and RCWD reported that they are always motivated by creative writers of a higher proficiency level. They learn from them and see examples of good writing. Some members from RWC and KWC indicated that the Facebook writing page/club/community is a fast way of publishing their creative work. That is why each blog is connected to a Facebook page containing samples of their work and links referring to their full blog posts. Unlike Chinese students in Ruan's study, anxiety does not seem to be a factor in creative writers in the present study even in less-proficient writers, probably because no marks and grades and no pass/fail are involved.

Another important personal factor is the centrality of emotions in the observed creative text and expression. All creative writers produce poems, reflections and short stories that are meaningful and original. They tend to express their feelings and emotional involvement with a personal experience that they had or their views of an issue or philosophy about something (See Examples 1, 6, 7, 10 to 14). Ali says:

Emotions need education. They need pacification and guidance. Emotions are so powerful that they can take the heart to extremes of despair and to the greatest heights of joy. Both resulting in destruction of the soul or even the body sometimes. These same emotions can be the source of good health and a good life if nurtured by wisdom. Wisdom, which is acquired from reflections and realizations of existence...

Ali also wrote:

What is expressed reveals what one may seek to discover. I have ventured into my own soul and mind to discover my real self. My inner self. Is any knowledge in this world sufficient to define to me myself? And, is there anything that deserves to be defined more?

No. It is only my own expression of my thoughts and Realities that will reveal to me what I am and in acquiring this knowledge I shall know mankind...and you...

These findings are also consistent with findings of studies by Vass (2007) and Vass, Littleton, Miell and Jones (2008) in which they conducted longitudinal observations of ongoing classroom-based collaborative writing activities conducted with 24 British children in grades 3 and 4, aged 7-9. A functional model was developed to analyze the cognitive processes associated with creative text composition (engagement and reflection) via an in-depth study of collaborative discourse. A major finding was the centrality of emotions in the observed creative writing sessions. Their results also highlighted the role of emotion-driven thinking in phases of shared engagement and creative-thinking.

As for social factors affecting creativity, results of the questionnaire-survey showed that most of the creative writers in the present study like to share their feelings and thoughts with readers and other members of the Facebook club and other Facebook users. They also reported that the Facebook creative writing community, as a social network, has a positive effect on their creativity and on their attitude towards writing, as it creates an open and supportive environment, where creative writing is appreciated, and authors are encouraged to trust their own linguistic ability. The Facebook writing community nurtures their creativity in every way possible. It makes writing an enjoyable task. They dive into the writing task because it is exciting, challenging and fun. They feel comfortable and unthreatened to reach maximal creativity. As a result, a positive personal relationship with other members of the Facebook writing club is fostered, although the members reported that they did not know each other before joining the Facebook club. Here is what Hassan says about the club:

i enjoy collaborating; it is what we do in EIC; different ideas make me pause; i practice English while decorating my mind; i enjoy UNDERSTANDING.

Peer support, interaction and feedback among members of the creative writing Facebook community were also found to be important. Creative writers in EIC, RWRD and CR, in particular, receive *"likes,"* positive comments, encouraging remarks and words of admiration from other club members such as *"I am proud of you," "perfect," "Mashallah," "Go," "well said,"* as shown in Figures 3 and 4. Such positive comments and remarks make them feel good about themselves and enhance their motivation. Analysis of the members' comments has shown that more proficient members do not correct grammatical, spelling or punctuation mistakes of less proficient writers. No matter how poor the writing is, no negative comments are given. Comments are usually given on the content, not the form of what is posted. An example of the comments that Emi received on her poem is:

EIC: *great massage Emi...and nice advice .. keep it up.*

Aya: *like this.*

Osome!

Other members show their admiration and appreciation of a post by requesting to share the post with others as in Figure 3.

On the other hand, interaction and feedback do not seem to be a factor in members' creative writing activities in RWC and KWC, because "likes" and "comments" are almost lacking in these two clubs. It seems that their sophisticated writing products are a result of their feeling of self-efficacy and self-motivation, as in Ruan's study, in addition to off-line support and encouragement, as members are friends

Figure 3. Post and comments, i.e., interaction from RCWD

Figure 4. Post and Comments, i.e., Interaction from RCWD

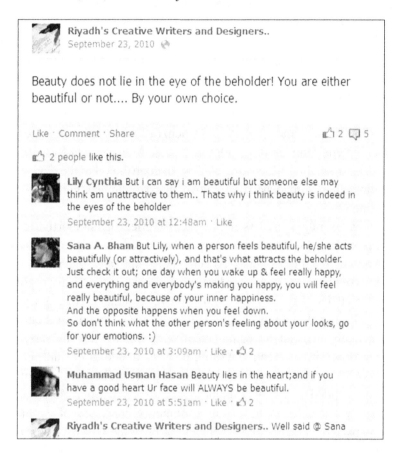

and meet face-to-face occasionally to discuss their projects and creative activities. Their published work in the blogs seems to give them a sense of accomplishment.

In all clubs, Facebook provides members with a non-threatening environment for trying out new ways of expressing themselves in English. Environmental friendliness prevails, especially in the EIC, RCWD and CW clubs. Writers of lower proficiency level, such as Emi and Zahra feel free to make mistakes without being afraid of losing marks or receiving negative comments regarding their mistakes as in a formal classroom setting. Creative writers in EIC, RCWD and CW sounded joyful and more relaxed in the Facebook writing environment, as it is the case in Gonçalves, Campos and Garg's (2015) study in which 10-16-year old students felt better, happier, more relaxed and somewhat empowered and creative while using OmniWriter, which positively influenced their productivity and the mental well-being and increased their desire to write more.

Furthermore, analysis of the members' comments has shown that more proficient members in RCWD and EIC give more sophisticated comments; some respond creatively to the theme and view it from a different angle. For example, they reflect upon the ideas of the post as in Figure 4. Such comments help expand the theme of the original post and enrich the writer's outlook and creative thinking. Here again, as Vass (2007) indicated, peer interaction and collaboration can resource, stimulate and enhance creative writing. Also, it shows writers' reliance on the collaborative floor which was indicative of joint focus and intense sharing, thus facilitating mutual inspiration in the content generation phases of the writers' writing activities (Vass, 2007). The Facebook creative writing environment in this study, as Bruno (2002) indicated, seems to serve as a source of constant evolving inspiration and a place where there is trust, no matter what the students write and whether they are native speakers, as in Bruno's study, or non-native speakers as in the present study. Involvement in the Facebook writing activity fostered participants' creative freedom in producing a well-formed piece of writing showing appropriate control of tone, style, and register (Howarth, 2007; Anae, 2014). Findings of the presents study are also consistent with findings of Al-Jarf's (2007) and Yunus, Salehi and Chenzi (2012) studies in which the collaborative, supportive and motivating environment of the Online Course Management System and the integration of an online discussion board in the ESL writing classroom contributed to the development of freshman students' creative writing ability, broadened their knowledge, heightened their motivation, and felt more confident in their writing ability.

As in the present study, several researchers emphasized the effect of a supportive learning environment and of peer feedback provided to student creative writers using other forms of technology or receiving traditional classroom instruction. For example, Hyland (1993) indicated that use of word processors for developing writing skills of foreign language students created unrealistically high expectations regarding learning gains and indicated that only teachers could improve the situation through a supportive learning environment. Essex (1996) noted that most children enter school with a natural interest in writing, and teachers can become actively involved in teaching creative writing to their students, and highlighted the effectiveness of peer feedback in the creative writing process. In addition, Kaufman, Gentile and Baer (2005) and Wichadee's (2013) studies supported the use of peer feedback among gifted novice creative writers. They also recommended the use of collaborative feedback in gifted classrooms as in Mohapatra & Mohanty's (2017) study where the online writing community, consisting of an amalgamation of writers, has been greatly impacted by technology. Members of this kind of online writing community play a vital role in reading, sharing and voting for the appropriateness of the content.

CONCLUSION AND RECOMMENDATIONS

The present study examined a sample of Facebook creative writing texts in English, created by Arab Facebook users to find out the kinds of creative writing forms and themes posted, the characteristics of the Facebook creative discourse in English and creative writing environment, the Facebook activities that initiated creative writing by non-native speakers of English, and the personal and social factors that impacted the Facebook creative texts by non-native speakers of English. It was found that creative writers in the present study, even those who are 14 and 15 years old and are still in high school, like to share their feelings and thoughts about themes of their choice with the other Facebook creative writing club members. They find the Facebook environment supportive as they receive positive "comments" and "likes" from them. Thus, Facebook nurtures their creativity and makes writing an enjoyable task.

Findings of the present study show, as James (2008) indicated, that all people have a creative potential, and that the right environment with prompts and encouragement can elicit creative work to a degree. Although talent, motivation and desire play an important role in creative writing, creative writing also involves tools, techniques and concepts. Since creative writing is subjective and personal, novice writers need to acquire those tools and techniques in order to experiment, practice, and master them.

Based on the findings of the present study, writing instructors can do a lot to nurture creative writing skills in their students. They can encourage them to get involved in writing-based extracurricular activities that take place outside the formal college or classroom setting, such as arranging writing contests to develop students' writing skills, spending more time on interactive writing rather than independent and solitary writing, and creating a Facebook writing page/club as a supplement to in-class instruction. To encourage students to write freely, enjoy writing and share thoughts and experiences, a Facebook environment that is supportive and safe for trial and error is necessary. All an instructor needs to do is to convince the students to get in touch with their inner voices, and encourage them to write for communication, rather than focusing on grammatical and spelling correctness, as it is usually the case in the classroom. "Likes" and positive "comments" are essential. Students need to feel free to express themselves, and need to feel good about themselves and what they can do and achieve. A writing group, such as a Facebook writing class page, can be a source of constant evolving inspiration and a place where there is trust (Bruno, 2002).

Developing creative writing skills via a Facebook writing class page can go through graded steps depending on the students' proficiency level and writing ability: (i) posting a picture, painting or logo and asking the students to describe it; (ii) posting an inspirational quote and having the students comment on it; (iii) posting story starters such as *"once upon a time ..., why do you think that ..., have you ever wondered ..., imagine that..., pretend that ..., what if ..., a funny/sad thing happened ..."* and asking the students to finish the story. Such an activity will help them use their imagination and express their feelings; (iv) Working on a collaborative project or theme of their choice; and (v) free writing and journaling. Students' poems and short stories can be published on the Facebook timeline and the number of "likes" on each poem or story can be tracked and comments responded to.

Simple poetry selected from the five Facebook writing pages/clubs described in the present study may be posted on the writing class Facebook page and/or in class. It can be used with all learners, even those with limited literacy and proficiency in English (Peyton and Rigg, 1999). Teaching great poetry to students can enhance their perceptions, improve their writing, challenge their minds, and enrich their lives as well (Certo, 2004). It provides rich learning opportunities in language, content, and community building. Repetition of words and structures encourages language play with rhythmic and rhyming devices. Poetic themes are often universal. When teachers and students write and read poetry together, they connect with texts and with one another in powerful ways as well. Such activities are believed to help develop L2 students writing skills, in particular, and creative writing skills, in particular.

Future research may compare the forms, themes and discoursal features of creative texts posted on Facebook by native and non-native speakers, and identify the factors that affect the creativity of each group. Since Twitter does not allow the posting of tweets longer than 140 characters, the forms, themes and discoursal features of English creative tweets by native and non-native speakers of English may be subject to further investigation by future research. The discoursal features of Arabic creative writing on social media and comparisons of creative writing skills in both English (L2) and Arabic (L1), by the same authors, are still open for further investigation. In addition, future research may investigate the effects of using a writing class Facebook page/club by EFL students as a supplement to in-class college instruction on writing achievement, and/or the effects of integrating creative texts from the Facebook creative writing pages in classroom reading and writing instruction.

To encourage and produce more research in the area of discourse and social media, Facebook and Twitter need to add an analytics tool or App whereby researchers can get statistics about posts and tweets according to the subject and author. Tags can be added to the Facebook timeline. Those would facilitate browsing and sorting out of posts and tweets and make the processes less tedious and less time-consuming.

REFERENCES

Abdallah, M. (2013). A Community of practice facilitated by Facebook for integrating new online EFL writing forms into Assiut university college of education. *Journal of New Valley Faculty of Education, 12*(1). ERIC Document No. ED545728.

Ahn, J. (2010). *The influence of social networking sites on high school students' social and academic development* (Unpublished doctoral dissertation). University of Southern California, Los Angeles, CA.

Al-Jarf, R. (2007). Online instruction and creative writing by Saudi EFL freshman students. *The Asian EFL Journal Professional Teaching Articles, 22.*

Al-Jarf, R. (2012). *Learning English on Facebook.* Paper presented at the 11th Asia CALL Conference, Ho Chi Minh City Open University, Vietnam.

Al-Jarf, R. (2014). Integrating ethnic culture Facebook pages in EFL instruction. In M. V. Makarych (Ed.), *Proceedings of the International Scientific Conference on Ethnology: Genesis of Hereditary Customs* (41-43). Minsk, Belarus: Belarusian National Technical University (BNTU).

Al-Sayed, R. K., Abdel-Haq, E. M., El-Deeb, M. A., & Ali, M. A. (2016). *Fostering the memoir writing skills as a creative non-fiction genre using a WebQuest model.* ERIC Document No. ED565329.

Alon, I., & Herath, R. K. (2014). Teaching international business via social media projects. *Journal of Teaching in International Business, 25*(1), 44–59. doi:10.1080/08975930.2013.847814

Anae, N. (2014). Creative writing as freedom, education as exploration: Creative writing as literary and visual arts pedagogy in the first-year teacher-education experience. *Australian Journal of Teacher Education, 39*(8). doi:10.14221/ajte.2014v39n8.8

Arshavskaya, E. (2015). Creative writing assignments in a second language course: A way to engage less motivated students. *InSight: A Journal of Scholarly Teaching, 10,* 68-78.

Awada, G. (2016). Effect of WhatsApp on critique writing proficiency and perceptions toward learning. *Cogent Education, 3*(1).

Barrs, M., & Horrocks, S. (2014). *Educational blogs and their effects on pupils' writing.* CFBT Education Trust. ERIC Document No. ED546797.

Bartscher, M., Lawler, K., Ramirez, A. & Schinault, K. (2001). *Improving student's writing ability through journals and creative writing exercises.* ERIC Document No. ED455525.

Bayat, S. (2016). The effectiveness of the creative writing instruction program based on speaking activities (CWIPSA). *International Electronic Journal of Elementary Education, 8*(4), 617–628.

Bentley, J. & Bourret, R. (1991). *Using emerging technology to improve instruction in college transfer.* ERIC Document No. ED405011.

Biesenbach-Lucas, S., & Weasenforth, D. (2001). E-Mail and word processing in the ESL classroom: How the medium affects the message. *Language Learning & Technology*, *5*(1), 135–165.

Bikowski, D., & Vithanage, R. (2016). Effects of web-based collaborative writing on individual l2 writing development. *Language Learning & Technology*, *20*(1), 79–99.

Bintz, W. P. (2017). Writing etheree poems across the curriculum. *The Reading Teacher*, *70*(5), 605–609. doi:10.1002/trtr.1544

Bremner, S., Peirson-Smith, A., Jones, R., & Bhatia, V. (2014). Task design and interaction in collaborative writing: The students' story. *Business and Professional Communication Quarterly*, *77*(2), 150–168. doi:10.1177/2329490613514598

Bruno, M. (2002). *Creative writing: The warm-up*. ERIC Document No. ED464335.

Casella, V. (1989). Poetry and word processing inspire good writing. *Instructor*, *98*(9), 28.

Cheung, W., Tse, S., & Tsang, H. (2003). Teaching creative writing skills to primary school children in Hong Kong: Discordance between the views and practices of language teachers. *The Journal of Creative Behavior*, *37*(2), 77–98. doi:10.1002/j.2162-6057.2003.tb00827.x

Creto, J. (2004). Cold plums and the old men in the water: Let children read and write great poetry. *The Reading Teacher*, *58*(3), 266–271. doi:10.1598/RT.58.3.4

Del Nero, J. R. (2017). Fun while showing, not telling: Crafting vivid detail in writing. *The Reading Teacher*, *71*(1), 83–87. doi:10.1002/trtr.1575

Dizon, G. (2016). A comparative study of Facebook vs. paper-and-pencil writing to improve L2 writing skills. *Computer Assisted Language Learning*, *29*(8), 1249–1258. doi:10.1080/09588221.2016.1266369

Dollins, C. A. (2016). Crafting creative nonfiction: From close reading to close writing. *The Reading Teacher*, *70*(1), 49–58. doi:10.1002/trtr.1465

Donovan, M. (2015). *14 types of creative writing*. Retrieved August 15, 2017 from https://www.writingforward.com/creative-writing/types-of-creative-writing

Dzekoe, R. (2017). Computer-based multimodal composing activities, self-revision, and L2 acquisition through writing. *Language Learning & Technology*, *21*(2), 73–95.

Elgort, I. (2017). Blog posts and traditional assignments by first- and second-language writers. *Language Learning & Technology*, *21*(2), 52–72.

Essex, C. (1996). *Teaching creative writing in the elementary school*. ERIC Document No. ED391182.

Ferguson, C. (2010). *Online social networking goes to college: Two case studies of higher education institutions that implemented college-created social networking sites for recruiting undergraduate students*. ERIC Document No. ED516904.

Feuer, A. (2011). Developing foreign language skills, competence and identity through a collaborative creative writing project. *Language, Culture and Curriculum*, *24*(2), 125–139. doi:10.1080/07908318.2011.582873

Fraser, D. (2006). The creative potential of metaphorical writing in the literacy classroom. *English Teaching*, *5*(2), 93–108.

Gammon, G. (1989). You won't lay an egg with the bald-headed chicken. *B. C. The Journal of Special Education*, *13*(2), 183–187.

Gonçalves, F., Campos, P., & Garg, A. (2015). Understanding UI design for creative writing: A pilot evaluation. In *Adjunct Proceedings of the INTERACT 2015 Conference* (179-186). University of Bamberg Press.

Guzzetti, B., & Gamboa, M. (2005). Online journaling: The informal writings of two adolescent girls. *Research in the Teaching of English*, *40*(2), 168–206.

Haines, S. (2015). Picturing words: Using photographs and fiction to enliven writing for ELL students. *Schools: Studies in Education*, *12*(1), 9–32. doi:10.1086/680692

Henry, L. (2003). *Creative writing through wordless picture books*. ERIC Document No. ED477997.

Hodges, B. (1999). Electronic books: Presentation software makes writing more fun. *Learning and Leading with Technology*, *27*(1), 18–21.

Hong, E., Peng, Y., & O'Neil, H. Jr. (2014). Activities and accomplishments in various domains: Relationships with creative personality and creative motivation in adolescence. *Roeper Review*, *36*(2), 92–103. doi:10.1080/02783193.2014.884199

Howarth, P. (2007). Creative writing and Schiller's aesthetic education. *Journal of Aesthetic Education*, *41*(3), 41–58. doi:10.1353/jae.2007.0025

Hung, H., & Yuen, S. (2010). Educational use of social networking technology in higher education. *Teaching in Higher Education*, *15*(6), 703–714. doi:10.1080/13562517.2010.507307

Hyland, K. (1993). ESL computer writers: What can we do to help? *System*, *2*(1), 21–30. doi:10.1016/0346-251X(93)90004-Z

James, D. (2008). A short take on evaluation and creative writing. *Community College Enterprise*, *14*(1), 79–82.

Jayaron, J., & Abidin, M. J. (2016). A pedagogical perspective on promoting English as a foreign language writing through online forum discussions. *English Language Teaching*, *9*(2), 84–101. doi:10.5539/elt.v9n2p84

Jonesa, A., & Issroffb, K. (2005). Learning technologies: Affective and social issues in computer-supported collaborative learning. *Computers & Education*, *44*(4), 395–408. doi:10.1016/j.compedu.2004.04.004

Kaufman, J., Gentile, C., & Baer, J. (2005). Do gifted student writers and creative writing experts rate creativity the same way? *Gifted Child Quarterly*, *49*(3), 260–265. doi:10.1177/001698620504900307

Keiner, J. (1996). *Real audiences-worldwide: A case study of the impact of WWW publication on a child writer's development*. ERIC Document No. ED427664.

Khalaf, R. S. (2014). Lebanese youth narratives: A bleak post-war landscape. *Compare: A Journal of Comparative Education*, *44*(1), 97–116. doi:10.1080/03057925.2013.859899

Lee, K., & Ranta, L. (2014). Facebook: Facilitating social access and language acquisition for international students? *TESL Canada Journal*, *31*(2), 22–50. doi:10.18806/tesl.v31i2.1175

Malesky, L., & Peters, C. (2012). Defining appropriate professional behavior for faculty and university students on social networking websites. *Higher Education: The International Journal of Higher Education and Educational Planning*, *63*(1), 135–151. doi:10.1007/s10734-011-9451-x

Mohapatra, S., & Mohanty, S. (2017). Assessing the overall value of an online writing community. *Education and Information Technologies*, *22*(3), 985–1003. doi:10.1007/s10639-016-9468-y

Morgan, W. (2006). Poetry makes nothing happen: Creative writing and the English classroom. *English Teaching*, *5*(2), 17–33.

O'Day, S. (2006). *Setting the stage for creative writing: Plot scaffolds for beginning and intermediate writers.* ERIC Document No. ED493378.

Owen, T. (1995). Poems that change the world: Canada's wired writers. *English Journal*, *84*(6), 48–52. doi:10.2307/820891

Peyton, J., & Rigg, P. (1999). *Poetry in the adult ESL classroom.* ERIC Document No. ED439626.

Pifarré, M., Marti, L., & Guijosa, A. (2014). *Collaborative creativity processes in a wiki: A study in secondary education.* Paper presented at the Conference on Cognition and Exploratory Learning in Digital Age (CELDA), Porto, Portugal. Retrieved August 15, 2017 from http://www.scopus.com/record/display.uri?eid=2-s2.0-84925220858&origin=inward&t xGid=28E2C4265FE0D85B8C5C3946BE70 3BAB.WlW7NKKC52nnQNxjqAQrA %3a2

Racco, R. G. (2010). Creative writing: An instructional strategy to improve literacy. Attitudes of the intermediate English student. *Journal of Classroom Research in Literacy*, *3*, 3–9.

Rambe, P. (2011). Exploring the impacts of social networking sites on academic relations in the university. *Journal of Information Technology Education*, *10*, 271–293.

Reinhardt, J., & Zander, V. (2011). Social networking in an intensive English program classroom: A language socialization perspective. *CALICO Journal*, *28*(2), 326–344. doi:10.11139/cj.28.2.326-344

Ruan, Z. (2014). Metacognitive awareness of EFL student writers in a Chinese ELT context. *Language Awareness*, *23*(1-2), 76–90. doi:10.1080/09658416.2013.863901

Ryan, M. (2014). Writers as performers: Developing reflexive and creative writing identities. *English Teaching*, *13*(3), 130–148.

Sahbaz, N., & Duran, G. (2011). The efficiency of cluster method in improving the creative writing skill of 6th grade students of primary school. *Educational Research Review*, *6*(11), 702–709.

Salcito, A. (2012). *Exploring creative writing through technology.* Retrieved August 15, 2017 from http://dailyedventures.com/index.php/2012/01/30/1369/

Sarar Kuzu, T. (2016). The impact of a semiotic analysis theory-based writing activity on students' writing skills. *Eurasian Journal of Educational Research*, *63*, 37–54.

Scott, V. (1990). Task-oriented creative writing with system-D. *CALICO Journal*, *7*(3), 58–67.

Sessions, L., Kang, M. O., & Womack, S. (2016). The neglected "R": Improving writing instruction through iPad apps. *TechTrends*, *60*(3), 218–225. doi:10.1007/s11528-016-0041-8

Shafie, L. A., Yaacob, A., & Singh, P. K. (2016). Facebook activities and the investment of L2 learners. *English Language Teaching*, *9*(8), 53–61. doi:10.5539/elt.v9n8p53

Stillar, S. (2013). Raising critical consciousness via creative writing in the EFL classroom. *TESOL Journal*, *4*(1), 164–174. doi:10.1002/tesj.67

Stoddart, A., Chan, J. Y., & Liu, G. (2016). Enhancing successful outcomes of wiki-based collaborative writing: A state-of-the-art. *Review of Facilitation Frameworks. Interactive Learning Environments*, *24*(1), 142–157. doi:10.1080/10494820.2013.825810

Tin, T. (2011). Language creativity and co-emergence of form and meaning in creative writing tasks. *Applied Linguistics*, *32*(2), 215–235. doi:10.1093/applin/amq050

Vass, E. (2002). Friendship and collaborative creative writing in the primary classroom. *Journal of Computer Assisted Learning*, *18*(1), 102–110. doi:10.1046/j.0266-4909.2001.00216.x

Vass, E. (2007). Exploring processes of collaborative creativity--The role of emotions in children's joint creative writing. *Thinking Skills and Creativity*, *2*(2), 107–117. doi:10.1016/j.tsc.2007.06.001

Vass, E., Littleton, K., Miell, D., & Jones, A. (2008). The discourse of collaborative creative writing: Peer collaboration as a context for mutual inspiration. *Thinking Skills and Creativity*, *3*(3), 192–202. doi:10.1016/j.tsc.2008.09.001

Wang, S., & Vásquez, C. (2014). The effect of target language use in social media on intermediate-level Chinese language learners' writing performance. *CALICO Journal*, *31*(1), 78–102. doi:10.11139/cj.31.1.78-102

Wichadee, S. (2013). Peer feedback on Facebook: The use of social networking websites to develop writing ability of undergraduate students. *Turkish Online Journal of Distance Education*, *14*(4), 260–270.

Wilson, P. (2011). Creative writing and critical response in the university literature class. *Innovations in Education and Teaching International*, *48*(4), 439–446. doi:10.1080/14703297.2011.617091

Witty, P., & Labrant, L. (1946). *Teaching the people's language*. New York: Hinds, Hayden & Eldredge. Retrieved August 15, 2017 from http://archive.org/stream/teachingpeoplesl00witt#page/n3/mode/2up

Your Dictionary. (2015). Retrieved August 15, 2017 from http://reference.yourdictionary. com/word-definitions/definition-of-creativewriting. html

Yunus, M., & Salehi, H. (2012). Integrating social networking tools into ESL writing classroom: Strengths and weaknesses. *English Language Teaching*, *5*(8), 42–48. doi:10.5539/elt.v5n8p42

APPENDIX

Examples of Creative Texts

Example 1: Ali Qrote in the RCWD

Sorrow flows out from your eyes when your heart is hurt and tears fill your eyes.. let them flow.. let your sorrow go...make room in your heart.. for joy shall come .. to stay..

Example 2: Ali Wrote in the RCWD

I find you wandering in the desert of your mind looking for an oasis. Seeking that drop of water that would make the desert a garden of beauty. I see you looking and I see that you are tired. In your search, you forgot to look somewhere... in the well of wisdom - your heart. Look there, you will find your beloved ready to make your mind a place of peace and security...

Example 3: A Poem Written by Sameera in RCWD

If I feel
If I know
If I see
A Wrong...
YET...
I turn away
I ignore
I cover up
WHAT AM I ?
A coward
A hypocrite
A liar
SORRY...
Must Stop
Must Speak
Must Resist
OR I'll BE JUST AS...
Evil
Dishonest
Mischievous
AS
I feel
I know
I see
THE OTHER PERSON TO BE!!!

Example 4: "A Reflection" Written by Sameera in RCWD

God created Human Beings as the most superior of his creations but they insist on acting like the lowest of creatures - deceiving, lying and cheating & God lets them be till all limits have been crossed and then reckoning.....

Example 5: "The Unforgettable Past" by Eman from RCWD

The ghost of memories is haunting my mind
I'm trying to escape but its pulling me behind
my body is aching peace I want to find
my soul is bleeding, for life it has declined...

The ocean of the past is filled with sin
the waves are crashing and I'm drowning in
my thoughts are chasing me, away far away i want to swim
I'm breaking up, i wish i couldn't remember how my life has been...

I'm pouring out my heart its filled with regret
its color is black another memory is a threat
a threat to spill the black blood on my last sunset
then my life will come to an end, is this the only way to forget?

Example 6: "Coffee" by Mimi from RWC

I remember when we laid in bed You wore nothing but your skin, And the blanket I slid on you. You said you'd rather wear nothing but me, all day Feel nothing but the tips of my fingers On every inch of your body. I said I'd always love you, you said you would too...

Example 7: Abdulaziz from RWC Reflects on the Project "Conspiracy"

I, one of the many citizens of the current world, have asked questions that were too uninspiring to be asked; why has the world lost its sweet flavor? And the bitter taste never left? Why are the sweat drops that we suffer to make fall into other men's pockets?

The thoughts we proudly claim are only meant to be thought, the tears that we shed fall into the never ending river of sadness, every time I look into a pure soul I find the same black dots that tarnish their beauty.

Yes, we have been alienated; we are now the property of the strongest, we have lost every single characteristic that define us as human beings. We might as well strongly oppose to the idea but let's face it, we are now pieces of a puzzle, and when our roles end, we get nuclear bombs flying towards us.

I, one of the many citizens of the current world, am full of laughter for our sad misfortune.

Example 8: "Coffee" by Hala from RWC

I want you fire-eyed, helpless, and passion-driven. I want you irrational and guilt-stricken. I want you spiteful, conflicted, loving, I want you going, coming, and running. I want you heaving, shoving, screaming, sighing, pulling, pushing, laughing, lying. I want you holding my face in the palms of your hands and crying I don't want…

Example 9: "Coffee" by Meshael from RWC

I will write my way home tonight; I will write my way to you. Love, you may ask: But where do you find me? I find you in music. Every sweet song, and every strum of the guitar that comes through my headphones brings you out. I find you in literature. In every word…

Example 10: "Coffee" by Nouf from RWC

The elevator door opened as she strutted out in all her glory and shame. The strappy heals that were flung across the room the night before were barely clinging on to her feet. Huge Victoria Beckham-like sunglasses were perched on her perfectly altered nose, hiding the puffy eyes and the smudged mascara. Click-clack. She walked…

Example 11: "Coffee" by Albutol from RWC

"One more," he says.
Over steaming mugs,
in musty rooms,
amongst nicotine-clouded heads, –
one more.

Stay, he pleads
without words,
the hanging promise of mahogany tinted liquid
of wood roasted beans.
Stay.

"One more," he tells the waiter.
The empty chair a shade of the drink,
a shade of his skin,
a shade of his eyelashes.
She breathes deep.

Her thumb scratching at the table,
her fingers gripping the corner,
her blood pulsating caffeine

and half-intended goodbyes,
her knees succumb.

Wrinkles etch his kerosene eyes,
his teeth wary of an audience.
She sits apprehensive.
The promise of maybe, of again, of possibility
dwelling in a cup of joe.

Example 12: "Blood" by Zoha

False,
False dreams, false people, false world
Each hand is dirty with the blood
My utmost desire to again reveal
The wonders that die long ago
The treasures about thy, regale
Fluttered in mind, in heart grow
But the desires are ^ never ending flood
Each hand is dirty with the blood
My corniest, attributes, prevails
The boat on ^ sea, with ^ wave it sails
Mysteries thy, told, stories thy tell
To make me determined, they always fails
No one is pure as a beautiful bud
Each hand is dirty with the blood
Growing in the massive, disastrous way
Even no time to settle and pray
Walk or run, but be the best
No one cares, they always say
After every thunder rain, I am covered with mud
Each hand is dirty with the blood

Example 13: "Wake Up !!!" by Emi Zen Wrote from EIC

Please Wake Up Before It's Too Late
If You Aren't Praying, Please Pray.
If You Aren't Lowering Your Gaze, Please Do.
If You Are Mistreating Your Parents, Please Stop.
If You Are Listening To Music, Please Replace It With Qur'an.
If You Are Swearing A Lot, Talking Behind People's Back, Please Fix This And Stop.
If You Aren't Wearing Hijab, Please Do.
If Your Cloths Are Tight, Revealing, Non-Islamic, Please Winder And More Moddest Cloths.

If You Are Making "Gheebah" And Talk Behind Your Sisters Back, Please Stop.
What Are You Waiting For?? Sudden Death??
Where's NO WAY TO REPENT ?!!!
Fear ALLAH !
Wake UP ... REPENT !

Example 14: By Mamadou from EIC

It's good to dream....but better to be realistic
It's good to harry..... but better to go safe and slow
It's easy to start.......but hard to keep it up
It's better not to start something that you * not gonna End in a good way...

WITH YU GUYS.....JUST SOME WORDS ...UP FRV

Example 15: "Birth" by Hawra'a

She smiles at the man sitting across from her at the café.
Oh, what a beautiful man, she thinks.
Hunting down her next prey gives an exuberant feeling,
She examines him, to see if he fits the code.
Tall,
Muscular,
Handsome,
Is that a dimple?
A black haired, dark bearded creature, the perfect prey.
His big chest calls for her.
Yum, he should be a tasty one.
She goes over to talk to him,
And sooner rather than later, he is devoured.
The creases on her forehead tell the unsaid
Blood dripping
Love no longer matters
Life no longer matters
All she wants to do is rip his heart out and feed on it
Enjoying the taste of his blood, his flesh.
As tough as it is to chew on a muscle, she has managed with exaggerated movements of her jaw.
She chews and chews, then aches for more.
She licks her blood-covered lips as she smiles and thinks about how her plan never fails her,
Step one
Study him
Step two
Trap him using the one thing she will ultimately feast upon,

And then finally,
It's dinner time.
Oops,
She has devoured yet another one.
The taste of his blood
The texture of his heart on her tongue,
He was okay, next time with a side of veggies, though.

She moves on,
And on,
And on.
Her heart? Once as holy as the Black Stone, as sacred as its home.
Medusa's eyes got to it, though.
She would be proud.

A smile creeps on her lips as she envisions the next creature that will belong to her
The next person she is going to give the gift of life.
This is her way of giving Birth to these lifeless creatures.
This is her way of making their deaths meaningful.

Example 16: "Birth" by Shahd

I remember our first firsts
And that day you said "I can't love you" as easily as you said hello.
And I echoed that love was just another way teenagers labeled the bulges in their trousers and the spilled
    secrets under their t-shirts
I do not love you (because that's what you need to hear)
But I fell in love with you with the same intensity I fell in love with Bronte and Plath.
Those two madwomen bled life through me (and others, I know)
Just like you, you with your Inconsistency, one day breathing and the other bleeding into me.
You have a way with perfecting the plot- just like my dead writers.
You build me up and that makes me want to travel halfway across the world, just to kiss your vocal chords.
But you break me when you say "I can't love you, because there's no life here, only Death."

You say you are dead inside. That sentence stretches across my brain corners,
And I find the solution: that heart of yours must shed layers for me.
We put our hands together, my love and your patience, and sculpted you a new heart.
It beat slowly, tentatively at first.
I glued my head to your chest and heard it's first rhythmless beats.
And as I looked up into your eyes, they held me in place and asked me to stay.
This time, a new you was born
I have missed this you since day one.

Example 17: "Honesty" by Zahraa

Facing a problem is not that easy like the sea waves that can overcome all the difficulties. However, it is difficult like finding a treasure who will lend you a shoulder to cry on and listen to you. The sun of honesty got down when I was accused of being a cheater during the exam. One day, we were having exam. My partner who was sitting beside me was cheating when he has finished. After that, he threw the paper under my desk. The teacher saw it, and she thought that it was for me. Without listening me, she prohibited me from a dazzling activity and put for me zero. My friends left me alone whenever they knew that I am innocent . That time I felt like a sword who cried for revenge, and my dream of being the best of the best was shattered into pieces. I felt like an abandoned place having imbalanced life. After the long cherished path of that was full of difficulties, I decided to let the volcano that was in my heart erupt, and to take off my innocent face and then to shout and say: "Why you are punishing me taking in consideration that I can't do that ?" I started complaining and complaining, so the teacher suggested that she is going to recording taken by the camera, and if I were mistaken she would fire me from the class. I was lucky, for she found that I was innocent and not mistaken, and she gave me all my rights. To sum up, we must know that "HONESTY IS THE BEST POLICY," and we should tell the truth in order to be trusted.

Example 18: Are They Receiving Their Rights by Zahraa

The problem of pollution and global warming is increasing and there are several solutions to overcome them and give our planet its rights, for scientists are going into more depth in such problems; but think for a while Are they thinking how to make life easier for disabled people and accept them?

Now, as a responsible person, I will stand in front of the cage of disabilities and open the lock to let them fly freely in the wide sky. Just think! Why are not we accepting them? Look around and witness! Are all our schools accessible for such people? It's sad to say that we have been like animals living in the jungle . The strong one is taking his rights, but the weak on is being neglected. Let us take them from darkness into light.

Dear ladies and gentlemen, we were the ones who damaged their self-confidence, and we must be the ones who are going to rebuild it.

Chapter 2

Not Just a Pretty Face:
The Cost of Performative Beauty and Visual Appeal in Beauty Pageants

Mariana Bockarova
Harvard University, USA

ABSTRACT

The primary focus of the studies on adult beauty pageants involves their creation, negotiation, and implication vis-à-vis national and/or political identity within the pageant industry; or else they examine national pageants which are primarily scholarship-based. The present chapter is an attempt to understand the many psychological costs that come from practicing beauty within the realm of pageantry, and the rationale behind entering into an expensive venture for which there is little to gain, but much to lose emotionally. It will address two main questions: What are the physical and emotional (or metaphysical) costs of entering into and, later, winning, a beauty pageant? Who enters into beauty pageants and why? The objective is to examine the incentive to publicly parade oneself against dozens of other women, at the risk of simply being dismissed at the hands of quasi-objective opinion.

INTRODUCTION

You have to have a certain aesthetic for pageantry... it's so much more than the 'whole you' [that] you put on stage; [it's] what kind of work do you do philosophically, emotionally. It's a whole package that makes you a title holder...

— Denise, International Title Holder

Beauty pageants constitute an interesting phenomenon, both through a cultural and academic lens: Though they have been popularly dismissed as recursively archaic, barbaric, or, when observed with analytic scrutiny, are often slotted into 'sex work' or 'fashion modeling' for their obvious emphasis on exploiting their physical appearance (and therefore seemingly unworthy of further inquiry due to their apparent saturated nature of analysis), they are, in effect, ripe with opportunity for study beyond the

DOI: 10.4018/978-1-5225-5622-0.ch002

expected discourses of aesthetic capital (Latham, 1995; Brenner & Cunningham, 1992). In fact, unlike other industries which focus on the aesthetic appearance of the human form, beauty pageant participants often engage in a costly venture while attempting to win the short-lived fame, intensely laboring both their physical bodies and beyond, and creating a purposeful reflection of archetypal femininity, carefully negotiated toward an 'idealized femininity' in a modern time (Balogun, 2012).

Unlike the limited previous research conducted on adult beauty pageants, which focuses on the creation, negotiation and implication of national and/or political identity within the international pageant industry (Barnes, 1994; Rogers, 1998; Wu, 1997), or analyses which examine national pageants that are primarily scholarship-based (Banet-Weiser, 1999; Banet-Weiser & Portwood-Stacer, 2006; Gundle, 1997), the present paper is an attempt to understand the figurative costs associated with both defining and practicing beauty within the realm of international pageantry, and the rationale behind entering into an expensive venture for which there is exceedingly more to lose and little to gain. Specifically, the present paper will address two main questions: First, what are the physical and meta-physical costs of entering into and, later, winning, a beauty pageant? Second, who enters into beauty pageants and why? In other words, what incentive is there to publically parade oneself against dozens of other women to likely be dismissed at the hands of quasi-objective opinion, at one's own expense.

The basic plot of any international beauty pageant begins with a group of generally self-selected young women (under age 27) who have undergone a screening process by gatekeepers (pageant managers, talent agents, etc.) in order to enter into their local beauty competitions. Commonly, and as discovered through the present research, at this level, a so-called 'pageant girl' will depend on her own community for support in the form of: (a) finances (wherein she must gather sponsors to help her pay for her expenses, including travel, hotel stay, make-up and costume); (b) social upholding (as she must win over her community, generally via philanthropic efforts, in order to obtain popularity and incentivize a following to vote for her to win the 'popular vote,' which will fast-track her to a later round in the competition); and (c) emotional support (as she looks towards her community to provide her with encouragement and incentive to go on during particularly stressful and turbulent emotional experience, testing one's ability to manage the 'self'). If successful in the local competition, the typical 'pageant girl' will then progress to a larger pageant, usually at the state level, competing against winners of other local competitions, respectively. If successful yet again, she will compete bearing the name of her state, "Miss X", with 'X' denoting the name of the state, into a national competition, vying for her country's title, "Miss Y,' with 'Y' denoting the name of her country, alongside other state titleholders. With each subsequent win, the pageant girl receives greater attention, greater praise, and greater reward, with the ultimate reward being the crowned winner of an international title" 'Miss World,' 'Miss Universe,' 'Miss Earth,' or 'Miss International'. In this sense, beauty pageants are a superb example of the "winner-take-all" model often seen in sports and art, designed such that one clear-cut winner in a concentrated market receives disproportionate benefit from the win in comparison to all others involved (Frank and Cook, 1995).

Despite the categorical name of 'beauty,' however, modern pageants purport to judge their contestants on more than physical appearance, suggesting that beauty pageants form a 'third space' mediating judging criteria from what is biological fact (the objective value of physical appearance) with social approval (the status of women and what it means to be a woman beyond, or, in conjunction with, physical appeal). The judging criteria, which largely remains amorphous and not defined in particular detail on any official publically accessible websites, nevertheless includes some form of culturally established feminine values such as confidence, ambition, and philanthropic aptitude. To illustrate, for instance, the following is noted on the Miss Universe website:

The winner of the competition must be confident. She must understand the values of our brand and the responsibilities of the title. She must have the ability to articulate her ambition. A contestant should demonstrate authenticity, credibility and exhibit grace under pressure. The women who compete embody the modern, global aspiration for the potential within all women.

As such, beauty pageants, at least according to their mandate, are not simply a matter of outward appearance, but that of intense and focused performative "inner and outer" beauty and thus requiring an entire set of skill involved in optimizing their aesthetic capital, including: body work (needing to constantly be in intense scrutiny of their physical appearance, and working toward a certain end goal in terms of appearance); emotional labor (embodying the emotional work necessary to win over fellow contestants and judges); cultural labor (needing to identify and successfully perform current cultural expectations such as espousing values of confidence and success and new age femininity in order to further help produce the cultural work which is the pageant); and what I have dubbed 'archetypal labor' (working towards defining and embodying the ultimate figure of femininity) which amalgamates and forms the ideological bedrock from which all other 'work' can be performed.

THEORETICAL BACKGROUND: LABOR AND WORK

In her review of the sociological literature, Gimlin (2007), notes that 'body work' can be defined as "(i) the work performed on one's own body, (ii) paid labor carried out on the bodies of others, (iii) the management of embodied emotional experience and display, and (iv) the production or modification of bodies through work". Certainly, pageantry involves and comes into contact with all definitions of body work: pageant girls are expected to be of a certain aesthetic, which requires physical training, the application of make-up and false eye lashes in order to modify the face, the removal of body hair to modify a natural bodily appearance, the addition of hair extensions to modify the appearance of her hair, and, in some instances, plastic surgery to permanently modify herself as a whole for the purposes of winning a prize, usually involving a salary of sorts (in the form of sponsorship pay, prize money, and gifts). Lastly, it is of prime importance for the pageant girl to be cognizant of her public display of emotions, requiring constant maintenance whenever in the public sphere.

Popularized by Hochschild (1983), emotional labor can be defined as the regulation of one's emotions within the workplace, requiring face-to-face contact with a member of members of the public, in efforts to change his/her/their feeling state. While pageantry does not constitute 'work' in the traditional sense of the word (as there is no steady and set pay associated with entering into or winning a pageant, but, instead, prizes which could constitute as a delayed salary, psychologically), the term 'emotional labor' will be used for the purposes of this paper. Within emotional labor, Hochschild identified three strategies in which one can regulate his or her emotions, namely via cognitive, bodily, and expressive means. Cognitive regulation involves the changing of thoughts in order to alter associated feelings. For instance, within pageantry, while one may not be the ultimate winner of an international title, with a disappointing emotion ensuring, the pageant girl may alter her thought such that she is thankful to have won her country's title, in any case. Bodily regulation involves altering physical symptoms in order to change emotions, such as deep breathing when feeling nervous. Expressive regulation involves changing visible and expressive gestures in order to change one's feelings. Within pageantry, an example may be continuing to smile on stage, while not being called to the next round.

In a similar vein, cultural labor, while described by Hesmondhalgh (2013) as "primarily with the industrial production and circulation of texts," with texts understood as broadcasting, film, music, print and electronic publishing, video and computer games, advertising, marketing and public relations, and web design, certainly encompasses the experience of pageantry which involves broadcasting (as major pageants are often broadcast) and the marketing of both pageant and girl, this paper will use the definition given by Mosco and McKercher (2009), in which "anyone in the chain of producing and distributing knowledge products", which constitutes the pageant girl, not only the producers of the pageant, is in the process of creating and utilizing previous cultural work.

Lastly, it can be conceived that pageant girls participate in archetypal labor, which I will define as the 'work' involved in the identification and expression of an archetype, in the Jungian sense of the word. Jung believed that archetypes were primordial concepts which reside in the collective unconscious. Archetypes are often repeated in universal stories which exist without specific time or place. As Yuhua (2002) notes, archetypes "represent specific forms existing in chronological and/or geographical space, but are both direct projections of the human psychological landscape and images instrumental to its construction: they invariably manifest themselves in mythological form." Bolen (1984) defines seven feminine archetypes based on Greek mythology: Aphrodite (a seductive, erotic woman, using charm and obvious beauty to seduce); Persephone (a maiden archetype, who emphasizes compliance and passivity); Demeter (the mother archetype, representing nurturance and compassion); Hera (as a 'queen' represents loyalty and regality); Artemis (a hunter within mythology, represents strength, courage, and self-reliance); Athena (representing the pursuit of knowledge, level-headedness, and intelligence); and Hestia (who, as the protector of the hearth, represents wisdom and confidence). In this sense, pageant girls work, on both a conscious and unconscious level (inherent in archetypes) to identify and express one or more of these feminine archetypes, sometimes oscillating from one archetype to the next, as a strategy to increase her odds of winning.

WHY STUDY 'PAGEANT GIRLS'?

While fashion modeling has been extensively explored either through qualitative research on the fashion models themselves (Mears, X) or through the critical dissection of the static images within which fashion models appear (and then further scrutiny of the resultant commentary), beauty pageants and 'pageant girls,' overshadowed by the surface similarities of the two industries, rarely receive the critical attention that is warranted. Unlike modeling, pageantry involves a carefully constructed extremely public dominance hierarchy wherein which there can be one and only one 'winner' who reaps the greatest reward. Therefore, pageantry forms its own category within the intersection of beauty and sport; cultural competition which changes the expectations of womanhood through competitive femininity, all seemingly based on visual appeal.

METHODS

This study was conducted in attempt to understand pageant contestants in an explorative way. The present sample consisted of 45 pageant contestants, with 34 national (country) titleholders, 8 state titleholders and 2 international titleholders. The participants were aged between 19 and 25, and were from the United

States of America, Canada, Italy, Honduras, Spain, Nicaragua, Philippines, Venezuela, Nigeria, Ghana, Albania, Haiti, Curacao, Ethiopia, Ukraine, Russia, Gabon, Egypt, and Hungary. Some countries had multiple participants, as pageant winners from previous years. Once consent was obtained, all participants were assured of their anonymity and interviewed separately. Once each interview was recorded and transcribed, the original recording was destroyed.

Semi-structured, open-ended interview questions were employed, and follow-up questions were asked to encourage participants to further the initial data they provided, in order to ensure an accurate understanding of their lived experiences. The 'core' questions asked were as follows:

1. Why did you enter into beauty pageants?
2. How important are beauty pageants to you?
3. Can you describe the mental and/or physical preparation you have put into preparing for pageants?
4. How has winning your (local, national, international) title impacted your life?
5. How do you think winning your (national, international) title will impact your future?

COST OF BEAUTY: BODY WORK

The amount of work needed in order to 'transform' oneself into a pageant girl is often immense: She must undergo significant, and often, painful changes to her body, including daily physical exercise and training, the removal of body hair, eyebrow maintenance, engaging in spray tanning or bed tanning for a prolonged period of time, the maintenance of manicures and pedicures, the application of makeup, the application of hair extensions, the physical alteration of hair style via curling irons and straightening irons, and/or plastic surgery including breast implants, buttock augmentation, liposuction, rhinoplasty, jaw restructuring, skin resurfacing using lasers, botox, and fillers.

As Melissa, 24, national titleholder, explains:

Well, I had a personal trainer and I would be working out, lifting weights 4 days a week…. But now, I'm lifting weights to kind of create this image and tone up the body… I'm eating vegetables, eggs, fish, and chicken, and protein. It is actually the healthiest that I've ever been! And it's not—I am not depriving myself… You're bringing your body back into this mode of what is naturally—what it's supposed to be. We're working so hard, and its hours upon hours before our next meal and we're starving and our bodies start feeling depleted or our blood sugar level is so low or we are lacking energy.

As pageant girls tend to compete in multiple pageants for multiple years, their first year of competition is often entered into with little experience of body work, and they return to competitions with increased labor. As Denise, 25, an international title holder, explains:

…to compete again and again and eventually win the crown, girls improve themselves each time around. …with the [modeling] industry, it just—it wasn't the same kind of growth opportunity. I was personally turned off of modeling [for that reason] and my talent agent was like, "Oh! You should go back and do modeling!" And I'm like, "No." … I'm more of a pageant girl. I like bettering myself and going on stage and promoting who I have made myself to be instead of just promoting a product.

As such, the pressure on girls to pursue greater and greater means of modification, including plastic surgery, grows with experience. Natalie, a national titleholder, noted:

Some people [get plastic surgery]. Yup. That's something also interesting. I mean, Brazil—if you wanna get anything plastic surgery operated—go to Brazil. Like it's the best country for plastic surgery. But I think that's also like—you can tell. You know looking at pictures of contestants whose had augmentations, and corrections, face lifts and all that. It does happen. It's very frequent because everyone is trying to attain that like perfect Barbie image, right? I mean our director said that, the girl who did win Miss World last year is not a Barbie in her form and in her body and I think that's something interesting to look at...there's more to her. And that's true, but not really. You still have a lot more of aesthetic changes [compared to other contestants]. But there is an image that—I guess—you're trying to get to. Everyone wants to have that certain look that is that popular, nice look. But yeah, I don't know. I think—for me— the idea of [getting plastic surgery] is different because I only competed in that one competition but it can be a slippery slope. You can end up spending a lot of money on those types of things, which—at the end of the day—if you don't win, there's a lot of money down the drain. Really.

ARCHETYPAL LABOR

Interestingly, most contestants in beauty pageants look very similar, encompassing characteristics of feminine idealization including being slender yet still amplifying features of sexual dimorphism; displaying ample cleavage, with tight fitting clothing to draw attention to the waist. Pageant girls also disproportionately display their hair in a long and wavy manner, and are trained to walk and dance gracefully and swing their hips. In this sense, as this look is pervasive within the pageant industry, they employ an archetypal identity, borrowed from myth and previous pageant winners (which could constitute a 'cinderella' myth, of sorts) to further their chances of winning. As Amelia, 19, state titleholder explains:

Well, there were certain elements that I would take from previous title-holders when crafting my look. What I specifically look at are title-holders that I felt had certain features like mine and had certain qualities and characteristics like mine. And then I would use them as an example as well to kind of learn from and to emulate a certain look. I would look at a certain photo of a former title-holder that I felt had certain features like mine. I'm like, "She looks really great in the photo. So, what is the color combination that she is wearing because it obviously works?" And then, I kind of work towards myself. So, like, I know that would probably work for me because there are similar traits and [...] with the color of her skin and her eyes and her hair, and how similar it is to mine, I would take the look because that would also work on me. So, I use former title-holders as inspiration because that was like the overall by the look that girl who won, so I get the ideal look for pageantry but, I mean, it's not the only look. That's why I look at photos of title-holders that I felt had certain characteristics similar to mine and I would kind of use those as lessons for myself.

As Antonella, 23, national title holder, further explains:

Everyone in the pageants has long porn-star hair, super long eyelashes, extensions, big blinged out earrings. That's what we perceive as beautiful so we all try to look like that and end up in a straight line looking the same. On the world stage, you get a lot of that as well but now you're bringing someone from Kenya who is very different from the girl from Belgium. So you do get some cultural difference but it's very much the same at that point, the way the girls look. We all look like the perfect woman, except for Kenya... but it's not like she's going to win because she just doesn't look like what you expect from a pageant winner, that perfect woman. Maybe she's an interesting contestant, but not winner. She just doesn't look like one.

Similarly, Britney, 25, national title holder, also found that there was a certain look which was expected of the pageant girls:

I think for the most part that a lot of people looked similar to me in many ways and I was thinking "How are [the judges] going to decide?" because all of these girls look the same. That's a problem that we have in society too. I'm a walking example of that. I'm wearing a maxi dress and how many people wear maxi dresses? My hair is in a sock-bun; how many people—do you know what I mean? We all have iPhones, we're all at Starbucks. It's a reflection of who we are and I think beauty pageants are an extension of that, but, somehow more. Every [pageant girl] sees what the media portrays in magazines is what's good-looking and what's not, or who won last year and that's how we decide what we should look like.

Apart from the physical appearance of the feminine ideal, pageant girls also draw from archetypes in order to craft their image, whether fundraising at a local level and needing to win over sponsors, or creating public profiles for themselves on pageant websites:

As Amy, 19, state title holder explains:

So, there's girls that fundraise their butts off and only make a small amount of money but it's probably because no one believes in them and what they are about. Like, you can't say you care about kids and that your platform is going to be sick kids if you go on Facebook and call your younger brother stupid or say kids are dumb. It means you don't really care about kids. You have to be like what the image of someone who cares about kids is like for people to believe you and sponsor you...you know, like warm and friendly and nice and kind.

CULTURAL LABOR

Participating in the pageant industry as a pageant girl means not only utilizing previously defined cultural content in order to maximize the chances of winning, but producing cultural content as a key and necessary figure in order for the pageant itself to come to realization, which can be a point of pride for pageant contestants:

As Melissa, 24, national titleholder, explains:

Coming from me, personally, and I have this internal struggle all the time coming from the educational background that I do have, entering into [a beauty pageant] like this, you know and I think, "That's kinda off!" like you know—"it's kinda weird you did that". But I guess I get to be a part of something bigger, like my face will always be Miss [X]. It feels good.

As Denise, 25, international titleholder, explains:

I get to do something that so few women get to do with their lives and be sort of the unifying force for all of these different countries and cultures. Like anyone can look up who Miss [Y] was and admire her as someone greater than everyone else, even herself, and I get to be that, forever. It's crazy!

Beyond the participation in cultural production, pageant girls also utilize traits of femininity that are embedded within the current culture climate, that are then identified, defined, analyzed, utilized and amplified in the pageant world:

As Samantha, 20, national titleholder, explains:

Thankfully, with Miss [Y], I didn't need to be too sexy in that competition. Partly because I don't think it's necessarily part of their image they want to portray. We were in Indonesia, which is a Muslim country. So going there was interesting and of course beauty pageants are famous for their swimsuit competition and there was no swimsuit competition. We did a sarong competition. So they wrapped up in five layers of sarong. So I knew for the whole competition, because of where we were and who the judges were, they sort of wanted a more modest girl to win. And so that's what I became. Like I changed my walk, the way I flirted, everything…When they picked the top ten, they were very modest, like how they acted.

Similarly, Rebecca, 23, national titleholder, felt that she had to mold herself into what the pageant culture demanded:

If I were a judge—and it's so hard, like you said they have a short glimpse of a second—I would look for someone whose first impression to me is strong. But I don't think that's what like Miss [Y] for example is sorta what they're looking for. I don't think they're looking for someone who is very strong. I think they are look for someone who is more docile, a little more bendable, malleable is a good word. So that they can mold them into the figure that they want them to be….Yes, that's what I tried to be.

Amelia further noted:

The interesting thing about [girls in the competition] is that they all portray this image that they're all strong individuals that they have this confidence, they have that and when they're trashed—for the lack of a better word—everything falls apart and it's like, "What happened to that strong independent person that you told me you were". So, again, it's those characteristics or things they think people want to hear that make them seem like the ideal candidate to win that is what's being portrayed—that's not what's being portrayed!

EMOTIONAL LABOR

Throughout the pageant competition, which generally begins 1-2 weeks before the pageant, there is little time to be spent on activities done in private. Usually, pageant girls share their bedrooms with other girls, at the local or even state-level. They usually eat breakfast with other contestants first thing in the morning, do public appearances or rehearse for the competition, which usually involves a number of dances and stage positions, and meet with media for press interviews, all which require an intense amount of emotional labor. Contestants conduct archetypal labor (for instance, if she is attempting to embody the Aphrodite archetype, she will be flirtatious and seductive throughout all non-private engagements), and also cultural labor (meeting the expectations of what it means to be a beauty queen) throughout all public appearances, requiring her to maintain an emotional image, despite the grueling hours of daily work which goes into pageant activities and the exhaustion that is felt.

Christine, 23, national titleholder explains:

The whole thing is so tiring but you can never show that. You can absolutely never show how tired you are or how upset you are, or that you just want to sleep in and eat something. You also are evaluating yourself against everyone else all the time but you can't show it. Like, I think blonde hair and blue eyes is very ideal, so I thought that Miss [X] was going to win but I couldn't show that at all. I just had to look so fierce, like no one could touch me, like I'm the best.

Cindy, 23, international titleholder, adds:

With the ideal pageant girl...there's a confidence when she walks on stage and it's that charisma that will then cause a shift in the judge's mind. If you're torn between two girls because they both have this look based on your preference on what beauty is but then if the girl that has that extra confidence and that connection with the judges ...that's where I think the scale is then tipped The audience member feels it too; they see it, they watch it on television. Sometimes [the audience] thinks they have a favorite based on her photo, but once they see these girls in action, like on the stage, it's like, "Oh, I really liked that girl but watching her on stage, she just didn't have it or she was very dull! But this girl—I never thought I saw it coming! I really loved her because it's the personality and that confidence." And that's when the scale gets tipped. That's what you have to aim for, to be the best and brightest, the one they will see from a mile away.

Jasmine, 19, state titleholder, further explains:

You always have to smile. People said to me, "Your smile! Your smile! Everyone seems to notice your smile!" and I'm like, "Okay! I'll play up my smile!" ...When you have a smile, people tend to be a lot more attracted to you because it's a friendly expression and it shows you're congenial and welcoming... so you always have to have it on...It's hard and tiring, but you can't stop.

REASONS FOR ENTERING INTO A PAGEANT

While this section warrants further exploration from a psychological viewpoint, three main themes were synthesized based on the participants; responses to why they had entered into the pageant, and further probing. The responses include either winning the pageant being the only chance for a monumental shift in their current life situation, via (a) in the self, by wanting to prove to themselves that they are of a certain appeal through mass social approval), or (b) to improve their financial situation (which is exemplified by the 'Cinderella' myth, whereby a poor girl in dire economic situation becomes a princess based on her beauty and natural goodwill. Lastly, some girls simply wanted to compete in pageants as a new opportunity which would beget further opportunities in the beauty industry.

In terms of a shift of the self via mass approval, several girls noted being bullied in their younger years, and wanting to prove to others that they were beautiful. Denise, 25, international titleholder, explains:

[I differentiated myself on stage] by being proud of who I was. I was bullied a lot as a kid and so I worked really hard to change. I could've allowed external voices that were telling me, "Denise, you're too big, your knees look too muscular, it makes you look too athletic," to allow me to put myself down and then to lose my confidence and the moment you start losing it, it's a slippery slope. [Voice breaking] I just have to proud. I can't change who I am anymore. This is what I am, so I have to just let it shine and I'll have the judges know that I'm proud of it and I love it as well, and then the audience will see that too. So, and that's what I did.

With regard to pageants being the only change to change one's financial situation, Julia, 18, noted:

I live in a poor country. With pageants, I meet with a lot of interesting and beautiful people, with whom I became friends with. I also got amazing prices which included a car, which I gave to my dad and $100,000 which I spent to help my parents buy a new apartment. I couldn't do that if I wasn't Miss [Y] competing in this pageant.

Lastly, some pageant girls simply use beauty pageants as a way to further showcase themselves, such as the case of Rachel, 20, state titleholder:

Why pageants? Okay. So I never competed in a pageant before. This was my first one. And I decided to enter it because when I read it online, it did seem like interesting to do. I've been a dancer my whole life and competed in dance competitions. So, to me, it was almost an extension of that. In the sense that I got to draw all of the fun things that you see about beauty pageants—like wear pretty dresses and all that kind of stuff. But I also thought that the mantra of it—the mandate—seemed up my alley because I was doing a degree in human justice and social justice change and all that kind of stuff was kinda something I was interested in and could maybe use the title for, but also, really, I could find an avenue to pursue dance if I did win that competition.

DISCUSSION AND CONCLUSION

As international phenomena, beauty pageants mark an important cultural experience, which is projected onto society in multiple levels. Nevertheless, few studies have researched the pageant contestant themselves, apart from political motive or that of charity. The reason for this study was to explore the cost of being beautiful, in a sense, within the pageant industry. As has been found in this study, pageant girls typically embark on heavy body work, including the achievement and maintenance of an ideal form. This has also been found in Mears' (2009) work fashion models, wherein she notes:

At a minimum, models need to conform to general norms of conventional attractiveness, such as symmetrical features, clear skin, and healthy teeth, as well as to the modeling industry's specific requirements for height, weight, and bodily shape. Beyond these standards, however, what makes a model's appearance right for a particular advertising campaign or a particular client becomes somewhat variable. It depends on current fashion, the market that the advertiser has targeted, and the client's individual taste and preferences (p. 327).

Unlike models, pageant girls, as discovered, do not target their look according to the client's individual taste, but rather, they use an idealized feminine form of physical appearance, which is naturally occurring but culturally mediated within the expected look of a pageant winner, in order to comply with the desired appeal. This typically includes a slender figure, exaggerated eye lashes, and long hair.

Further, as discovered through this study, pageant girls employ significant techniques of emotional labor in order to increase their chances of winning. Unlike what typically defines labor, in that the laborer is paid a salary, the pageant girl embarks on emotional labor in order to have the chance of winning a prize which generally equates to financial gains. Despite negative feelings, contestants nevertheless continuously feel the need to smile despite facing rejection (not being chosen to compete in the next round), or nervousness (in the face of competition). This is similar to the experience of fashion models (Mears, 2009), in needing to manage "their own feelings to create a desired facial and bodily display for those watching them".

Lastly, and perhaps most intriguingly, pageant contestants employ archetypal labor, often attempting to embody one archetype, or multiple at different times, in order to maximally influence their chances of winning the pageant based on the desires of the specific audience at hand. This then manifests itself via both emotional and cultural labor, aside from body work. In sum, pageants are an important cultural experience which require significant effort on the part of the contestant, from which general sociological and social psychological knowledge can be extrapolated and warrants further exploration.

REFERENCES

Balogun, O. M. (2012). Cultural and cosmopolitan: Idealized femininity and embodied nationalism in Nigerian beauty pageants. *Gender & Society*, *26*(3), 357–381. doi:10.1177/0891243212438958

Banet-Weiser, S. (1999). *The most beautiful girl in the world: Beauty pageants and national identity*. Berkeley, CA: University of California Press.

Banet-Weiser, S., & Portwood-Stacer, L. (2006). 'I just want to be me again!' Beauty pageants, reality television and post-feminism. *Feminist Theory*, *7*(2), 255–272. doi:10.1177/1464700106064423

Barnes, N. B. (1994). Face of the nation: Race, nationalisms and identities in Jamaican beauty pageants. *The Massachusetts Review*, *35*(3/4), 471–492.

Brenner, J. B., & Cunningham, J. G. (1992). Gender differences in eating attitudes, body concept, and self-esteem among models. *Sex Roles*, *27*(7), 413–437. doi:10.1007/BF00289949

Frank, R. H., & Cook, P. J. (2010). *The winner-take-all society: Why the few at the top get so much more than the rest of us*. New York: Random House.

Gimlin, D. (2007). What is 'body work'? A review of the literature. *Sociology Compass, 1*(1), 353–370. doi:10.1111/j.1751-9020.2007.00015.x

Gundle, S. (1997). *Bellissima: Feminine beauty and the idea of Italy*. New Haven, CT: Yale University Press.

Hesmondhalgh, D. (2013). *Why music matters*. John Wiley & Sons.

Hochschild, A. (1983). *The managed heart*. Berkeley, CA: University of California Press.

Latham, A. J. (1995). Packaging woman: The concurrent rise of beauty pageants, public bathing, and other performances of female "nudity.". *Journal of Popular Culture, 29*(3), 149–167. doi:10.1111/j.0022-3840.1995.00149.x

Mosco, V., & McKercher, C. (2009). *The Laboring of communication: Will knowledge workers of the world unite?* Lanham, MD: Rowman & Littlefield.

Rogers, M. (1998). Spectacular bodies: Folklorization and the politics of identity in Ecuadorian beauty pageants. *Journal of Latin American and Caribbean Anthropology, 3*(2), 54–85. doi:10.1525/jlat.1998.3.2.54

Wu, J. T. C. (1997). "Loveliest daughter of our ancient Cathay!:" Representations of ethnic and gender identity in the Miss Chinatown USA beauty pageant. *Journal of Social History, 31*(1), 5–31. doi:10.1353/jsh/31.1.5

Yuhua, X. (2002). Shu: Naxi nature goddess archetype. *Gender, Technology and Development, 6*(3), 409–426. doi:10.1177/097185240200600305

Chapter 3
Riddle of the Sphinx Revisited:
Self-Referential Twists to an Ancient Myth

Terry Marks-Tarlow
Insight Center, USA & Italian Universita Niccolo Cusano London, UK

ABSTRACT

Myth is a universal conveyor of culture whose stories capture the human heart and whose embodied set of guidelines serve to conduct everyday life. When Freud added the Oedipus myth to his theory of psychosexual development, his method of psychoanalysis subsequently launched worldwide. Whereas Freud viewed the myth of Oedipus quite literally as a prohibition against infanticide, patricide, and incest, this chapter views the myth more metaphorically to examine how the riddle of the Sphinx informs self-referential thinking as a collective stage of human consciousness. Two contemporary theoretical lenses are adopted: 1) interpersonal neurobiology, which proposes that mind, brain, and body develop from relational origins, and 2) second-order cybernetics, which examines how observers become entangled in their very processes under observation. From within these perspectives, the Sphinx's riddle appears as a paradox of self-reference whose solution requires humankind to leap from concrete to metaphorical thinking. Only upon retaining recursive loops in consciousness can humans attain full self-reflection as a beacon towards full actualization.

INTRODUCTION

It is all imaginary and only the imaginary is real! (Louis Kauffman)

During ancient times, myths were passed along as teaching tales told from generation to generation. Yet, for most of contemporary Western society, it is not ancient tales but rather modern science and math that predominantly guide the way. Although ancient tales—of Greek heroes and Gods, of Buddha, Arjuna, and the Ramayana—are still around and sometimes make their mark as animation features, these figures have largely fallen into collective shadows. Especially in written form, the classics easily lose their luster compared with the bright icons and shiny features of computers, iPads, tablets, and other digital devices.

DOI: 10.4018/978-1-5225-5622-0.ch003

Our collective excitement has been drawn more toward science, perhaps because of its concrete power to transform information, communication, and the general quality of life. Science and especially physics comprise our culture's contemporary creation mythology (Marks-Tarlow, 2003). Whereas the 19th century Newtonian model of physics separated observers cleanly from the realm of the observed, 20th and 21st century models offer inner and outer worlds more fully and reflexively blended (see Orsucci & Sala, 2008; 2012). In gaming technologies, virtual avatars take the place of real bodies, while in medical research, thoughts drive prosthetic limbs (Peck, 2012).

Of all forms of contemporary science, inner and outer worlds appear blend in fantastic, even surreal ways, within quantum physics. Quantum entanglement, nonlocality, and the uncertainty paradox are just a few ideas that shake our sense of ordinary reality to the core. This is the stuff of modern fairytales, a good example of which is the book, *Alice and the Quantum Cat* (Shanley, 2011). Written in the tradition of Martin Gardner (1999), author of *The Annotated Alice*, Shanley introduces his book as "A Twenty-First Century Myth." Its chapters are written by physicists, e.g., Amit Goswami (e.g., 1995) and Fred Alan Wolf (e.g., 1995), and chaos and complexity theorists, e.g., John Briggs and David Peat (e.g., 2000), who regularly popularize physics in service of new ways to think, see, and be in the world. The book's main character, the Quantum Cat, is a blend of the Cheshire Cat, whose smile appears out of nowhere, and Schrödinger's Cat, who embodies the quantum paradox of existing and not existing simultaneously. With Alice as his sidekick, the Quantum Cat battles a sterile, Newtonian, mechanistic world, where observers and observed are so antiseptically separated as to threaten their very aliveness:

In Newton's world, ambiguity was the enemy—mechanism stresses the absolute, the unchanging and the certain—things are 'either/ or,' 'good/bad.' In the quantum world reality is 'both/and'—a coexistence of mutually contradictory possibilities, all equally true, each one a potentially possible constituent of reality. Acausal, non-local synchronicities can give rise to events that seem to 'pop-up' out of thin air. There are no isolated, separate, closed systems in Nature. In this universe of wholeness, everything affects everything else, from the most fundamental particles to faraway galaxies at the edge of the universe.

The central theme of Shanley's quantum tales is the Observer Effect, through which the awareness of observers forms deep, invisible foundations for material existence. With observers and observed intertwined to the point of full interpenetration, this world view implies a radically relational perspective. Here it becomes absurd to try to parse out isolated elements, people, or traditional concepts of cause and effect. Much akin to the worldviews revealed by Maya's veil within Hinduism or the Indra's net within Buddhism, the appearance of observers as separated from observed is mere illusion, born of evolutionary needs for survival. And so mythology of contemporary science dovetails with ancient mystical and spiritual traditions the world over (Marks-Tarlow, 2003; 2008a).

Just as The Quantum Cat uses science to illustrate microscopic truths, so too does this paper use science to reveal deep truths implicit within the neurobiological weave of our social and relational worlds. After a section on myth broadly, I review the myth of Oedipus by exploring the Riddle of the Sphinx as a paradox of self-reference. I claim Oedipus is uniquely positioned to answer the riddle because of his own traumatic origins. Within Oedipus's relentless search for truth, we can see how recursive, self-referential loops in consciousness increase cognitive capacity, enabling the leap from concrete to metaphorical thought. By using second-order, cybernetics to explore the dynamic, embodied unconscious of Oedipus, observer and observed remain hopelessly entangled, here at macroscopic levels of body, brain, and relationship.

THE SELF-REFERENTIAL ROLE OF MYTH

Throughout history, mythology has inspired the psychology of everyday life at implicit levels. Myths help to organize cultural categories and mores by supplying archetypal stories with roles, rules, and relationships that are prescriptive. Speculation exists that ancient and traditional peoples experienced myths quite literally (e.g., Jaynes, 1976), with people hearing the voices of Gods as if from the outside, speaking to them personally from above in order to guide the behavior of mortals below. Over time, people have come to hold myths more metaphorically, where they serve the role of "as-if" tales that point toward universal themes, predicaments, and solutions. Finally, over recent decades, the social sciences, particularly psychology, have shifted focus to view myths in increasingly symbolic and self-referential terms. Especially since Jung's ground breaking work (e.g., 1961), contemporary analyses examine myths as they illuminate the inner world and culture of the mythmakers themselves.

One myth rises above all others to signal entry into modern consciousness, the ancient Greek tale of Oedipus. This story has been analyzed throughout the millennia by well-known thinkers, such as Aristotle, Socrates, Nietzsche, Lévi-Strauss, Lacan and Ricoeur. Some (e.g., Lévi-Strauss, 1977; Ricoeur, 1970) have understood the myth as the individual quest for personal origins or identity. Others (e.g., Aristotle, 1982, Nietzsche, 1871/1999) have used sociopolitical and cultural lenses to focus on the tale's prohibitions against the very taboos it illustrates. Indeed, this cautionary tale's prohibitions against infanticide, patricide and incest helped to establish the modern day state. This was accomplished partly by erecting boundaries to protect society's youngest and most vulnerable members, and partly by prohibitions serving as a kind of social glue to bind individuals into larger collective units. Evolutionarily, these prohibitions have prevented inbreeding, while maximizing chances for survival and healthy propagation within the collective gene pool.

Perhaps the most prominent analyst of the Oedipus myth has been Sigmund Freud. At the inception of psychoanalysis, this myth proved central to Freud's psychosexual developmental theory as well as to his topographical map of the psyche. That this tragic hero killed his father and then married and seduced his mother occupied the psychological lay of the land, so to speak, immortalized as the "Oedipal complex." Whereas Freud (1900) viewed the myth quite literally, in terms of impulses and fantasies towards real people, his successor Jung (1956) interpreted it more symbolically, in terms of intrapsychic aspects of healthy individuation.

My main purpose in revisiting early origins of psychoanalysis that pivot around the Oedipus myth is to re-examine the narrative from a second-order cybernetic point of view. Cybernetics is the study of information; second-order cybernetics views information science self-referentially by implicating the observer within the observed (see Heims, 1991). From the vantage point of self-reference, the Oedipus story yields important clues about how the modern psyche became more complex via recursive loops in consciousness. Such internal feedback loops in body, brain, and mind allow implicit memories to become explicit, leading to an increased, more complex capacity for self-reflection.

In sections to follow, I refresh the reader's memory first by briefly reviewing the Oedipus myth. Then I reason to a new level of abstraction, by applying the approach of Lévi-Strauss to treat the myth structurally. I view the Sphinx's riddle as a paradox of self-reference and argue that both the riddle of the Sphinx and the life course of Oedipus bear structural similarities that signify the self-reflective search for origins. I examine the shift from a literal Freudian interpretation to a more symbolic Jungian one within the early history of psychoanalysis and then show how Freud's interest in the Oedipus myth was itself self-referentially re-enacted in real life through his struggles for authority with Carl Jung. Next I follow Feder

(1974) to examine Oedipus clinically. Oedipus' relentless search for the truth of his origins, combined with his ultimate difficulty accepting what he learns, appears at least partly driven by psychobiological symptoms of separation and adoption trauma combined with the physical abuse of attempted murder by his biological father. In the process, I link contemporary research on the psychoneurobiology of implicit versus explicit memory with a cybernetic perspective and the power of Universal Turing Machines able with full access to implicit and explicit memory. Finally, I claim that affective, imagistic, and cognitive skills necessary to move developmentally from concrete to metaphorical thinking, and eventually to full self-actualization, relate to implicit cognition within Lakoff's (1999) embodied philosophy as well as to mature, abstract cognition within Piaget's (e.g., Flavell, 1963) developmental psychology. Recursive loops in consciousness, by which the observer can be detected within the observed, signal enhanced internal complexity (Marks-Tarlow, 2008a, 2012, 2015) and the power of self-reflection to break inter-generational chains of abusers unwittingly begetting abusers.

Please note that although I refer to Sigmund Freud amply throughout this paper, my purpose is primarily historical and contextual. I do not intend to appeal to Freud as the ultimate authority so much as the originator of psychoanalysis and precursor to contemporary thought and practice. Especially since Jeffrey Masson (1985) documented Freud's projection of his own neuroses onto his historical and mythological analyses, including the invention of patients to justify his theories, Freud largely has been de-centered, if not dethroned, within most contemporary psychoanalytic communities. Yet, contemporary neuropsychoanalysis reinstates some of Freud's early claims about the nature of the human unconscious (Schore, 2011; 2012; Solms, 2004; Solms & Turnbull, 2002). Meanwhile, through lens of interpersonal neurobiology, one of the implicit themes drawn out by the myth of Oedipus highlights intersubjectivity as adopted by more present day forms of relational psychoanalysts (e.g., Bromberg, 1998; Mitchell, 1988; Stern, 1983). Along with revealing roots of these contemporary trends, I hope my reading of Oedipus helps to reinstate the majesty of this myth to the human plight, without sacrificing the many gains and insights gleaned by psychoanalysts and other psychotherapists since Freud's time.

THE MYTH OF OEDIPUS

There is an ancient folk belief that a wise magus can be born only from incest; our immediate interpretation of this, in terms of Oedipus the riddle solver and suitor of his own mother, is that for clairvoyant and magical powers to have broken the spell of the present and the future, the rigid law of individuation and the true magic nature itself, the cause must have been monstrous crime against nature—incest in this case, for how could nature be forced to offer up her secrets if not by being triumphantly resisted—by unnatural acts? (from Frederick Nietzsche's The Birth of Tragedy)

In the myth of Oedipus, which dates back to Greek antiquity, King Laius of Thebes was married to Queen Jocasta, but the marriage was barren. Desperate to conceive an heir, King Laius consulted the oracle of Apollo at Delphi, only to receive a shocking prophecy—the couple should remain childless. Any offspring of this union would grow up to murder his father and marry his mother. Laius ordered Jocasta confined within a small palace room and placed under strict prohibitions against sleeping with him.

But Jocasta was not to be stopped. She conceived a plot to intoxicate and mate with her husband. The plot worked, and a son was born. Desperate once again to prevent fulfillment of the oracle, Laius ordered that the boy's ankles be pinned together and that he be left upon a mountain slope to die. But the shepherd who was earmarked to carry out this order took pity on the boy and delivered him instead to yet another shepherd. This second shepherd brought the wounded boy to King Polybus in the neighboring realm of Corinth. Polybus, who suffered from a barren marriage, promptly adopted the boy as his own. Due to his pierced ankles, the child was called "Oedipus." This name, which translates either to mean "swollen foot" or "know-where," is telling, given Oedipus' life-long limp plus his relentless search to "know-where" he came from. I return to the self-referential quality of Oedipus' name in a later section.

As Oedipus matured, he overheard rumors that King Polybus was not his real father. Oedipus was eager to investigate his true heritage, and unwittingly following in the footsteps of his biological father, he visited the oracle at Delphi. The oracle grimly prophesized that Oedipus would murder his father and marry his mother. Oedipus was horrified by the prophecy; much like his biological father before him, he attempted to avoid this fate. Still believing Polybus his real father, Oedipus decided not to return home. Instead, he took the road from Delphi towards Thebes, rather than back toward Corinth.

Unaware of the underlying situation, Oedipus met his biological father, who appeared to him as a stranger at the narrow crossroads of the three paths both separating and connecting the cities of Delphi, Corinth and Thebes. King Laius ordered the boy out of the way in order that royalty may pass. Oedipus responded that he himself was a royal prince of superior status. Laius ordered his charioteer to advance in order to strike Oedipus with his goad. Enraged, Oedipus grabbed the goad, in the process striking and killing Laius plus four of his five retainers. This left a single witness to tell the tale.

Upon Laius' death appeared the Sphinx, a lithe monster perched high on the mountain. This creature possessed the body of a dog, the claws of a lion, the tail of a dragon, the wings of a bird and the breasts and head of a woman. The Sphinx began to ravage Thebes, stopping every mountain traveler attempting to enter the city unless they solved her riddle:

What goes on four feet in the morning, two at midday and three in the evening?

Whereas the priestess of the Oracle at Delphi revealed a glimpse of the future to her visitors, often concealed in the form of a riddle, the Sphinx, by contrast, killed anyone unable to answer her riddle correctly. The Sphinx either ate or hurled her victims to their death on the rocks below. Until the arrival of Oedipus, the riddle remained unsolved. With no visitors able to enter the city, trade in Thebes had become strangled and the treasury depleted. Confronted by the Sphinx's riddle, Oedipus responded correctly and without hesitation, to indicate that it is "mankind" who crawls on four legs in the morning, stands on two in midday and leans on a cane as a third in the twilight of life. The Sphinx was horrified at being outwitted, and responded by self-referentially applying the punishment she had meted out to others to herself: she cast herself to her death on the rocks far below. As a consequence, Thebes was freed. As reward for saving the city, Oedipus was offered its throne plus the hand of the deceased king's widow Jocasta. Still unaware of his true origins, Oedipus accepted both honors. He ruled Thebes and married his mother, with whom he multiplied fruitfully. In this manner, Oedipus fulfilled the second part of the oracle.

But the city of Thebes was not finished suffering and soon was stricken with a horrible plague and famine, rendering all production barren. Out of eagerness to end the affliction, Oedipus once again consulted the oracle. This time, he was told that in order to release Thebes from its current plight, the murderer of Laius must be found. Because he wanted only what was best for the city, Oedipus relentlessly pursued a quest for truth. He made an important declaration: whenever Laius' murderer was found, the offender would be banished forever from Thebes.

In line with his blind search for Laius' murderer, Oedipus called in the blind prophet Tiresias to help. But Tiresias refused to reveal what he knew. In the meantime, Jocasta intuited the truth and dreaded the horror of her sins exposed. Unable to bear what she saw, Jocasta committed suicide by hanging herself. Soon Oedipus discovered that the one he sought was none other than himself. After learning that he had indeed fulfilled the Oracle by murdering his father and marrying his mother, Oedipus was also unable to bear what he saw. Tearing off a brooch from Jocasta's hanging body, Oedipus blinded himself. He then faced the consequence that he himself had determined most just. As Laius' banished murderer, Oedipus was led into exile by his sister/daughter Antigone.

Here ended the first of Sophocles' tragedies, "King Oedipus." The second and third of this ancient Greek trilogy, "Antigone" and "Oedipus at Colonus," detail Oedipus' and his sister/daughter's extensive wanderings. Oedipus' tragic insight into unwittingly having committed these crimes of passion brought the mature man eventually out of suffering and into wisdom. In later years, Oedipus reached a mysterious end in Colonus, near Athens, amidst the utmost respect from his countrymen. Despite his sins, Oedipus ended his life with the blessings of the Gods. In completion of one more self-referential loop, Oedipus' personal insight in-formed the very land itself, as Colonus became an oracular center and source of wisdom for others.

NEW TWISTS TO AN ANCIENT MYTH

To Freud, the tale of Oedipus was initially conceived in terms of real sexual and aggressive impulses towards real parents. Later, he revised his seduction theory, by downplaying incestuous desires to the level of fantasy and imaginary impulses. Within Freud's three-part, structural model of the psyche, the Id was the container for unbridled, unconscious, sexual and aggressive impulses; the Super-Ego was a repository for social and societal norms; and the Ego was assigned the difficult task of straddling these two inner, warring factions, by mounting defenses and mediating the demands and restrictions of outside reality. We easily detect the influence of Freud's military background within metaphors he chooses to detail his conflict model of the psyche (Berkower, 1970).

According to Freud, symptoms formed out of the tension between conscious and unconscious factors, including conflicting needs both to repress and express. Among many different kinds of anxiety Freud highlighted, an important symptom was castration anxiety. This was the fear that one's incestuous desire for one's mother would be discovered by the father and punished by him with castration. Both desire for the mother and fear of castration became sources of murderous impulses towards the father. Working through these feelings and symptoms consisted of lifting the repression barrier and thereby gaining insight into the unconscious origins of the conflict.

Note that Freud's developmental model of the psyche was primarily intrapsychic. Because he emphasized the Oedipal complex as a Universal struggle within the internal landscape of all (the adaptation for girls became known as the "Electra" complex, in honor of another famous Greek tragedy), it mattered little how good or bad a child's parenting. Most contemporary psychoanalytic theories, such as object relations (e.g., Klein, 1932), self-psychology (e.g., Kohut, 1971), intersubjectivity theory (e.g., Stolorow, Brandchaft & Atwood, 1987), and relational psychoanalysts (e.g., Bromberg, 1998; Mitchell, 1988; Stern, 1983), have abandoned the importance of the Oedipus myth partly by adopting a more interpersonal focus. Within each of these newer therapies, psychopathology is believed to develop out of real emotional exchanges (or the absence of them) between infants and their caregivers. Symptoms are maintained, re-enacted, and ideally altered within the relational context of the analyst/patient dyad.

LIFE IMITATING THEORY

Prior to these relational theories, near the origins of psychoanalysis, the myth of Oedipus took on an ironic, self-referential twist when it became embodied in real life. Carl Jung, a brilliant follower of Freud, had been earmarked as the "royal son" and "crown prince" slated to inherit Freud's psychoanalytic empire (see Jung, 1961; Kerr, 1995; Monte & Sollod, 2003). The early intimacy and intellectual passion between these two men gave way to great bitterness and struggle surrounding Jung's creative and spiritual ideas. In his autobiography, Jung (1961, p. 150) describes Freud as imploring: "My dear Jung, promise me never to abandon the sexual theory. This is the most essential thing of all. You see, we must make a dogma of it, an unshakable bulwark…against the black tide of mud…of occultism."

For Jung, Freud's topography of the psyche maps only the most superficial level, the "personal unconscious," which contains personal memories and impulses towards specific people. Partly on the basis of a dream, Jung excavated another, even deeper, stratum he called the "collective unconscious." This level had a transpersonal flavor, containing archetypal patterns common in peoples of all cultures and ages. By acting as if there was room only for what Jung called the "personal unconscious" within the psyche's subterranean zone, Freud appeared compelled to re-enact the Oedipus struggle in real life. He responded to Jung as if to a son attempting to murder his symbolic father. This dynamic was complicated by yet another, even more concrete, level of enactment: both men reputedly competed over the loyalties of the same woman, initially Jung's patient and lover, later Freud's confident, Sabina Speilrein, (see Kerr, 1995).

Freud and Jung acted out the classic Oedipal myth at multiple levels, with Jung displacing Freud both professionally (vanquishing the King) and sexually (stealing the Queen). An explosion ensued when the conflict could no longer be contained or resolved. As a result, the relationship between Freud and Jung severed permanently. Jung suffered what some believe was a psychotic break (see Hayman, 1999), while others termed it a "creative illness" (see Ellenberger, 1981), from which he recovered to mine the symbolic wealth of his own unconscious.

Jung overcame his symbolic father partly by rejecting the Oedipus myth in favor of Faust's tale. "Jung meant to make a descent into the depths of the soul, there to find the roots of man's being in the symbols of the libido which had been handed down from ancient times, and so to find redemption despite his own genial psychoanalytic pact with the devil" (Kerr, 1995, p. 326). After his break with Freud, Jung self-referentially embodied his own theories about individuation taking the form of the hero's journey. Whereas Jung underscored the sun-hero's motif and role of mythical symbols, mythologist Joseph Campbell (1949/1973) differentiated three phases of the hero's journey: separation (from ordinary consciousness), initiation (into the night journey of the soul) and return (integration back into consciousness and community). This description certainly fits Jung's departure from ordinary sanity, his nightmarish descent into haunting symbols, if not hallucinations, and his professional return to create depth psychology. Jung's interior descent and journey is chronicled in writing and pictures in the Red Book. Although this 2005 journal was written between 1914 and 1930, following Jung's fallout with Freud, it was released publically only in 2009, due to decades of suppression by Jung's heirs.

Jung and his followers have regarded the Oedipus myth less literally than Freud. In hero mythology, as explicated by one of Jung's most celebrated followers, Eric Neumann (1954/93), to murder the father generally and the King in particular, was seen as symbolic separation from an external source of authority, in order to discover and be initiated into one's own internal source of guidance and wisdom. Whereas Freud viewed the unconscious primarily in terms of its negative, conflict-ridden potential, Jung recognized the underlying universal and positive potential of the fertile feminine. But in order to uncover this positive side, one first had to differentiate and confront the destructive shadow of the feminine. At the archetypal level, some aspects of the feminine can feel life threatening. To defeat the Sphinx was seen as conquering the Terrible Mother. In her worst incarnation, the Terrible Mother reflected the potential for deprivation and destructive narcissism within the real mother. In some cultures, e.g., the Germanic fairytale of Hansel and Gretel, the Terrible Mother appeared as the Vagina Dentate, or toothed vagina, a cannibalistic allusion not to the Freudian fear of castration by the father, but rather to the Jungian anxiety about emasculation by the mother.

Symbolically, once the dark side of the Terrible Mother was vanquished, her positive potential could be harvested. To have incest and fertilize the mother represented overcoming fear of the feminine, of her dark chaotic womb, in order to tap into riches of the unconscious and bring new life to the psyche. Psychologically we can see how Sphinx and incest fit together for Neumann (1954/93): The hero killed the Mother's terrible female side so as to liberate her fruitful and bountiful aspect. For Jung, to truly individuate was to rule the kingdom of our own psyche, by overthrowing the father's masculine influence of power, the ultimate authority of consciousness, while fertilizing and pillaging the mother's feminine territory, that of the unconscious. By breaking with Freud and finding his way through his psychosis, Jung killed the King and overcame the Terrible Mother to harvest her symbolism for his own creative development, both in theory and self.

Judging from the drama of real life, both Freud and Jung arrived at their ideas at least partly self-referentially by living them out. Along with affirming Ellenberger's (1981) notion of "creative illness," this coincides with Stolorow's thesis that all significant psychological theory derives from the personal experience and worldview of its originators (Atwood & Stolorow, 1979/1993). It also dovetails with contemporary relational perspectives that emphasize the importance of enactments within psychotherapy (Ginot, 2007). Enactments bring the underlying emotional dynamics alive in the room, yet carry the possibility of a less traumatic resolution.

RIDDLE AS PARADOX

As mentioned, in the last several decades, the Freudian interpretation of the Oedipus story largely has been laid aside. With the early advent of feminism, the significance of the tale to a woman's psyche was challenged. With the recognition that sexual abuse was often real and not just fantasy, later feminist thought challenged Freud's early abandonment of his seduction theory. As knowledge about the neurophysiology of the posttraumatic stress condition increased, so has clinical interest in "horizontal," dissociative splits between cortical and subcortical aspects of the brain (e.g., Lanius, Vermetten & Pain, C., 2010; Rothschild, 2000; Schore, 2007; 2012), versus the "vertical" splits maintaining Freud's repression barrier (see Kohut, 1977). Greater relational emphasis among contemporary psychoanalysts shifts interest towards early mother/infant attachment dynamics, as well as toward here-and-now relations between psychotherapist and patient. Finally, the current climate of multiculturalism disfavors any single theory, especially one universalizing development.

In the spirit of Levi-Strauss, I propose a different way of looking at the Oedipus myth. I aim to harvest meaning primarily by sidestepping narrative content to derive an alternative interpretation both structural and cybernetic in nature. When understood literally, both the "improbable" form the Sphinx embodies plus her impossible-seeming riddle present paradoxes that appear to contradict all known laws of science. Surely no creature on earth can literally walk on four, two and then three limbs during the very same day. With the possible exception of the slime mold, no animal changes its form of locomotion this radically; and not even the slime mold undergoes such complete metamorphosis in the course of a single day.

The Sphinx's riddle presents the type of "ordinary" paradox that science faces all the time. Here, paradox is loosely conceptualized as a set of facts that contradicts current scientific theory. Just as Darwin's embodied evolution proceeds in fits and starts (e.g., Gould, 1977), so too does the abstract progression of scientific theory. Kuhn (1962) described the erratic evolution of scientific theory, when resolution of ordinary contradiction leads to abrupt paradigm shifts that offer wider, more inclusive contexts in which to incorporate previously discrepant facts.

Beyond this type of "ordinary" scientific paradox, the Sphinx's riddle was essentially a paradox of self-reference (Marks-Tarlow, 2008a, 2008b, and 2008c). Within the history of mathematics, paradoxes of self-reference have arisen since ancient Greek times. A good example is The Liar: "This sentence is a lie," which is true only if false, and false only if true. Paradoxes of self-reference ultimately destroyed all hopes of mathematics supplying a logical foundation that is entirely complete and consistent. Instead, paradoxes of self-reference require creative leaps outside of their normal parameters, which is exactly what Oedipus accomplished by solving the Sphinx's riddle. The solution—humanity—required a leap under the surface to deep understanding of the nature of being human, including knowledge of self. In order to know what crawls on four legs in the morning, walks on two in midday and hobbles on three in the evening, Oedipus had to understand the entire human life cycle. He needed to possess intimate familiarity with physical changes in the body, ranging from the dependency of infancy, through the glory of maturity, to the waning powers of old age.

To approach the riddle without self-reference was to look outwards, to use a literal understanding, and to miss a metaphorical interpretation. To approach the riddle with self-reference was to seek knowledge partly by becoming introspective. At a deep, somatic level, Oedipus was uniquely positioned to apply the riddle to himself. Almost killed at birth and still physically handicapped, he harbored virtual, vestigial memories of death in life. His limp and cane were whispers of a helpless past and harbingers of a shattered future.

Self-referentially, Oedipus' own life trajectory showed the same three parts as the Sphinx's riddle. Through the kindness of others Oedipus survived the traumatized helplessness of infancy. In his prime, he proved more than able to stand on his own two feet—strong enough to kill a king, clever enough to slay the proverbial monster, and potent enough marry a queen and spawn a covey of offspring. Ironically, in the case of our tragic hero, it was Oedipus' very in-sight into his own origins that led to the loss of his kingdom and wife/mother, leaving him to hobble around blindly in old age, physically leaning on his cane, and emotionally learning upon the goodness of others, primarily his daughter/sister, Antigone. The namesake and body memories of Oedipus connected him with chance and destiny, past and future, infancy and old age. Recall that the name Oedipus means both "swollen foot" and "know-where." Feder (1974/1988) analyzed the Oedipus myth in terms of the clinical reality of adoption trauma. Like many adopted children, Oedipus was relentlessly driven to seek his own origins in order to "know where" he came from both genetically and socially.

Taking this approach a step further, we can see the impact of early physical abuse—attempted infanticide—on the neurobiology of different memory systems. Oedipus "knows where" he came from implicitly in his body due to his "swollen foot," even while ignorant of traumatic origins explicitly in his mind. This kind of implicit memory has gained much attention in recent clinical lore (e.g., Bucci, 2011; Cortina & Liotti, 2007; Fosshage, 2011; Mancia, 2006; Marks-Tarlow, 2011, 2012, 2013; Rothschild, 2000; Ruth-Lyons, 1998; Schore, 2010, 2011, 2012; Siegel, 2001). In early infant development, implicit memory is the first kind to develop. Implicit learning includes unconscious processing of exteroceptive information from the outer world as well as interoceptive information from the inner world. Such information helps tune ongoing perception and emotional self-regulation in the nonverbal context of relationships with others. In this way contingent versus non-contingent responses of caretakers become hardwired into the brain and body via particular neural pathways. While alluded to by others, e.g., Ornstein (1973), Allan Schore (2001; 2010; 2011; 2012) specifically proposed that implicit memory exists within the right, nonverbal, hemisphere of the human cerebral cortex, which I suggest constitutes the biological substrate for Freud's unconscious instincts and memories. Although hotly contested, neurobiological evidence mounts for Freud's repression barrier as hardwired into the brain (e.g., Solms, 2004).

Schore proposed a vertical model of the psyche, where the conscious, verbal, mind is localized in the left hemisphere of the brain, while the unconscious and body memory is mediated by the nonverbal right hemisphere (for most right handed people). The hemispheres of the brain and these different modes of processing are conjoined as well as separated by the corpus callosum, such that the perspective of only one hemisphere comes forward at any given time, while the perspective of the other recedes into the background (McGilchrist, 2009). Early trauma plus his secret origins caused a haunting and widening of the gap between what Oedipus' body knew versus what Oedipus' mind knew. Oedipus' implicit memory of his early abandonment and abuse became the invisible thread that provided deep continuity despite abrupt life changes. His implicit memory offered a clue to the commonality beneath the apparent disparity in the Sphinx's three-part riddle.

Structurally, to solve the riddle became equivalent to Oedipus' self-referential quest for explicit memory of his own origins. This interpretation meshes with anthropologist Lévi-Strauss' (1977) emphasis on structural similarities within and between myths, plus the near universal concern with human origins. It also dovetails with Bion's (1983, p. 46) self-referential understanding of the Sphinx's riddle as "man's curiosity turned upon himself." In the form of self-conscious examination of the personality by the personality, Bion used the Oedipus myth to illuminate ancient origins of psychoanalytic investigation.

METAPHORICAL THINKING IN COGNITIVE DEVELOPMENT

In order to solve both the riddle of the Sphinx as well as that of his own origins, Oedipus had to delve beneath the concrete level of surface appearances. Here he'd lived happily, but in ignorance, as children and innocents are reputed to do. Ignorance may have been bliss, but it did not necessarily lead to maturity. Prior to Oedipus solving the riddle, humankind lived in an immature state, an idea supported by the work of Julian Jaynes (1976). Writing about the "bicameral mind," as mentioned earlier, Jaynes speculated that ancient humanity hallucinated gods as living in their midst. Here myths were concretely embodied, serving as externals sources of authority before such executive functions became internalized within the cerebral cortex of the modern psyche, including our increased capacities for self-reflection, inner guidance and self-control, all functions of the frontal lobes.

The Sphinx's riddle was a self-referential mirror reflecting and later enabling explicit memory and knowledge of Oedipus' traumatic origins. Upon successfully answering the riddle, Oedipus bridged the earlier, developmental territory of the "right mind" with the evolutionarily and developmentally later left-brain (see Schore, 2001). In the process, Oedipus healed and matured on many levels. Not only did he address his castration fears by conquering the Terrible Mother in the form of the Sphinx after killing the Terrible Father, but also and perhaps more significantly, Oedipus made the leap from concrete to metaphorical thinking. By understanding "morning," "midday" and "evening" as stages of life, he demonstrated creativity and mental flexibility characteristic of internal complexity.

Cognitive linguists Lakoff and Johnson (1980) have suggested that metaphor serves as the embodied basis for all abstract thinking. More recently, Lakoff and Johnson (1999) argued that metaphor forms part of the implicit memory of the cognitive unconscious, where its immediate conceptual mapping is hard-wired into the brain. These researchers speculate that all cognitive activity is embodied, because it derives from a primary set of metaphors that surround how the body moves, functions and interacts in the physical and social world in which we are embedded. Verticality and balance are among Lakoff and Johnson's primary metaphors. This makes a great deal of sense given early developmental milestones. Babies universally shift from the horizontal posture of lying down to more vertical postures by first rolling over, then sitting upright, crawling and eventually rising up to balance and walk on two legs. Each shift is associated with increased mobility, agency and potency in the world. Psychoanalyst Arnold Modell (2003) has picked up on the relevance of Lakoff and Johnson by writing extensively on how the body uses metaphor to bridge disconnected experience, create somatic templates, and weave the illusion of constancy amidst continual change. Meanwhile, I have emphasized the role of spontaneous, embodied metaphors during psychotherapy as portmanteaus, or double signs that represent the core problems to be addressed in therapy, while simultaneously pointing toward their solutions.

Lakoff and Johnson's notions of embodied metaphors also dovetail with Piaget's developmental epistemology (e.g., Flavell, 1963). Though many details are still disputed, overall Piaget's theory has remained one of the most important and universal accounts of intellectual development to date (see Sternberg, 1990). Using careful observation and empirical studies, Piaget mapped the shift from a sensorimotor period of infancy, through the pre- and concrete operations of early childhood, into a formal operations stage of later childhood characterizing the adult, "mature" mind. Piaget's hallmark of maturity involved freedom from the particulars of concrete situations. This grants cognitive flexibility necessary for both abstract and metaphorical thinking. In sum, Oedipus's leap from concrete to metaphorical thinking can be understood both as an important developmental step for the individual as well as an important historical leap in the history of collective consciousness.

SELF-REFERENCE AND UNIVERSAL TURING MACHINES

So far, I have suggested that self-reference is central to a metaphorical solution of the Sphinx's riddle. But self-reference also proves to be an essential part of cybernetics, the sciences of information. A computational model views the human psyche as a recursive system, where present behavior depends upon how it has processed its past behavior. Within abstract machines, different computational powers depend deterministically upon a system's retrospective access to memory.

In computational science, power is ranked according to "Chomsky's hierarchy." At the bottom of the hierarchy lies the finite state automaton. This machine possesses only implicit memory for its current state. In the middle lies the push-down automaton. This machine possesses explicit memory, but with only temporary access to the past. At the top of Chomsky's hierarchy lies the Universal Turing Machine. This abstract machine possesses unrestricted, permanent and explicit memory for all past states.

Cyberneticist Ron Eglash (1999) provides a text analogy to contrast these differences: The least powerful machine is like a person who accomplishes all tasks instinctively, without the use of any books; in the middle is a person limited by books removed once they've been read; at the top is a person who collects and recollects all books read, in any order. The power of the Universal Turing Machine at the top is its capacity to recognize all computable functions. The point at which complete memory of past actions is achieved marks a critical shift in computational power. It is the point when full self-reference is achieved, which brings about the second-order, cybernetic capacity of a system to analyze its own programs. My reading of the Oedipus myth illustrates this very same point—that powerful instant when full access to memory dovetailed with self-reference to signal another step in the "complexification" of human consciousness (Marks-Tarlow, 2008).

THE RIDDLE AS MIRROR

Just as the Sphinx presented a paradigm of self-reference to hold a mirror up to Oedipus, the myth of Oedipus also holds a mirror up to us as witnesses. The story of Oedipus reflects our own stories in yet another self-referential loop. Like Oedipus, each one of us is a riddle to him or herself. The tale rocks generation after generation so powerfully partly because of this self-referential quality, which forces each one of us to reflect upon our own lives mythically.

Throughout the tale, there is dynamic tension between knowing and not-knowing—in Oedipus and in us. Oedipus starts out naïvely not-knowing who he is or where he came from. We start out knowing who Oedipus really is, but blissfully unaware of the truth in ourselves. By the end of the tale, the situation reverses: Oedipus solves all three riddles, that of the Oracle of Delphi, that of the Sphinx and that of his origins, while ironically, we participant/observers are left not-knowing. We harbor a gnawing feeling of uncertainty—almost as if another riddle has invisibly materialized, as if we face the very Sphinx herself, whose enigma must be answered upon threat of our own death.

Eglash (1999) notes that the power of the Universal Turing Machine lies in its ability not to know how many transformations, or applications of an algorithm a system would need ahead of time, before the program could be terminated. Paradoxically, to achieve full uncertainty about the future and its relationship to the past is symptomatic of increasing computational power. This kind of fundamental uncertainty is evident collectively within the modern sciences and mathematics of chaos theory, stochastic analyses, and various forms of indeterminacy. For example, Heisenberg's uncertainty principle states the impossibility of precisely determining both a quantum particle's speed as well as its location at the same time. Meanwhile, chaos theory warns of the impossibility of precisely predicting the long-term future of highly complex systems, no matter how precise our formulas or capacity to model their past behavior.

Experientially, we must deal with fundamental uncertainty with respect to the riddle of our own lives (see Marks-Tarlow, 2003), leaving us ultimately responsible to glean meaning from this self-reflective search. The Oedipus myth presents a self-referential mirror through which each one of us individually enters the modern stage of self-reflective consciousness. Capabilities for full memory, to consider the past and future, to contemplate death, to confront paradox, to self-reflect and to consider self-reference all represent critical levels of inner complexity that separate human from animal intelligence, the infant from the mature individual, plus the weakest from the most powerful computing machines.

CONCLUSION

I end this paper by speculating how this complex state of full self-reference serves as a prerequisite to a fully self-actualized human being. To have thorough access to memory for the past plus the cognitive flexibility not to have to know the future represents a state of high integration between left and right brain hemispheres, between body and mind, and between implicit, procedural memory versus explicit memory for events and facts. Such integration maximizes our potential for spontaneous response and creative self-expression that is the hallmark of successful individuation.

Furthermore, I argue that this complex state of "good-enough" self-reflective awareness is necessary to break the tragic intergenerational chain of fate and trauma symbolized by Greek tragedy in general and the Oedipus myth in particular. At the heart of the Oedipus myth lies the observation, echoed by a Greek chorus, that those born into abuse unwittingly grow up to become abusers. Laius' unsuccessful attempt to kill his son all but sealed Oedipus' fate to escalate this loop of violence by successfully killing his father. Meanwhile, the intergenerational transmission of incest within families is readily spread by seemingly innocuous mechanisms, such as forms of play (Marks-Tarlow, 2017).

The only way out of the fatalistic tragedy of abusers begetting abusers is to possess enough insight to unearth violent instincts before the deed is done, to exert sufficient self-control to resist and transcend such instincts, plus to tell a cohesive, self-reflective narrative. Multigenerational, prospective research within the field of attachment (e.g., Siegel, 1999) suggests that the best predictor of healthy, secure attachment in children remains the capacity for their parents to tell a cohesive narrative about their early childhood. It matters little whether the quality of this narrative is idyllic or horrific. What counts instead is whether parents possess the self-reflective insight to hold onto specific memories concerning their origins, which can be cohesively woven into the fabric of current life without disruption. This kind of self-referential reflection carries the full computational power of Universal Turing Machine. This provides the necessary mental faculties to break intergenerational chains of emotional, sexual, and physical abuse. It also allows for creative self-actualization, without a pre-determined script, set upon the stage of an open future.

In life, people gain self-awareness by looking into the mirror of experience self-referentially. By looking backwards toward past experience, we glean knowledge and meaning for dealing with present circumstances and moving toward the future wisely. Interestingly, within the field of neurobiology, there is current speculation that the brain itself is deeply intentional and forward looking (e.g., Van Boven & Ashworth, 2007), right down to the level of single cells (Freeman, 1999). Rather than representing static pictures from the past, memory is dynamically reconfigured according to ever-shifting present contexts (see Marks-Tarlow, 2008), as it serves the primary function of navigating through uncertainty toward the future. Indeed, the hippocampus, which is the main structure in the mammalian brain devoted to encoding long term memories, evolved out of the part of the brain that depends upon place cells to record the what and where of current environmental context. A current candidate for a universal mechanism of change within psychotherapy is the notion of memory reconsolidation (Lane, Ryan, Nadel, & Greenberg, 2015).

This theory proposes that psychotherapy works via retrieval and updating of prior emotional memories through reconsolidation that incorporates new emotionally charged experiences.

Whereas a traumatic mindset anticipates future experiences based on traumatic events of the past, a non-traumatic mindset is open to a future that could be different from what took place in the past. The role of uncertainty in moving with fullest complexity and computational power into the future is also evident within new data mining techniques. Without knowing what is sought, these computational algorithms can sift through a mountain of material until clear patterns emerge. Whether such techniques are used wisely in ways consistent with humanistic values, increased self-awareness, and humanitarian aims, or whether they are abused in service of decreasing personal freedoms and further eroding the environment, only time will tell. Either way, both in the wetware of the human brain and in software of the computer, recursive loops continue to ensure that past, present, and future lose distinctive orientations, while observers continue to blend ever more seamlessly with the observed.

ACKNOWLEDGMENT

The first version of this paper appeared under the title, "Riddle of the Sphinx Revisited," in the Electronic Conference of the Foundations of Information Sciences, May 6-10, 2002. The second version of this paper appeared in 2008 under the title. "Alan Turing meets the sphinx: Some new and old riddles" in *Chaos & Complexity Letters, 3*(1), 83–95. A third version of this paper also appeared in 2008 under the title "Riddle of the sphinx: A paradox of self-reference revealed and reviled," in *Reflecting interfaces: The complex coevolution of information technology ecosystems*. Hershey, PA: Idea Group. A fourth version of this paper appeared in 2014 under the title "Myth, metaphor, and the evolution of self-awareness" in the International Journal of Signs and Semiotic Systems, 3, 1, 46-60.

REFERENCES

Aristotle. (1982). Poetics: Vol. 23. *Loeb Classical Library*. Cambridge, MA: Harvard University Press.

Atwood, G., & Stolorow, R. (1979/1993). *Faces in a cloud: Intersubjectivity in personality theory*. Northvale, NJ: Jason Aronson.

Berkower, L. (1970). The military influence among Freud's dynamic psychiatry. *The American Journal of Psychiatry, 127*(2), 85–92. doi:10.1176/ajp.127.2.167 PMID:4919635

Bion, W. (1983). *Elements of psycho-analysis*. Northvale, NJ: Jason Aronson.

Briggs, J., & Peat, D. (2000). *Seven life lessons of chaos: Spiritual wisdom from the science of change*. New York, NY: Harper Perennial.

Bromberg, P. (1998). *Standing in the spaces*. Hillsdale, NJ: Analytic Press.

Bucci. (2011). The interplay of subsymbolic and symbolic processes in psychoanalytic treatment: It takes two to tango—But who knows the steps, who's the leader? The choreography of the psychoanalytic interchange. *Psychoanalytic Dialogues, 21*, 45-54.

Campbell, J. (1949/1973). *The hero with a thousand faces*. Princeton, NJ: Bollingen Series, Princeton University.

Cortina, M., & Liotti, G. (2007). New approaches to understanding unconscious processes: Implicit and explicit memory systems. *International Forum of Psychoanalysis*, *16*(4), 204–212. doi:10.1080/08037060701676326

Eglash, R. (1999). *African fractals: Modern computing and indigenous design*. Rutgers University Press.

Ellenberger, H. (1981). *The discovery of the unconscious*. New York, NY: Basic Books.

Feder, L. (1974). Adoption trauma: Oedipus myth/clinical reality. In G. Pollock & J. Ross (Eds.), *The Oedipus papers*. Madison, CT: International Universities Press.

Flavell, J. H. (1963). *The developmental psychology of Jean Piaget*. New York, NY: Van Nostrand; doi:10.1037/11449-000

Fosshage. (2011). How do we "know" what we "know?" and change what we "know?" *Psychoanalytic Dialogues*, *21*, 55-74.

Freeman, W. (1999). Consciousness, intentionality, and causality. *Journal of Consciousness Studies*, *6*, 143–172.

Freud, S. (1900). *The interpretation of dreams* (J. Strachey, Trans.). New York, NY: Basic Books.

Gardner, M. (1999). *The annotated Alice*. New York, NY: Norton.

Ginot, E. (2007). Intersubjectivity and neuroscience: Understanding enactments and their therapeutic significance within emerging paradigms. *Psychoanalytic Psychology*, *24*(2), 317–332. doi:10.1037/0736-9735.24.2.317

Goswami, A. (1995). *The self-aware universe*. Los Angeles, CA: Tarcher.

Gould, S. J., & Eldredge, N. (1977). Punctuated equilibria: The tempo and mode of evolution reconsidered. *Paleobiology*, *3*(02), 115–151. doi:10.1017/S0094837300005224

Hayman, D. (1999). *The life of Jung*. New York, NY: W.W. Norton.

Heims, S. (1991). *The cybernetics group*. Cambridge, MA: The MIT Press.

Jaynes, J. (1976). *The origin of consciousness in the breakdown of the bicameral mind*. Boston, MA: Houghton Mifflin.

Jung, C. (1956). Symbols of transformation. In *Collected works*. London, UK: Routledge & Kegan Paul.

Jung, C. (1961). *Memories, dreams, reflections*. New York, NY: Random House.

Jung, C. (2009). *The red book*. New York, NY: Norton.

Kerr, J. (1995). *A most dangerous method*. New York, NY: Vintage Books/Random House.

Klein, M. (1932). *The psycho-analysis of children*. London, UK: Hogarth.

Kohut, H. (1971). *The analysis of the self*. New York, NY: International Universities Press.

Kohut, H. (1977). *The restoration of the self*. New York, NY: International Universities Press.

Kuhn, T. (1962). *The structure of scientific revolutions*. Chicago, IL: University of Chicago Press.

Lakoff, G., & Johnson, M. (1980). *Metaphors we live by*. Chicago, IL: University of Chicago Press.

Lakoff, G., & Johnson, M. (1999). *Philosophy in the flesh: The embodied mind and its challenge to Western thought*. New York, NY: Basic Books.

Lane, R. D., Ryan, L., Nadel, L., & Greenberg, L. (2015). Memory reconsolidation, emotional arousal, and the process of change in psychotherapy: New insights from brain science. *Behavioral and Brain Sciences*, 38. PMID:24827452

Lanius, R., Vermetten, E., & Pain, C. (2010). *The impact of early life trauma on health and disease: The hidden epidemic*. Cambridge, UK: Cambridge University Press; doi:10.1017/CBO9780511777042

Lévi-Strauss, C. (1977). *Structural anthropology* (C. Jacobson & B. G. Schoepf, Trans.). Harmondsworth, UK: Penguin.

Lyons-Ruth, K., Bruschweiler-Stern, N., Harrison, A. M., Morgan, A. C., Nahum, J. P., Sander, L., & Tronick, E. Z. et al. (1998). Implicit relational knowing: Its role in development and psychoanalytic treatment. *Infant Mental Health Journal*, *19*(3), 282–289. doi:10.1002/(SICI)1097-0355(199823)19:3<282::AID-IMHJ3>3.0.CO;2-O

Mancia, M. (2006). Implicit memory and early unrepressed unconscious: Their role in the therapeutic process (How the neurosciences can contribute to psychoanalysis). *The International Journal of Psycho-Analysis*, *87*, 83–103. PMID:16635862

Marks-Tarlow, T. (2003). The certainty of uncertainty. *Psychological Perspectives*, *45*(1), 118–130. doi:10.1080/00332920308403045

Marks-Tarlow, T. (2008a). *Psyche's veil: Psychotherapy, fractals and complexity*. London, UK: Routledge.

Marks-Tarlow, T. (2008b). Alan Turing meets the sphinx: Some new and old riddles. *Chaos & Complexity Letters*, *3*(1), 83–95.

Marks-Tarlow, T. (2008c). Riddle of the sphinx: A paradox of self-reference revealed and reveiled. In *Reflecting interfaces: The complex coevolution of information technology ecosystems*. Hershey, PA: Idea Group; doi:10.4018/978-1-59904-627-3.ch002

Marks-Tarlow, T. (2012). *Clinical intuition in psychotherapy: The neurobiology of embodied response*. New York, NY: Norton.

Marks-Tarlow, T. (2013). *Awakening clinical intuition*. New York, NY: Norton.

Marks-Tarlow, T. (2015). The nonlinear dynamics of clinical intuition. *Chaos & Complexity Letters*, *8*(2-3), 1–24.

Marks-Tarlow, T. (2017). I am an avatar of myself: Fantasy, trauma and self deception. *American Journal of Play*, *9*(2), 169–201.

Masson, J. (1985). *The assault on truth: Freud's suppression of the seduction theory*. New York: Penguin Press.

McGilchrist, I. (2009). The Master and his emissary: The divided brain and the making of the Western world. New Haven, CT: Yale University Press.

Michell, S. (1988). *Relational concepts in psychoanalysis: An integration*. Cambridge, MA: Harvard University Press.

Modell, A. (2003). *Imagination and the meaningful brain*. Cambridge, MA: MIT Press.

Monte, C., & Sollod, R. (2003). *Beneath the mask: An introduction to theories of personality*. New York: John Wiley & Sons.

Neumann, E. (1954/1993). *The origins and history of consciousness*. Princeton, NJ: Princeton.

Nietzsche, F. (1999). *The birth of tragedy and other writings (Cambridge texts in the history of philosophy)*. Cambridge, UK: Cambridge University Press. (Original work published 1871)

Ornstein, R. (Ed.). (1973). The nature of human consciousness. San Francisco: W.H. Freeman.

Orsucci, F., & Sala, N. (2008). *Reflexing interfaces: The complex coevolution of information technology ecosystems*. Hershey, PA: IGI Global; doi:10.4018/978-1-59904-627-3

Orsucci, F., & Sala, N. (2012). *Complexity science, living systems, and reflexing interfaces: New models and perspectives*. Hershey, PA: IGI Global; doi:10.4018/978-1-4666-2077-3

Peck, M. (2012). *Prosthetics of the future: Driven by thoughts, powered by bodily fluids*. Retrieved from http://spectrum.ieee.org/tech-talk/biomedical/devices/prosthetics-of-the-future-driven-by-thoughts-powered-by-bodily-fluids

Pollock, G., & Ross, J. (1988). *The Oedipus papers*. Madison, CT: International Universities Press.

Ricoeur, P. (1970). *Freud and philosophy*. Cambridge, MA: Yale University Press.

Rothschild, B. (2000). *The body remembers: The psychophysiology of trauma and trauma treatment*. New York, NY: W. W. Norton.

Schore, A. (2001). Minds in the making: Attachment, the self-organizing brain, and developmentally-oriented psychoanalytic psychotherapy. *British Journal of Psychotherapy*, *17*(3), 299–328. doi:10.1111/j.1752-0118.2001.tb00593.x

Schore, A. (2010). The right-brain implicit self: A central mechanism of the psychotherapy change process. In J. Petrucelli (Ed.), *Knowing, not-knowing and sort of knowing: Psychoanalysis and the experience of uncertainty* (pp. 177–202). London, UK: Karnac.

Schore, A. (2011). The right brain implicit self lies at the core of psychoanalytic psychotherapy. *Psychoanalytic Dialogues, 21,* 75–100. doi:0481885.2011.54532910.1080/1

Schore, A. (2012). *The science of the art of psychotherapy*. New York, NY: Norton.

Shanley, W. (Ed.). (2011). *Alice and the quantum cat*. Pari, Italy: Pari Publishing.

Siegel, D. (2001). Memory: An overview, with emphasis on developmental, interpersonal, and neurobiological aspects. *Journal of the Academy of Child & Adolescent Psychiatry, 40*(9), 997–1011. doi:10.1097/00004583-200109000-00008 PMID:11556645

Solms, M. (2004). Freud returns. *Scientific American*, 83–89. PMID:15127665

Solms, M., & Turnbull, O. (2002). *The brain and the inner world: An introduction to the neuroscience of subjective experience*. London, UK: Karnac Books.

Stern, D. (1983). Unformulated experience: From familiar chaos to creative disorder. *Contemporary Psychoanalysis, 19*, 71–99. doi:10.1080/00107530.1983.10746593

Sternberg, R. (1990). *Metaphors of mind: Conceptions of the nature of intelligence*. Cambridge, UK: Cambridge University Press.

Stolorow, R., Brandchaft, B., & Atwood, G. (1987). *Psychoanalytic treatment: An intersubjective approach*. Hillsdale, NJ: The Analytic Press.

Van Boven, L., & Ashworth, L. (2007). Looking forward, looking back: Anticipation is more evocative than retrospection. *Journal of Experimental Psychology. General, 136*(2), 289–300. doi:10.1037/0096-3445.136.2.289 PMID:17500652

Wolf, F. (1995). *The dreaming universe: A mind-expanding journey into the realm where psyche and physics meet*. New York, NY: Touchstone.

Chapter 4
Existential Graphs and Cognition

Caterina Clivio
Columbia University, USA

Marcel Danesi
University of Toronto, Canada

ABSTRACT

When looked at cumulatively, it can be said that American pragmatist philosopher Charles Sanders Peirce strove to understand cognition via his sign theory and especially his notion of existential graphs. Peirce put forth ideas for a discipline that would incorporate notions of psychology and semiotics into a unified ontological and epistemological theory of mind. The connecting link was his system of dia-grammatic logic, called "existential graphs." For Peirce a graph was more powerful than language as a means of understanding because it showed how its parts resembled relations among the parts of cognitive acts. Existential graphs show that cognition cannot be extracted from a linear or hierarchical succession of structures, but the very process of thinking itself in actu. In fact, Peirce called his graphs "moving pictures of thought" because they allow us to see how are thoughts are unfolding. In short, as Kiryuschenko (2012) puts it, "Graphic language allows us to experience a meaning visually as a set of transitional states, where the meaning is accessible in its entirety at any given here and now during its transformation" (p. 122).

INTRODUCTION

Diagrams are keys to understanding real-world phenomena, as scientists and mathematicians certainly know, allowing them to both represent their theoretical hunches and then to use these very representations to derive further ideas and to conduct experiments. Peirce certainly understood this, inventing Existential Graphs (EGs) to explain not only the nature of discovery in science but also how the visual imagery of the mind generates ideas. The focus in this paper will be on the relevance of EGs in mathematics (Danesi 2013), but the concept of EGs as "discovery devices" applies to all domains of knowledge. EGs mirror how the mind synthesizes into abstract images of real-world objects, rearranging parts in various ways

DOI: 10.4018/978-1-5225-5622-0.ch004

to see what the blending yields. EGs offered Peirce the possibility of linking semiotics and psychology into a model of how thought and discovery unfold. As he points out in the following excerpt, diagrams are maps of thought, which may be used "to stick pins into" in order to mark anticipated changes.

But why do that [use maps] when the thought itself is present to us? Such, substantially, has been the interrogative objection raised by an eminent and glorious General. Recluse that I am, I was not ready with the counter-question, which should have run, "General, you make use of maps during a campaign, I believe. But why should you do so, when the country they represent is right there?" Thereupon, had he replied that he found details in the maps that were so far from being "right there," that they were within the enemy's lines, I ought to have pressed the question, "Am I right, then, in understanding that, if you were thoroughly and perfectly familiar with the country, no map of it would then be of the smallest use to you in laying out your detailed plans?" No, I do not say that, since I might probably desire the maps to stick pins into, so as to mark each anticipated day's change in the situations of the two armies." "Well, General, that precisely corresponds to the advantages of a diagram of the course of a discussion. Namely, if I may try to state the matter after you, one can make exact experiments upon uniform diagrams; and when one does so, one must keep a bright lookout for unintended and unexpected changes thereby brought about in the relations of different significant parts of the diagram to one another. Such operations upon diagrams, whether external or imaginary, take the place of the experiments upon real things that one performs in chemical and physical research. (CP4: 530)

The study of diagrammatic representations is a productive area of investigation in various domains of semiotics and cognitive science (Shin 1994, Chandrasekaran, Glasgow, & Narayanan 1995, Hammer 1995, Hammer & Shin 1996, 1998, Allwein & Barwise 1996, Barker-Plummer & Bailin 1997, 2001, Kulpa 2004, Stjernfelt 2007, Roberts 2009, Kiryushchenko 2012). EGs are consistent with two prominent trends in these field—namely, phenomenology and blending theory in cognitive science (for example, Lakoff & Núñez 2000)—trends that were prefigured by Peirce's notion of "phaneroscopy," which he described as the formal analysis of appearances apart from how they appear to interpreters and of their actual material content.

EXISTENTIAL GRAPHS

Peirce argued that discoveries in chemistry were phaneroscopic, because chemical compounds could be studied not as mixtures of actual substances but as diagrammatic structures. Chemists discovered that the structure of a molecule and transformations of chemical compounds themselves gave birth to the scientific language that explained them though the diagrams used to represent them. Diagrams contain within them "virtual objects," which are like real objects and can thus be used to experiment cognitively with the latter. Peirce wrote an entry on the concept of "virtual" for Baldwin's (1902, p. 763) *Dictionary of Philosophy and Psychology*, which is of relevance here:

A virtual X is something, not an X, which has the efficiency (virtus) of an X. This is the proper meaning of the word; but it has been seriously confounded with "potential," which is almost its contrary. For the potential X is of the nature of X, but is without actual efficiency. A virtual velocity is something, not a velocity, but a displacement; [it is] equivalent to a velocity in the formula, "what is gained in velocity is lost in power." (3) Virtual is sometimes used to mean pertaining to virtue in the sense of an ethical habit.

According to this definition, any virtual object is not a mental copy of its real object, but a portrayal of its practical applications, predicting how other real objects are connected to it. Thus, the virtuality of diagrams generally is what leads to discoveries. He also saw logic as a form of diagrammatic thinking which superseded sentential logic to explain phenomena. So, an EG can be defined as a virtual map of something in the mind that takes on form and shape as it is drawn to represent something. The notion of EG extends to equations and other mathematical notions. In effect, algebraic notation is a kind of diagrammatic strategy for compressing information, much like pictography does for representing referents. An equation is an EG. As Kauffman (2001, p. 80) states, EGs are powerful cognitively because they contain arithmetical information in an economical, and thus, structurally-expository form:

Peirce's Existential Graphs are an economical way to write first order logic in diagrams on a plane, by using a combination of alphabetical symbols and circles and ovals. Existential graphs grow from these beginnings and become a well-formed two dimensional algebra. It is a calculus about the properties of the distinction made by any circle or oval in the plane, and by abduction it is about the properties of any distinction.

Consider the Pythagorean equation ($c^2 = a^2 + b^2$) as an EG; as such it is a visual-virtual portrait of the relations among the variables (originally standing the sides of the triangle). But, being a EG, it also tells us that the variables relate to each other in many ways other than geometrically. Expressed in language, we would not be able to *see* the possibilities that the equation presents to us. To use Susan Langer's (1948) concept of discursive-versus-presentational representation, the equation tells us much more than the statement ("the square on the hypotenuse is equal to the sum of the squares on the other two sides") because it literally "presents" the structure inherent in the linguistic version, fleshing it out as an abstract form. Describing it in language (with sentences) is a *discursive* process, forcing us to think of the information in a different, cognitively-constrained way. In effect, an equation is an EG that represents real-world objects in a holistic revelatory way. Further mathematical knowledge occurs by unpacking the inherent suggestive information from virtual forms such as equations to literally see what is in them. In a way, all mathematical notations and forms are EGs, allowing mathematicians to experiment with their own observations and thus to advance their work. They use language to explain their EGs and to contextualize them in the real world. Mathematics is thus both a presentational and discursive craft. What an EG does, like a map, is turn a real-world problem into a paper-and-pencil one and then suggests language for explicating it.

This line of reasoning raises deep philosophical issues. Although the structures of the cosmos certainly predate the human mind, they are not understood nor do they exist outside of human minds. As Bergin and Fisch (1984, p. xlv) have perceptively pointed out, in reference to the philosophy of Neapolitan philosopher Giambattista Vico, human beings "have themselves made this world of nations, but it was not without drafting, it was even without seeing the plan that they did just what the plan called for." As Peirce (volume 6, 1931-1958, p. 478) similarly put it, the human mind has "a natural bent in accordance with nature" (CP 6.478). This blending of mind and nature trough visual diagramming becomes perception, which Peirce called the "outward clash" of the physical world on the senses. As neuroscientific research has shown rather convincingly, mental imagery and its expression in diagrammatic form seems to be a more fundamental form of cognition, probably predating the advent of vocal language in carrying out counting and measurement tasks (Cummins 1996, Chandrasekaran et al. 1995). Even sentences, as Peirce often argued, hide within their logical structure a visual form of understanding that can be easily rendered diagrammatically.

In sum, diagrams show relations that are not apparent in linguistic or in other symbolic forms (Barwise and Etchemendy 1994, Allen and Barwise 1996). As Radford (2010: 4) puts it, they present information to us by means of "ways of appearance." They constitute a veritable explanatory system of logic of their own. Diagrams are inferences (informed guesses) that translate hunches (raw guesses) visually. These then lead to abductions (insights). The process of cognition is complete after the ideas produced in this way are organized logically (deduction). In fact, this suggests a model of cognition:

Hunches are the brain's attempts to understand what something means initially. These lead to inferences through a consideration of what the hunches suggest in terms of previous experience. So, the Pythagorean triangle leads to the previously-hidden concept of number triples. Eventually, this concept led to an hypothesis, namely that only when n = 2 does the generalized Pythagorean formula hold ($c^n = a^n + b^n$)—called Fermat's Last Theorem. This, in turn, led to many discoveries. It also led to a conclusive proof, which came, of course with the Taylor-Wiles (1995) proof.

As another example, consider imaginary numbers, which were discovered serendipitously. At first, it was not clear how they fit into the existing number system at the time or how they could be represented on the Cartesian plane. This conundrum led to the ingenious invention of a diagram, called the Argand diagram, that made it possible to show the relation of imaginary numbers to real ones. The diagram locates imaginary numbers (Im) on one axis and real ones (Re) on the other. The point $z = x + iy$ represents a complex number in the plane (called the Argand plane) and shows its vectorial features in terms of the angle θ that it forms. This is a geometric interpretation of complex numbers building on the previous diagrammatic system of Cartesian representation. The Argand plane allows for a visual interpretation of complex numbers, showing that they can be added like vectors and can be multiplied in terms of polar coordinates with the product of the two moduli (absolute values). The angle of the product is the sum of the two angles. Multiplication by a complex number of modulus 1 is a rotation—a discovery that has been incorporated into the theory of complex numbers. The invention of the Argand diagram turned out, therefore, to be not only a heuristic device, showing how addition, multiplication, and other operations of the complex numbers can be carried out systematically, but also a source of investigation of the structure of these numbers, having led to many discoveries in number theory.

Figure 1.

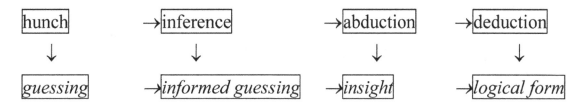

DIAGRAMS IN LOGICAL THINKING

Peirce saw his EGs as more powerful models of logic than sentential (syllogistic) logic. In fact, the whole field of set theory is fundamentally diagrammatic. Venn diagrams (1880, 1881)) are indispensable for deducing logical implications, since they allude to various features of sets by their simple configuration. The principle of the Venn diagram is to show relationships among sets and elements in them.

Diagrams permeate set theory, perhaps because they reveal intrinsic *image schemata* in cognition—an idea derived from the work of George Lakoff and his research associates (Lakoff and Johnson 1980, 1999, Lakoff 1987, Johnson 1987, Lakoff and Núñez 2000). These are defined as largely unconscious mental outlines of recurrent shapes, actions, dimensions, and so on that derive from perception and sensation. The world is made up of different kinds and levels of physical energy. Our knowledge of the world is filtered by our sense organs, which react to these energies. The patterns of energies become objects, events, people, and other aspects of the world through semiotic classification. However, some perceptions are not categorized because we lack the appropriate knowledge schemata to interpret them. But by actually drawing our intuitive or instinctual images in diagram form, we gain direct access to their hidden structure. In a word, diagrams are the externalizations of image schemata. They not only mirror other kinds of stored information, such as sentential information, they also bring out the unconscious image schemata inherent in it (up-versus-down, containment-versus-openness, and so on). In so doing, they excise irrelevant detail from incoming information leaving only the relevant features in it in schema form.

The translation of sentential logic to diagram logic started with Euler. Before the advent of Venn diagrams, Euler represented categorical sentences in terms of diagrams the prefigure the Venn ones (Hammer and Shin 1996, 1998):

These are, in effect, image schemata that cut across language (and languages) and allows us to see the logical structure involved in bare outline form. The power of the diagrams over the linguistic forms lies in the fact that no additional conventions, paraphrases, or elaborations are needed—the relationships holding among sets are shown by means of the same relationships holding among the circles representing them. Euler was aware, however, of both the strengths and weaknesses of his diagrammatic system. For instance, in the statement: "No A is B. Some C is A. Therefore, Some C is not B," no single diagram can represent the two premises, because the relationship between sets B and C cannot be fully specified in one single diagram. Instead, Euler suggested three possible cases:

Figure 2.

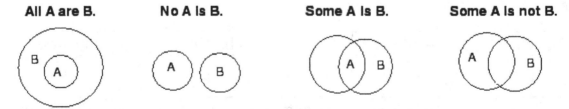

| **All A are B.** | **No A Is B.** | **Some A Is B.** | **Some A Is not B.** |

Figure 3.

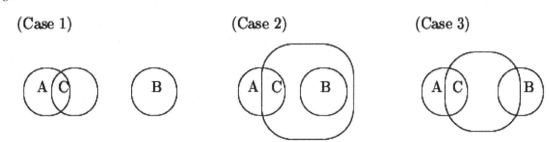

(Case 1)   (Case 2)   (Case 3)

He claimed that the proposition "Some C is not B" can be read from all these diagrams. But it is far from clear how this is so. Such anomalies have led some logicians to claim that diagrams are only ancillary devices, being ultimately incapable of representing all logical statements accurately. It was Venn (1881, p. 510) who tackled Euler's dilemma pointing out that the weakness lay in the fact that Euler's method did not show that imperfect knowledge exists. Venn aimed to overcome the weaknesses of Euler's diagrams by showing how partial information can be visualized. So, a diagram like the one above of three intersecting sets, A, B, C (which he called primary) does not convey specific information about the relationship between sets. So, for instance, the relations between two sets, A and B, can be shown as follows, by simply shading them (Venn 1881, 122). With this simple modification, we can draw diagrams for various premises and relations:

Figure 4.

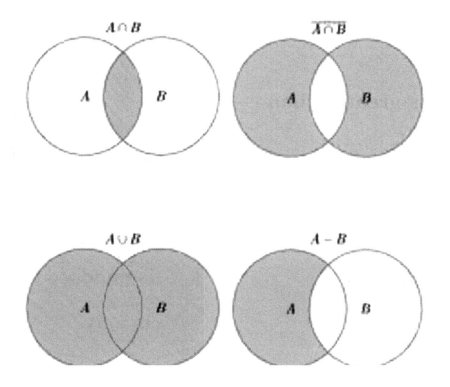

It was Peirce who pointed out that Venn's system had no way of representing existential statements, disjunctive information, probabilities, and relations. Peirce showed that "All A are B or some A is B" cannot be represented by neither the Euler or Venn systems in a single diagram. Like Euler, Peirce saw a graph as anything having its parts in relation to each other in such a way that they resemble relations among the parts of some different set of entities or referents. The relation was evident in the outline of the graph and thus showed in bare form how the thought process unfolded. The following EG is Peirce's brilliant solution:

The line is called a line of identity by Peirce. In any EG any line of identity whose outermost part is evenly enclosed refers to something, and any one whose outermost part is oddly enclosed refers to anything there may be (CP4.458). The following graph shows, essentially, how any EG can be used to represent logical statements (from Roberts 2009):

The first graph (where the outermost part of the line is evenly, zero, enclosed) says that something good is ugly, and the second graph (where the outermost part is enclosed once) says that everything good is ugly. The visual power of such a graph requires no comment (literally).

CONCLUSION

EGs are used extensively in philosophy and some areas of mathematics and logic, but they are relatively unknown in psychology and cognitive science generally. The dilemma of understanding cognition, like understanding an equation, is essentially a matter of understanding how we represent our thoughts. EGs are unconscious entities that can be formalized, as Peirce showed, to unravel the nature of thinking as a holistic activity. Peirce attached extreme importance to the task of making diagrams practicable and practical. EGs display not a linear succession of logical deductions, but how inference unfolds, thus conveying information and simultaneously explaining how it is being done (CP4. 619).

Figure 5.

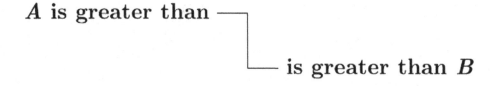

Figure 6.

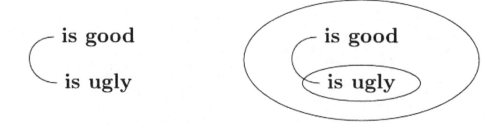

As is well known, in 1931 Kurt Gödel showed that there never can be a consistent system of statements that can capture all the truths of mathematics and logic. The makers of the statements could never extricate themselves from making them. Gödel made it obvious to mathematicians that mathematics was made by them, and that the exploration of "mathematical truth" would go on forever as long as humans were around. The final map of the mathematical realm will never be drawn. Like other products of the imagination, the world of mathematics lies within the minds of humans. In effect, all diagrams are theories of reality, evaluating it in their own particular ways.

REFERENCES

Allwein, G., & Barwise, J. (Eds.). (1996). *Logical reasoning with diagrams*. Oxford, UK: Oxford University Press.

Baldwin, J. M. (1902). *Dictionary of philosophy and psychology*. New York: Macmillan.

Barker-Plummer, D., & Bailin, S. C. (1997). The role of diagrams in mathematical proofs. *Machine Graphics and Vision, 8*, 25–58.

Barker-Plummer, D., & Bailin, S. C. (2001). On the practical semantics of mathematical diagrams. In M. Anderson (Ed.), *Reasoning with diagrammatic representations*. New York: Springer.

Bergin, T. G., & Fisch, M. (1984). *The New Science of Giambattista Vico*. Ithaca, NY: Cornell University Press.

Chandrasekaran, B., Glasgow, J., & Narayanan, N. H. (Eds.). (1995). *Diagrammatic reasoning: Cognitive and computational perspectives*. Cambridge, MA: MIT Press.

Cummins, R. (1996). *Representations, targets, and attitudes*. Cambridge, MA: MIT Press.

Danesi, M. (2013). *Discovery in mathematics: An interdisciplinary approach*. Munich: Lincom Europa.

Gödel, K. (1931). Über formal unentscheidbare Sätze der Principia Mathematica und verwandter Systeme, Teil I. *Monatshefte für Mathematik und Physik, 38*(38), 173–189. doi:10.1007/BF01700692

Hammer, E. (1995). Reasoning with sentences and diagrams. *Notre Dame Journal of Formal Logic, 35*, 73–87.

Hammer, E., & Shin, S. (1996). Euler and the role of visualization in logic. In J. Seligman & D. Westerståhl (Eds.), *Logic, language and computation* (Vol. 1). Stanford, CA: CSLI Publications.

Hammer, E., & Shin, S. (1998). Euler's visual logic. *History and Philosophy of Logic, 19*(1), 1–29. doi:10.1080/01445349808837293

Johnson, M. (1987). *The body in the mind: The bodily basis of meaning, imagination and reason*. Chicago: University of Chicago Press.

Kauffman, L. K. (2001). The mathematics of Charles Sanders Peirce. *Cybernetics & Human Knowing, 8*, 79–110.

Kiryushchenko, V. (2012). The visual and the virtual in theory, life and scientific practice: The case of Peirce's quincuncial map projection. In M. Bockarova, M. Danesi, & R. Núñez (Eds.), *Semiotic and cognitive science essays on the nature of mathematics*. Munich: Lincom Europa.

Kulpa, Z. (2004). On diagrammatic representation of mathematical knowledge. In A. Sperti, G. Bancerek, & A. Trybulec (Eds.), *Mathematical knowledge management*. New York: Springer. doi:10.1007/978-3-540-27818-4_14

Lakoff, G. (1987). *Women, fire and dangerous things: What categories reveal about the mind*. Chicago: University of Chicago Press. doi:10.7208/chicago/9780226471013.001.0001

Lakoff, G., & Johnson, M. (1980). *Metaphors we live by*. Chicago: Chicago University Press.

Lakoff, G., & Johnson, M. (1999). *Philosophy in flesh: The embodied mind and Its challenge to western thought*. New York: Basic.

Lakoff, G., & Núñez, R. (2000). *Where mathematics comes from: How the embodied mind brings mathematics into being*. New York: Basic Books.

Langer, S. K. (1948). *Philosophy in a new key*. New York: Mentor Books.

Peirce, C. S. (1931-1958). In C. Hartshorne, P. Weiss, & A. W. Burks (Eds.), *Collected papers of Charles Sanders Peirce* (Vols. 1-8). Cambridge, MA: Harvard University Press.

Radford, L. (2010). Algebraic thinking from a cultural semiotic perspective. *Research in Mathematics Education*, *12*(1), 1–19. doi:10.1080/14794800903569741

Roberts, D. D. (2009). *The Existential Graphs of Charles S. Peirce*. The Hague, The Netherlands: Mouton. doi:10.1515/9783110226225

Shin, S. (1994). *The logical status of diagrams*. Cambridge, UK: Cambridge University Press.

Stjernfelt, F. (2007). *Diagrammatology: An investigation on the borderlines of phenomenology, ontology, and semiotics*. New York: Springer. doi:10.1007/978-1-4020-5652-9

Taylor, R., & Wiles, A. (1995). Ring-theoretic properties of certain Hecke algebras. *Annals of Mathematics*, *141*(3), 553–572. doi:10.2307/2118560

Venn, J. (1880). On the employment of geometrical diagrams for the sensible representation of logical propositions. *Proceedings of the Cambridge Philosophical Society*, *4*, 47–59.

Venn, J. (1881). *Symbolic logic*. London: Macmillan. doi:10.1037/14127-000

Chapter 5
Framing in News Discourse:
The Case of the Charlie Hebdo Attack

Miriam Tribastone
University of Amsterdam, The Netherlands

Sara Greco
Università della Svizzera Italiana, Switzerland

ABSTRACT

By presenting the case study of the Charlie Hebdo attack in news discourse, this chapter combines a semantic analysis of the most frequent frame-activating words through text linguistics tools with frame analysis, developed according to the model proposed by Entman in the news making context. The linguistic perspective adopted in this chapter combines the works by Fillmore and Congruity Theory. As shown in the present work, both linguistics and news framing benefit from such integration.

INTRODUCTION

"The social world is … a chameleon, or, to suggest a better metaphor, a kaleidoscope of potential realities, any of which can be readily evoked by altering the ways in which observations are framed and categorized" (Edelman, 1993, p.232). These significant words illuminate the importance of framing in research and the fundamental connection between the social world and the phenomenon at the center of this research: Van Eemeren (2010, p.126) argues that "framing always involves an interpretation of reality that puts the facts or events referred to in a certain perspective." According to Entman (2004), no one can escape from framing: even journalists in news reporting, thought as objective, frame events, issues, and political actors. Indeed, as argued by Fowler (1996, p. 4), "news is a representation of the world in language" that reflects a fact and all the values and meanings attached to it. Not coincidentally, there is a lot of research on framing in different (and sometimes not compatible) disciplines and approaches (see for example the review in Dewulf et al., 2009).

More often than not, the different approaches to framing are not in dialogue one with another. This work has the methodological ambition to integrate different but complementary traditions on frames and framing in order to shed light on how frames are created in news discourse.

DOI: 10.4018/978-1-5225-5622-0.ch005

In this paper, the news framing tradition will be combined with the attention to words from a text linguistic perspective, with the purpose to combine framing in news discourse and highlight the role of frame-activating words. The Charlie Hebdo attack, which took place in the Parisian newsroom on the 7th January 2015, will be examined as a case study. Because this event has been interpreted in different ways on the media, it is particularly suitable for the studying of framing. The several frames will be investigated as different ways through which this event has been understood and interpreted; the analysis will be based on a combination of two complementary dimensions: a semantic analysis of the most frequent frame-activating words, as based on linguistic tools, and a news discourse analysis of frames. These two dimensions will be illustrated in the following section. In section 3, the method used for data collection and analysis will be explained. Then, the findings and discussion section will follow (section 4); finally, some conclusions will be drawn in section 5.

FRAMES AND FRAMING: A COMPLEX CONCEPTUAL PANORAMA

The concepts of *frame* and *framing* have interested researchers coming from different areas of research: from psychology, sociology and communication, to conflict resolution, artificial intelligence, and management (Dewulf et al., 2009; Goffman, 1974; Minsky, 1975). However, the dialogue between traditions is still not sufficiently developed: the varied of subjects, methodologies, and definitions scattered findings and increased the scepticism among disciplines (Entman, 1993).

Across the years, two different approaches have been forming in the frame tradition: the interactional or communicative perspective and the cognitive paradigm (Dewulf et al., 2009; Shmueli, 2008). The bipartite nature of framing is reflected in the use of the word "frame" as a noun (frame) or a verb (to frame or framing) (Shmueli, 2008). In a cognitive perspective, frame refers to the mental structure that helps individuals to reduce the complexity of reality through information selection, simplification, and categorization (Shmueli, 2008). In contrast, in a communicative perspective, framing is the act of frame's creation during social communicative interactions and the consequences that this phenomenon brings (Shmueli, 2008). Despite the ontological, epistemological and methodological differences between the two traditions (Dewulf et al., 2009), it is important to underline the complementarity and compatibility of the cognitive and communicative aspects (Shmueli, 2008). Linguistics and Discourse Analysis offer a set of theoretical tools that communication researchers can use to comprehensively analyse frames. In specific, the linguistic perspective enables the study of the lexicon level and the framing-activating power of words. This has the potential to integrate the news framing analysis by showing how frames are activated at the linguistic level.

FRAMING IN JOURNALISM

The importance of framing in journalism is hard to question. Journalists choose carefully words and mental images because by describing facts and issues, they can influence public opinion (Tewksbury & Scheufele, 2009). In fact, the power of communicative discourse is embodied by the concept of framing, which Entman (1993, p. 52) defines as:

To select some aspects of a perceived reality and make them more salient in a communicating text, in such a way as to promote a particular problem definition, causal interpretation, moral evaluation, and/ or treatment recommendation.

From the definition elaborated by Entman (1993), the two main characteristics of framing are *selection* and *salience*. Salience refers to the action of making information more relevant and meaningful for the public and is described as the "product of the interaction of texts and receivers" (Entman, 1993, p. 53); it could be reached with the repetition, placement, and association with culturally familiar symbols (Entman, 1993). In contrast, selection involves the inclusion and the omission of features of a situation (Entman, 1993). This process affects the way people perceive and process information (Entman, 1993). Moreover, for each frame, Entman (1993) individuates the focuses, i.e. issues, events, and/or political actors, and the four functions of framing in the news:

...define problems - determine what a causal agent is doing with what costs and benefits, usually measured in terms of common cultural values; diagnose causes - identify the forces creating the problem; make moral judgements - evaluate causal agents and their effects; and suggest remedies - offer and justify treatments for the problems and predict their likely effects (Entman 1993, p.52).

According to Entman (2004), in the news, frames are part of the reporting process and the three focuses by the four functions of framing create a matrix of twelve cells: a fully developed frame has all the twelve cells explicitly filled and consist of information about the event, related issues, and moral assessments reporting about political events, issues and actors such as individual leaders, nations or groups (Entman, 2004).

Furthermore, at least four locations take part in the communication process relative to a frame: the communicator, the text, the receiver and the culture (Entman, 1993). As reported by this author, frames are as powerful as language itself. Indeed, frames can promote analogies with previous stories and stimulate empathy through the use of words and images that allow the audience to assume the perspective of the subject (Entman, 2004). Furthermore, a successful frame has magnitude, i.e. favouring one side, while shrinking the elements of counter frames (Entman, 2004).

Further important contributors in the field of the news framing are Edelman (1993), who emphasizes the connection between public opinion and categories, and Pan and Kosicki (1993, p. 57) who define news media frames as "a cognitive device used in information encoding, interpreting, and retrieving ... related to journalistic professional routines and conventions." News media, in this process, play an important role, because they generate, organize, and transmit frames, and link social structure with the person (Baresch, Hsu, & Reese, 2009). Also, words choice and the sources selection are related to framing (Baresch et al., 2009). According to Kothari (2010), the typology of primary sources used in the story, journalists' location, and the subject of the story determine a media frame.

To build a coherent theory of framing, scholars such as Scheufele (1999) and de Vreese (2005) have developed process models of framing. Moreover, Tewksbury and Scheufele (2009) and Lecheler and de Vreese (2012) have suggested the influence and the effects of news framing on citizens' opinion. Another significant research area is news discourse, the application of some Discourse Analysis techniques to the field of news framing. Considering news text as a form of discourse, Pan & Kosicki (1993) conceptualize the news discourse as a sociocognitive cyclical process that involves sources, journalists, and audience members in a shared culture and that interact according to their socially defined roles (Pan & Kosicki,

1993). Also, van Dijk (1985, 1988) considers news as public discourses that have both cognitive and social aspects, being both text and social interaction.

However, in the field of news framing, tools from linguistic semantics are rarely integrated. This generates a lack of understanding of how frames are created using words. The integration of linguistic approaches to framing could help bridge the divide between cognitive and communicative approaches to framing by showing the frame-activating power of words.

LINGUISTIC APPROACHES TO FRAMES

Studies of framing in linguistics often rely on frame semantics and, in particular, on Fillmore's notion of frame (Rocci, 2009). According to Fillmore (1982/2006, p. 373), the semantics of frames gives a "way of looking at word meanings, … of characterizing principles for creating new words and phrases, for adding new meanings to words, and for assembling the meanings of elements in a text into the total meaning of the text." Frames are also (2003) defined as the lexical and grammatical set of choices in a given language for naming and describing the categories and connections in schemata; these collect and categorize different notions regarding actions, institutions, and objects (Fillmore, 2003). They are so conceptualized by Fillmore (2003) as tools that allow people reality comprehension and interpretation. In this way, words are chosen in accordance with a specific view of the world and have attached a value judgment (Bigi & Greco Morasso, 2012). As Fillmore (1982/2006) puts it:

By the term 'frame' I have in mind any system of concepts related in such a way that to understand any one of them you have to understand the whole structure in which it fits; when one of the things in such a structure is introduced into a text, or into a conversation, all of the others are automatically made available (p. 373).

Observing that more words may evoke the same scene, Fillmore (1982/2006) highlights that people not only choose a frame, but also the specific words within that frame. Every word has a certain focus on specific aspects and roles (Fillmore, 1982/2006). In other words, as Rocci (2009, p. 262) puts it, choosing one or the other word "*activates*, or highlights, certain elements of the schema leaving other elements unexpressed, and present the whole scene from a particular *perspective*." Thus, the framing process implies two levels of choice: the selection of a frame, and the choice of a specific word in the frame (Fillmore, 1982/2006).

A second and complementary instrument adopted in this paper for analysing framing in a linguistic perspective is Congruity Theory, an approach to the semantics and the pragmatics of texts developed by Rigotti (1993) and further elaborated in Rigotti and Rocci (2005). Congruity Theory is aimed to explain the meaning in terms of congruity between predicates and arguments at different levels (including the semantic and pragmatic levels). In Congruity Theory, the concept of frame is developed from the argument frame of a predicate and of the roles of each argument place:

Each predicate defines a set of argument slots (or semantic roles) corresponding to the participants in an event or action, or the elements of a state of affairs. Each role slot is defined by a set of features that the actual participant must fulfill (Rocci, 2009, p. 261).

METHODOLOGY

The methodology of this study has been developed to respond the following research questions:

What are the frames used in news discourse in order to talk about the Charlie Hebdo attack on the day after the event, the 8th January 2015? How is the event interpreted in different frames? What is the semantics of the most frequent frame-activating words?

The focus placed on a single event as a case study enables to investigate in depth the phenomenon of framing in news discourse. The Charlie Hebdo attack could be considered an important chapter of contemporary history and, for its multiple interpretations, it is particularly suited for the study of news framing. We applied tools developed in text linguistic, and, more specifically the models by Fillmore (1982/2006; 2003) and Congruity Theory, in the news context, where Entman (1993; 2004) has focused his work on framing. In particular, this paper sets out to show the frame-activating power of certain lexical items and how frames are created in media discourse.

In this work, the empirical data consists in a corpus of written articles from the press, which have been selected according to the following criteria. First, because of the huge number of articles about the attack, only on the day immediately after the attack, 8th January 2015 is considered; in fact, as argued by Entman (2004, p. 82), "early coverage is the most important because it shapes audience reactions to succeeding information." Furthermore, more for bad than for good news, "challenging first impression is difficult, particularly when they are vividly supported by emotional language and visual images of threat" (Entman, 2004, p. 43). Other criteria adopted for the selection of articles are: the frequency of publication, the language, the prestige of each newspaper, and the topic of the article were considered. Because the analysis is focused on a specific day, only articles published in daily newspapers were selected. Regarding the language of publication, only articles in Italian, English, and German were considered. Other criteria are the prestige, the relevance of the newspaper, and the number of copies sold daily: it was established that each newspaper should have listed in the main five newspapers of its country. Finally, the main topic of all articles was the Charlie Hebdo attack, irrespectively of the textual genre (news article, feature article or editorial). We retrieved articles published on the 8th January 2015 and marked with the keywords "Charlie Hebdo" using the electronic database *Factiva*.

The sample so constructed includes 100 articles, published on some of the major newspapers in the USA, the UK, Italy, Germany, and Switzerland, on the day after the Charlie Hebdo attack: 30 articles are in Italian, 50 in English, and 20 in German[1].

After a first reading of the articles, an a priori and concise operationalisation of frames, as recommended by de Vreese (2005), was made. Five different frames emerged. In a second reading of each article, the words associated to each frame were underlined and their frequency counted. As noted by Fillmore (1982/2006), a single word has the power to recall the entire scene in the mind of the receiver. For instance, Fillmore (1976) explains how the whole frame of commercial event, where the buyer, the seller, the goods, and the money are the fundamental roles, is activated in mind with just one word, such as "buy," "pay," "cost," "sell." Each frame-activating word has been semantically analysed in accordance with Congruity Theory. This kind of analysis, going beyond the shallow surface of texts, clarifies meaning, discloses ambiguities and evaluates congruity. In particular, it considers if arguments present in real utterances are congruous, i.e. if each argument satisfies *presuppositions*, i.e. the conditions that the predicate imposes on its arguments; then, *implications* show the predicate's semantic effects (Rigotti &

Cigada, 2013). In order to explain the analysis that will be done in the present work, the authors will use an example reported in Rocci (2005): in the simple sentence *Louis read a book*, *Louis* and *a book* are the two arguments of the predicate *to read* (x_1, x_2). The predicate imposes *presupposition*s onto its potential arguments; in particular, x_1 must be an alphabetized human being, whilst x_2 must be a written text. All the conditions are satisfied in this utterance; therefore, it is semantically congruous (Rocci, 2005).

In this work, the predicate-argument analysis is considered the first step for the reconstruction of how frames are activated. This analysis will then be integrated into frame analysis, mainly based on Entman's (2004) classification. The semantic and frame analysis are considered as two complementary elements that allow a complete comprehension of framing in news discourse and, particularly, on how framing is created using words. In this sense, the present work represents a methodological integration between the tradition of study of framing in news discourse and the linguistic tradition of frame analysis.

FINDINGS AND DISCUSSION

From the analysis of the corpus, five main frames have emerged according to which the Charlie Hebdo attack has been categorized: *act of the war on terror, attack on Western freedom, the price to pay for the offence, the result of disastrous integration policies,* and (result of) *Islam internal issues*. Depending on the frame, the meaning of this event changes substantially (c.f. the analysis in Greco Morasso, 2012). In each frame, one and the same event is understood as an effect – even a "symbolic effect" – due to a given cause. The event is seen as negative and unjustifiable if it is an attack on a positive value such as freedom; on the opposite, it is seen as a mere reaction if it is understood as a punishment for Charlie Hebdo's irreverent policies.

In what follows, it will be explained how these five frames are activated starting from lexical elements in newspaper discourse. Therefore, for each of these five frames, a frame-activating word will be semantically analysed; then, the authors will integrate the frame analysis in accordance with Entman (2004).

Frame 1: Act of War on Terrorism

The predicate *to be terrorist* activates the frame *act of the war on terror*. The word *terror** was counted 362 times in the whole sample and it is recorded as the most frequent word among the ones taken into account because of their relevance. In the present case study, the arguments of the predicate *to be terrorist* are the Charlie Hebdo shooting (terrorist event, terrorist act) and the Kouachi brothers (terrorists). The following example illustrates the first meaning: "François Hollande, who had rushed to the scene, called for national unity to combat what he described as 'a *terrorist* act'" (Thomson, 2015, p. 4).

To be terrorist hides two different predicates, depending on how it is used (in relation to acts or individuals). Congruity Theory enables to specify the semantic differences as in Table 1.

As it can be inferred from Table 1, the two meanings are interrelated, as someone is defined as a terrorist if he/she intends to commit terrorist acts. This word, as used in relation to Charlie Hebdo, activates the frame *act of war on terrorism,* which recalls the 9/11 attacks, when the attack was framed as a war, implying enemies, military action, and the state of war, e.g. the values of sacrifice and solidarity (Lakoff, 2001). At a metaphorical level, the frame of the war on terrorism is characterized by the eternal battle of good against evil, where the Western society represents the first pole (Lakoff, 2001). The application of this frame to the Charlie Hebdo attack means the opening of a new chapter or another war, which sees the

Table 1. Semantic analysis of to be terrorist A and B

To be terrorist - A(x1)	To be terrorist – B (x_1)
Presuppositions o ∃x1: an action or an event Implications o x1 is aimed to kill a large number of people in resounding way	Presuppositions o ∃ x_1: an individual or an organized group of individuals Implications o x_1 intends to commit terrorist acts

European nations at the frontline (see for example Reichelt, Wickert, Pointner, & Vehlewald, 2015). At the linguistic level, the analysis of "to be terrorist" shows its negative implications; saying that one fares war against a negative (unjustifiable) terroristic attack is to some extent a justification of the state of war.

According to Entman's model (2004), this frame is focused mainly on the event; the focus on issues and political actors is also developed. The Charlie Hebdo attack is explained as a reprisal attack for joining the USA front against terrorism (see for example Landauro, Bisserbe, & Gauthier-Villars, 2015). This condition of ongoing state of war, as a problematic effect for political actors, is defined as terror and fear for reprisal attack. The remedy on the short run is national unity and the enhancement of counterterrorism; whereas, in a bigger context, the ultimate remedy is to crush radical groups. In the fight "between evil and good," the terrorist groups of IS and Al Qaeda against Western countries, several moral judgments are conveyed. For example, Landauro et al. (2015) reported the quotation of Obama and Hollande, who respectively defined the shooting as an evil and barbarous attack. Furthermore, this frame highlights the military discipline of the perpetrators of the Charlie Hebdo attack that were not self-radicalized lone wolves but professional soldiers (Reichelt et al., 2015).

Frame 2: Attack on Western Freedom

Similar to the first frame for its defensive approach, the second frame identified is *attack on Western freedom*. The Charlie Hebdo attack is qualified as an attack to freedom, one specific value that the Western world treasures. Before moving to the discussion of this frame, the predicates *to defend* and *to attack,* activators of the frame, are semantically analyzed because their semantics offers different points of view on the same scene.

The word *attack** was counted 308 times in the entire sample. For example, a Daily Mail article presents in its title the predicate *to attack*: "A murderous *attack* on Western freedoms" (2015). In this example, the predicate *to attack* is used metaphorically with the following semantic structure:

To attack (x_1, x_2)
Presuppositions
o ∃ x_1: an institution, organization, or a person
o ∃ x_2: a value
Implications
o x_1, with a particular emphasis, assaults x_2
o x_2 is jeopardized

The predicate *to defend* considered in this work is illustrated in the following example:

Any society that's serious about liberty has to defend the free flow of ugly words, even ugly sentiments ("The killers in Paris trained their Kalashnikovs on free speech everywhere," 2015).

In this case, as the predicate *to attack, to defend* is metaphorically used in the sample and has the following semantic structure:

To defend (x_1, x_2)
Presuppositions
o $\exists\, x_1$: an institution, organization, or a person
o $\exists\, x_2$: a value that is dear to x_1 and that was attacked earlier
Implications
o x_1 protects x_2 from another attack
o x_2 has x_1 by its side

As we see from the semantics of "to defend," defense is meaningful if x_2 has been attacked earlier and if it is a value that is dear to x_1; by focusing on Western freedom as attacked by the Kouachi brothers, this frame justifies a reaction of defense.

At the center of this frame lies the concept of freedom. In the entire sample, *free** was counted 319 times, but, the number is even greater if we consider also the quasi-synonym *liberty* counted about ten times. As noted by Wierzbicka (1997), these two terms represent one of the most important values of Western societies. It is worth recalling Wierzbicka's semantic analysis of this term in order to give a clearer account of this frame. According to Wierzbicka (1997), *freedom* corresponds to the Latin *libertas* but has a more negative orientation: "it has to do with being able NOT TO DO things that one doesn't want to do, and ... with being able to do things that one wants to do WITHOUT INTERFERENCE from other people" (p.129). In the considered frame, *freedom* occupies a central position that reflects its importance for the Western society. Indeed, it is considered the pillar of democracy that protects the fundamental rights of freedom of thought, of expression, and of the press. In contrast, according to this frame, the Muslim fanatics respond to cartoons with a fatwa, according to the Islamic law, and in the worst cases, by killing journalists. The victims become martyrs and heroes, who had died for the defence of Western civilization. This frame is created by recalling also to the mind the idea of civil religion: the Charlie Hebdo attack is an attack on the heart of "our" civilization. For instance, Serra (2015) argues the incompatibility between the jihadists and the Western society by claiming extremists' inability and indifference to freedom.

According to Entman's (2004) model, the frame *attack on Western freedom* is focused on the *event* and on an *issue*. The defined problematic effect of the event is the martyrdom in the name of freedom; whereas, the problematic conditions of the issue are the differences between the Islamic dictatorship and the Western model of democracy. The major causes are, for the issue, the spread of Islamic Fundamentalism, and, for the event, fanatics' hate. In what is constructed as a fight of jihadists against Westerns, regarded as political actors, the endorsed remedy is to continue to follow and defend 'our' traditions and freedoms. Furthermore, this frame is characterized by Western pride and moral superiority.

Frames 3 and 4: The Price to Pay for the Offence and Result of Disastrous Integration policies

The third and fourth frames, *the price to pay for the offence* and *result of disastrous integration policies* present the same event from quite a different perspective. Both frames are mainly activated by three different predicates: *to provoke*, *to offend*, and *to be irreverent*.

Provoke is used 48 times in the sample. In reference to the Charlie Hebdo staff's behaviour, the predicate *to provoke* can be illustrated through the following example: "attack on the magazine that has *provoked* Muslims" (Paci, 2015, para. 4). The predicate *to provoke* has four arguments as in the following semantic analysis:

to provoke (x_1, x_2, x_3, x_4)
Presuppositions
o $\exists x_1$: an institution, organization, or a person
o $\exists x_2$: an institution, organization, or a person
o $\exists x_3$: an action or an intended action of x_1 towards x_2, negative for x_2
o $\exists x_4$: an action or a feeling of x_2 towards x_1
Implications
o with x_3, x_1 wants to trigger the reaction x_4 in x_2
o x_2 can decide to do or not x_4 as a response of x_3

Another predicate constituting this frame is *to offend*. In the sample offend* was counted 31 times, 28 of which in articles in English: an example is "If there is a right to free speech, implicit within it there has to be a right to *offend*" ("The killers in Paris trained their Kalashnikovs on free speech everywhere," 2015). *To offend* is a predicate with two arguments, as can be seen in this example: "expressions that appear intended to ... *offend* Muslims" ("The final word," 2015).

To offend (x_1, x_2)
Presuppositions
o $\exists x_1$: an action, a state of affairs, an item or an institution, organization, or a person that for its own characteristics can hurt x_2
o $\exists x_2$: an institution, organization, or a person that is emotionally involved in x_1
Implications
o x_2 is hurt in its values, feelings, and personality by x_1

To be irreverent, a one-argument predicate, is used 26 times in the whole sample for the description of the French satirists, in cases such as the following: "the irreverent French magazine" ("An irreverent French institution," 2015, para. 1).

To be irreverent (x_1)
Presuppositions
o ∃ x_1: an institution, organization, or a person that chooses to do not show respect to others
Implications
o deliberately x_1 with his/her behaviour does no pay attention to the sensibility of others, even if this
   could hurt other people

These three predicates *to provoke, to offend* and *to be irreverent* share the focus on the actions per-formed by Charlie Hebdo journalists. The attack is implicitly seen as a quasi-mechanical, unavoidable answer triggered by the journalists' behavior. Such behavior is qualified in different ways; in fact, each of the three predicates conveys different information and offers a unique point of view on the frame. In particular, *to be irreverent* is centred on Charlie Hebdo's respect-less approach to the society; someone who is irreverent might be naive, not calculating the consequences of his or her attitude. On the oppo-site, the predicate *to provoke* offers a more comprehensive perspective of the event and underlines the deliberate will to trigger a reaction in Muslims. Finally, the predicate *to offend* specifies that the action that is at the basis of the provocation is an offense which to some extent deserves punishment.

The *price to pay for the offence* frame is centred on the responsibility of the Charlie Hebdo staff in the continuous irreverent behaviour towards Muslims. We labeled this frame *the price to pay for the offence* because this image is recurring in the sample. In terms of news frame analysis, the focus of this frame is mainly on the attack as an event; and on an issue regarding the limits of satire and the differ-ences between cultures. The major agents involved in this the event are insolent journalists, on the one hand, and extremists, on the other hand.

An analysis of the identified frame-activating predicates permits to better understand how this frame is activated and why it conveys a reflection on the limits of satire and on the relation between cultures. Notably, an implication of right punishment is involved in the predicate *to offend*: by using this word the journalists imply that the French satirists have made some mistakes; offending is negative and, therefore, it deserves punishment. On 8th January 2015, several newspapers criticized the satirical magazine for its cruel attack on a religion with a different culture from the Western one. For example, Carvajal and Daley (2015) highlight the insolence and irreverence of the French institution highlighting the intentionality of the journalists' offensive behavior (e.g. "it featured a mock debate," "not food for thought," "proud to offend," and "intentionally offensive content"); whereas Häfliger (2015) reports the controversial comment of the Federal Council member Leuthard on the Charlie Hebdo attack "satire is not a freepass" (para. 3). According to this frame, the journalists of the satirical magazine were deliberately offensive in their cartoons, despite the awareness of the risks that they were facing. The *price to pay for the offence* frame does not justify the killing of the French satirists but is it critical about the (more or less) lack of Charlie Hebdo sensibility and caution toward the Islamic World.

A further frame activated mainly by the predicate *to provoke,* but also by *to offend* and *to be irreverent* is *the result of disastrous integration policies*. This frame contextualizes the Charlie Hebdo attack in a far bigger issue, the modern European society. The focus is more on the issue than on the event because the attack is regarded as the tip of an iceberg. The attack is seen as the result of disastrous integration policies; these articles underline European failures by conveying, at the same time, a moral judgment. The scaring scenario with marginalised young Muslims as agents but also victims encourages both the radicalization and the rise of right-wing parties, defined in the articles as problematic conditions. The only remedy, according to the frame, is the integration of migrants in the society, a solution that sees the cooperation of Muslims and Western countries, the two major political agents. The event, instead, is defined a merciless massacre; furthermore, a heavy moral judgment is expressed towards Charlie Hebdo, seen as a magazine run by old-fashioned satirists. For instance, Hussey (2015) defines the French magazine "a museum piece" (para. 9) and accuses the French satirists of unnecessary provocation towards immigrants.

This frame is similar to the previous one in the focus on the European attitudes as a possible cause of the attack; in this sense, both frames question the Europeans' (and the journalists') responsibility. However, the present frame sees the event in a broader political context, inserting Charlie Hebdo in the bigger picture of the Western attitude towards other cultures.

Frame 5: Islam Internal Issues

The two predicates *moderate and extremist* activate the fifth frame *Islam internal issues*. *Moderate* and *extremist*, in fact, are two examples of scalar predicates that Fillmore, Kay and O'Connor (1988, p. 525) explain as predicates that "require that some scalar array of the compared variable pairs be automatically set up as an initial step in interpreting the sentence." Indeed, if people think of something *moderate*, they position it in the middle, between the two extremities; in the case of *extreme*, they arrange it in the far end. The example of *moderate* and *extremist* is fascinating and shows as behind the apparent neutral lexicon choices there is much more. As underlined by the analysis of *moderate* and *extremist*, this distinction is already included in the presuppositions of each of these words.

The frame Islam internal issues has a strong focus on the event and political actors and deals with the controversies on the interpretation of the Koran among Muslims. The main agents of the Charlie Hebdo are intolerant extremists; however it is acknowledged that, from a political perspective, there are differences between moderate and extremist Muslims. As a remedy, it is said that the division should be officialised by Muslim authorities with a clear condemnation of ISIS and other terrorist groups. In this frame, it is remarked the urgent support from Muslim organizations in order to prevent this kind of episodes. An example of dissociation is the one of the Islamic thinker Tariq Ramadan, who commented that terrorists have betrayed and tainted Islam pillars (Black, 2015). From the analysis, the moral superiority of Westerns has emerged. As an example of the Islam incompatibility with the Western civilization, the issue of women's oppression was counted various times among the sample (see for example Kristof, 2015). By contrasting moderate and extremist Muslims, this frame creates a polarity that is implicitly seen as an internal conflict for Islam. Charlie Hebdo is portrayed as the consequence of this wider problem that allegedly characterizes Muslims. The semantic choice, in this sense, partially shifts the responsibility of the event from the perpetrators to a broader societal or cultural problem.

CONCLUSION

This paper has presented a novel combination of approaches for the study of framing in the news and its communicative functions. The work has applied some text linguistics tools, in particular, Fillmore (1982/2006) and Congruity Theory, to the news context. In this sense, an added value of the present work is methodological: the combination of linguistic tools with the studies on framing in news discourse covers a gap in literature and allows a better understanding of how frames are created in the news through the frame-activating power of words. This has given rise to a double-level analysis, in which the semantics of the main frame-activating words in the corpus have been associated with an analysis of the significance and value of frames in the news. Conversely, Entman's (2004) model for analyzing frames in media discourse is given more substance thanks to a linguistically grounded analysis of the frame-activating words, which explains why and how some frames are generated. The choice of the frame-activating words that we have analyzed is motivated on the basis of frequency. From the analysis of the 100 news articles, published on the 8th January 2015 in English, German, and Italian that make up the sample, five frames emerged: *act of war on terrorism, attack on Western freedom, the price to pay for the offence, the result of disastrous integration policies,* and *Islam internal issues*. Framing is activated by specific choices at the lexical level. Frame-activating words permit to understand one and the same event in very different ways, ranging from an unjustifiable terroristic attack against a Western value, to an unavoidable consequence of Western behaviors or cultural attitudes, to the sign of a surfacing problem that allegedly affects Muslims in general. It is clear that each of these frames triggers different questions at the level of responsibility and possible countermeasures. Following Greco Morasso (2012), we might say that this type of contextual framing of event in the news gives rise to different implicit argumentative and rhetorical discourses.

This study has analyzed a single case study in order to put the above described methodological integration into practice. This choice inevitably leads to limitations. First, the choice to focus on the day after the attack allowed exploring early coverage, which is the most powerful in terms of framing (Entman, 2004); however, we left out the changing nature of framing on a broader time-span. Secondly, despite the selection of representative European newspapers, other languages were excluded from this work. Also, it must be considered that the Factiva database does not offer access to all possible newspapers.

These limitations notwithstanding, this paper paves the way for a systematic integration of linguistic aspects into the analysis of frames in news discourse. This sheds light on how the interpretation of events is guided and constructed through news frames, thus providing a nuanced view of a complex communicative phenomenon.

NOTE

This paper has been written as a development of an unpublished bachelor thesis written in 2016 by M. Tribastone under the supervision of S. Greco at USI - Università della Svizzera italiana, Switzerland.

REFERENCES

A murderous attack on Western freedoms. (2015, January 8). *Daily Mail*, p.14.

An irreverent French institution (2015, January 8). *Financial Times*, p. 3.

Baresh, B., Hsu, S., & Reese, S. D. (2009). The power of framing: new challenges for researching the structure of meaning in news. In S. Allan (Ed.), *The Routledge Companion to News and Journalism* (pp. 637–647).

Bigi, S., & Greco Morasso, S. (2012). Keywords, frames and the reconstruction of material starting points in argumentation. *Journal of Pragmatics*, *44*(10), 1135–1149. doi:10.1016/j.pragma.2012.04.011

Black, I. (2015, January 8). Paris terror attack. Reaction: leaders condemn attack. *The Guardian*, p. 5.

Carvajal, D., & Daley, S. (2015, January 8). Proud to offend. Paper carries torch of political provocation. *The New York Times*.

De Vreese, C. H. (2005). News framing: theory and typology. *Information Design Journal, Document Design, 13*(1), 51-62.

Dewulf, A., Gray, B., Putnam, L., Lewicki, R., Aarts, N., Bouwen, R., & Van Woerkum, C. (2009). Disentangling approaches to framing in conflict and negotiation research: A meta-paradigmatic perspective. *Human Relations, 62*(2), 155–193. doi:10.1177/0018726708100356

Edelman, M. (1993). Contestable categories and public opinion. *Political Communication, 10*(3), 231–242. doi:10.1080/10584609.1993.9962981

Entman, R. M. (1993). Framing: Toward clarification of a fractured paradigm. *Journal of Communication, 43*(4), 51–58. doi:10.1111/j.1460-2466.1993.tb01304.x

Entman, R. M. (2004). *Projections of power: Framing news, public opinion, and US foreign policy.* Chicago: The University of Chicago Press.

Fillmore, C. J. (2003). *Form and meaning in language.* Leland, CA: CSLI Publications.

Fillmore, C. J. (2006). Frame semantics. In D. Geeraerts (Ed.), *Cognitive Linguistics: Basic readings* (pp. 373–400). Berlin: De Gruyter. (Original work published 1982) doi:10.1515/9783110199901.373

Fillmore, C. J., Kay, P., & O'Connor, M. C. (1988). Regularity and Idiomaticity in Grammatical Constructions: The Case of Let Alone. *Language, 64*(3), 501–538. doi:10.2307/414531

Fowler, R. (1996). *Language in the news: Discourse and ideology in the press.* London: Routledge.

Goffman, E. (1974). *Frame analysis: An essay on the organization of experience.* London: Harper and Row.

Greco Morasso, S. (2012). Contextual frames and their argumentative implications: A case study in media argumentation. *Discourse Studies, 14*(2), 197–216. doi:10.1177/1461445611433636

Häfliger, M. (2015, January 8). Satire-Freiheit relativiert; Leuthard erntet Empörung. *Neue Zürcher Zeitung*, 10.

Hussey, A. (2015, January 8). French humor, turned into tragedy. *The New York Times*, p. 23.

Kothari, A. (2010). The framing of the Darfur Conflict in *The New York Times*: 2003-2006. *Journalism Studies, 11*(2), 209-224.

Kristof, N. (2015, January 8). Satire, Terrorism and Islam. *The New York Times*, p. 23.

Lakoff, G. (2001). *Metaphors of terror.* Retrieved July 12, 2016, from http://www.press.uchicago.edu/sites/daysafter/911lakoff.html

Landauro, I., Bisserbe, N., & Gauthier-Villars, D. (2015, January 8). Terror strikes heart of Paris. Gunmen attack French magazine, killing at least 12. *The Wall Street Journal*.

Lecheler, S., & de Vreese, C. H. (2012). What a difference a day makes? The effects of repetitive and competitive news framing over time. *Communication Research*, *40*(2), 147–175. doi:10.1177/0093650212470688

Minsky, M. (1975). A framework for representing knowledge. In P. H. Winston (Ed.), *The psychology of computer vision* (pp. 211–277). New York: McGraw-Hill.

Paci, F. (2015, January 8). Islam diviso: le autorità condannano ma nel Web dilaga l'odio per Parigi. *La Stampa*, p. 5.

Pan, Z., & Kosicki, G. M. (1993). Framing analysis: An approach to news discourse. *Political Communication*, *10*(1), 55–75. doi:10.1080/10584609.1993.9962963

Reichelt, J., Wickert, U., Pointner, N., & Vehlewald, H. (2015, January 8). Blutiger Terroranschlag in Frankreich seit Jahrzehnten. *Bild*.

Rigotti, E. (1993). La sequenza testuale. Definizione e procedimenti di analisi con esemplificazione in lingue diverse. *L'analisi linguistica e letteraria, 1*(2), 43-148.

Rigotti, E., & Cigada, S. (2013). La comunicazione verbale (2nd ed.). Santarcangelo di Romagna: Maggioli.

Rigotti, E., & Rocci, A. (2001). Sens – non-sens – contresens. *Studies in Communication Sciences*, *2*, 45–80.

Rocci, A. (2005). Are manipulative texts 'coherent'? Manipulation, presuppositions and (in)congruity. In L. Saussure & P. Schulz (Eds.), *Manipulation and ideologies in the twentieth century: Discourse, language, mind* (pp. 85–112). Amsterdam: Benjamins. doi:10.1075/dapsac.17.06roc

Rosaspina, E. (2015, January 8). Liberi e irriverenti: la redazione che non c'è più. *Corriere della Sera*, p. 8.

Scheufele, D. A. (1999). Framing as a theory of media effects. *Journal of Communication*, *49*(1), 103–122. doi:10.1111/j.1460-2466.1999.tb02784.x

Serra, M. (2015, January 8). La brigata dei libertini. *La Stampa*.

Shmueli, D. F. (2008). Framing in geographical analysis of environmental conflicts: Theory, methodology and three case studies. *Geoforum*, *39*(6), 2048–2061. doi:10.1016/j.geoforum.2008.08.006

Tewksbury, D., & Scheufele, D. A. (2009). News framing theory and research. In J. Bryant & M. B. Oliver (Eds.), *Media effects: Advances in theory and research* (3rd ed.; pp. 17–33). London: Routledge.

The final word. (2015, January 8). *The Washington Post*.

Thomson, A. (2015, January 8). France grapples with state of shock. Terror in Paris. *The Financial Times*, p. 8.

van Dijk, T. A. (1985). Structure of news in the press. In T. A. van Dijk (Ed.), *Discourse and communication. New approaches to the analysis of mass media discourse and communication* (pp. 69–93). Berlin: De Gruyter. doi:10.1515/9783110852141.69

van Dijk, T. A. (1988). *News as discourse*. New York: Routledge.

van Eemeren, F. H. (2010). *Strategic maneuvering in argumentative discourse: extending the pragma-dialectical theory of argumentation*. Amsterdam: Benjamins. doi:10.1075/aic.2

Wierzbicka, A. (1997). *Understanding cultures through their keywords. English, Russian, Polish, German, and Japanese*. New York: Oxford University Press.

ENDNOTE

[1]   The choice of these languages depended on two aspects. First, we decided to verify how Charlie Hebdo was framed outside France (thus, French was excluded). Second, the choice of the other European languages depended on the first author's linguistic knowledge. The corpus was collected by the first author.

Chapter 6
Meaning in the Age of Big Data

Mihai Nadin
University of Texas at Dallas, USA

ABSTRACT

The most fascinating semiotic applications of recent years came not from semioticians but from those who practice semiotics without knowing they do so (what the author calls the Monsieur Jourdain syndrome). Military and surveillance applications, genome sequencing, and the practice of phenotyping are immediate examples. The entire domain of digital computation, now settled in the big data paradigm, provides further proof of this state of affairs. After everything was turned into a matter of gamification, it is now an exercise in data acquisition (as much as possible) and processing at a scale never before imagined. The argument made in this chapter is that semiotic awareness could give to science and technology, in the forefront of human activity today, a sense of direction. Moreover, meaning, which is the subject matter of semiotics, would ground the impressive achievements we are experiencing within a context of checks-and-balances. In the absence of such a critical context, the promising can easily become the menacing. To help avoid digital dystopia, semiotics itself will have to change.

PRELIMINARIES

Language interaction is the most definitory activity of the self-constitution of the species *homo sapiens*. Self-constitution—i.e., the making of ourselves through the activity in which we are involved (Nadin (1997))—takes place at all levels of life—in animals and even plants. However, the making and remaking of the human being under circumstances involving language associate the process of self-constitution with awareness. While it is true that a bacterium swimming upstream in a glucose gradient marks the beginning of goal-directed intentionality (Sowa 2017), it is only through language that purposiveness—a particular expression of anticipation—becomes possible, and indeed necessary (Brentano, 1874; Margulis, 1995). Of course, language-based human interaction is only one among the many sign systems through which self-constitution takes place. It became the focus of inquiry (philosophical, scientific, aesthetic, social, etc.) since all other forms of expression (images, sounds, odors, etc.) are, so to say, *more natural*, that is, they appear as extensions of the senses. Language conjures the association with thinking, and as such it is present even in sign processes transcending language. The abstraction of mathematical or chemical

DOI: 10.4018/978-1-5225-5622-0.ch006

formulae invites a language of explanations: what we would call *decoding*. Images, sounds, textures, rhythms, and whatever else are never language-free. Therefore, a re-examination of conceptions of language—the classic path from Aristotle to the computational theories of our days—is almost inevitable.

Today's ontology engineering, i.e., translating language into computable specifications of everything (for example, "Siri, what's the time?" new medical treatments, new materials, new forms of transactions) is nothing but the expression of how we can tame language so that machines (of today or of tomorrow) can "understand" what we want. Ideally, such machines would think the way we do. With this subject, we are moving from "What is X?" (any subject, such as what is matter, or sex, or justice) to how we make new entities, how we think, how we evaluate thinking.

Wittgenstein is laughing louder than ever (at least in spirit). In rejecting the *name theory of language* (associated with Socrates), he knew that words do not correspond to things. (By the way, Eco was a follower of Wittgenstein in this sense.) Although not a semiotician himself, Wittgenstein wrote in *Philosophical Investigations* (PI) what everyone active in semiotics should learn by heart: "Every sign by itself seems dead. What gives it life? In use, it lives" (Wittgenstein, 1953). In *On Interpretation*, Aristotle distinguished between *sêmeion*—natural sign, such as a symptom of disease—and *symbolon*—"casting together," adopted by convention, shared. But he remained pretty much captive to the idea that signs—and, by extension, words—correspond to objects, "same for everyone, and so are the objects of which they are likeness" (Aristotle, 350 BCE; see also Dewart, 2016). With Wittgenstein, we experience a change in perspective: signs, and especially language, which was his focus, are associated with tools. This translates as: language is associated with activities. This is exactly what ontology engineering means in our days: specify an object or process, and program the computer to produce it or recognize it. Make it even actionable: when risk is identified, a process affected by risk can be avoided or triggered. In Wittgenstein's words:

Think of the tools in a toolbox: there is a hammer, pliers, a saw, a screwdriver, a rule, a glue-pot, nails and screws. The functions of words are as diverse as the functions of these objects (PI, 11).

(Of course, the toolbox is now expressed virtually in one program or several, or in the ubiquitous apps.) It is no surprise then that Stuart Kauffmann (2011) adopts the screwdriver: his focus is on relational features (a subject I shall revisit in this study). He knows that no computer process can capture all functions of objects, as we use them, some according to their purpose, others according to purposes we make up. (Kauffman's take on the screwdriver is actually about the limits of algorithmic computation.)

But let's stay focused. What this study proclaims is the need to *rethink the foundations of semiotics*. In concrete terms: the sign as the knowledge domain of semiotics explains why semiotics entraps itself, as a discipline, in a dead-end street where all that can be expected from it are reflections in a house of mirrors, all showing the same image from many viewpoints, but none suggesting the path out of this self-delusional condition. Peirce—to whom we owe the modern foundation of semiotics—was aware of the danger of focusing on the sign. The interpretant, as part of the sign definition (uniting object, representamen, and interpretant) was meant to give a dynamic dimension to sign-based activities. But the notion of semiosis remained undefined; its nature as process was mostly ascertained, but not endorsed with an operational function. In a dictation for *Schlick* (December, 1932), Wittgenstein gave a convincing argument for the need, and indeed possibility, to transcend the sign as label, and word as name theory: "…if I were asked what knowledge is, I would enumerate instances of knowledge and add the words 'and similar things'." (Wittgenstein & Waisman, 2003). For describing the dynamics, how various instances of something (such as what is knowledge, what is life, what is justice) complete each other, he chose the metaphor of the game. Of course, there were no video or computer games to refer to (and even less to predict), but rather "board games, card games, bull-games, athletic-games" in which he discerned "similarities, affinities, like for family—build, features, color of eyes, gait, temperament...." Any choice is different, and any attempt at a new foundation is based on the narration of describing the life of signs, and thus of language characterized by their use. Narration is the record of the actions. With the risk of getting ahead of myself, I shall state here that the interpretation aspect in my conception of semiotics is expressed in stories associated with a narration: examine the record and interpret it (but I shall eventually return to this).

To take a sign out of context, i.e., out of the pragmatics in which it participates, is in my view not different from taking a pawn from the chessboard and asking someone who does not play the game what it is. Wittgenstein got it right: outside of the game of chess, the pawn is, for those not familiar with the game, a piece of wood, or metal, or plastic, a dead symbol. In my view, outside of the narration represented by the game (sequence of moves leading to the game's outcome), neither the pawn nor any other figure, not even the chessboard or the game itself, makes sense. Their meaning—the actual object of semiotics—is in the narration: the game played by the rules shared by those involved. The stories of particular games—e.g., Big Blue beating Kasparov, or some champion beaten by a less known player— are interpretations. Deep Learning (in some embodiment of algorithmic AI, such as AlphaGo beating Fan Hui, the champion at Go) are interpretations with an open-ended semiosis. By the way: Kasparov knows what a pawn is; Big Blue does not. Fan Hui knows what Go means, AlphaGo does not. The game was reduced to permutations within a large space of possibilities, and the winner is not intelligence but computational brute force! Kasparov as a winner, or Fan Hui as a winner would have enjoyed the meaning of the game; on the machine side, the engineers enjoy the success of computer performance, expressed in numbers (Big Data at work).

Having sketched here in the Preliminaries the path to the outcome of the study, I will revisit arguments leading to my attempt at a new foundation of semiotics.

CULTURE AS SIGN SYSTEM

The foundation of semiotics around the notion of the sign (shortly mentioned in the Preliminaries) explains its accomplishments. But it also suggests an answer to the question of what led to the failure of the discipline to become the backbone of modern sciences and humanities; or, alternatively, to ascertain its own pragmatic relevance. Indeed, not living up to its possibilities affected not just its own credibility as a specific knowledge domain, but also my claim that it might act as a useful participant in other endeavors. Relevant is the fact that the sciences and the humanities are becoming more and more fragmented in the absence of an integrating coherent semiotic theory. The necessity of such a theory is also highlighted by the extreme focus on quantitative aspects of reality, to the detriment of understanding qualitative aspects, in particular, the meaning of change. Physics, and even chemistry, economics, cognitive science, etc., without mathematics are not a conceivable alternative. But few scientists realize that only when semiotics might acquire the same degree of necessity will conditions be created for complementing the obsession with depth (specialized knowledge) with an understanding of breadth, corresponding to an integrated view of the world.

Many attempts have been made to write a history (or histories) of semiotics: biographies of semioticians, history of semantics, history of symptomatology, anthologies of texts relevant to semiotics, and the like. Few would argue against the perception that we have much better histories of semiotics (and semioticians) than contributions to semiotics as such. What can be learned from the ambitious projects of the past is that semiotic concerns can be identified along the entire history of human activity. In trying to define *The Subject of Semiotics*, Kaja Silverman (1984) correctly identifies authors (in particular, Eco 1976, as well as Lotman 1990) who considered culture as the subject matter of semiotics. Roland Posner (in Basic Tasks of Cultural Semiotics (2004)) correctly noticed that Cassirer (1923-1929) (to whose work we shall return) analyzed sign systems in culture, but also cultures as sign systems. Ana Maria Lorusso (2015), writing in the series *Semiotics and Popular Culture* (edited by Marcel Danesi) advanced a cultural perspective. Initially, semiotic activity was difficult to distinguish from actions and activities related to survival. Over time, semiotic concerns (especially related to language) constituted a distinct awareness of what is needed to succeed in what we do and, furthermore, to be successful.

The aim being the grounding of semiotics, we will examine the variety of angles from which knowledge was defined from its domain. In parallel to the criticism of conceptions that have led to the unsatisfactory condition of semiotics in our time, we will submit a hypothesis regarding a foundation different from that resulting from an agenda of inquiry limited to the sign. Finally, we will argue that the semiotics of semiotics (embodied in, for instance, the organization dedicated to its further development) deserves more attention, given the significance of "organized labor" to the success of the endeavor. Evidently, the American Academy of Arts and Sciences will continue to celebrate accomplishments in domains such as mathematics, physics, computer science, etc., but, so far, not semiotics, whose contributions to society are more difficult to assess. While the grounding of semiotics in the dynamics of phenomena characteristic of a threshold of complexity associated with the living (Figure 1) will be ascertained, the more elaborate grounding in anticipation, is a subject beyond our aims here (see Nadin 2012).

ZOON SEMIOTIKON

Paul Mongré (in 1897) knew more semiotics than Charles Morris (in 1938). Let me explain this statement. We don't really need an agreement on what the subject of semiotics is, or what a sign is, in order to realize that the underlying element of any human interaction, as well as interaction with the world, is semiotic in nature. Interaction takes place through an intermediary. Signs or not, semiotics is about the *in-between*, about *mediation*, about guessing what others do, how nature will behave. Even two human beings touching each other is more than the physical act. In addition to the immediate, material, energetic aspect, the gesture entails a sense of duration, immaterial suggestions, something that eventually will give it meaning. It is a selection (who/what is touched) in a given situation (context). And it prompts a continuation.

But there is more to this preliminary observation. Just as a detail, the following observation comes from brain imaging science: The three most developed active brain regions—one in the prefrontal cortex, one in the parietal and temporal cortices are specifically dedicated to the task of understanding the goings-on of other people's minds (Mitchell, De Houwer, & Lovibond, 2009). This in itself suggests semiotic activity related to anticipation. Actions, our own and of others, are "internalized," i.e., understood and represented in terms of what neurobiology calls *mental states*. So are intentions. In this respect, Gallese (2001, 2009) wrote about mind-reading and associated this faculty with mirror neurons.

Figure 1. Semiotics at the threshold of complexity defining the living

The Living: Complex Systems - escape prediction

The Nonliving: Complicated Systems - subject to prediction

From this perspective, the semiotics of intentions, desires, and beliefs no longer relies on signs, but on representations embodied in cognitive states.

It would be presumptuous, to say the least, to rehash here the detailed account of how the human species defined itself, in its own making, through the qualifier *zoon semiotikon* (Nadin, (1997, pp. 197, 226, 532, 805)), i.e., semiotic animal. Felix Hausdorff, concerned that his reputation as a mathematician would suffer, published, under the pseudonym Paul Mongré, a text entitled *Sant' Ilario. Thoughts from Zarathustra's Landscape* (1897). A short quote illustrates the idea:

The human being is a semiotic animal; his humanness consists of the fact that instead of a natural expression of his needs and gratification, he acquired a conventional, symbolic language that is understandable only through the intermediary of signs. He pays in nominal values, in paper, while the animal in real, direct values (...) The animal acts in Yes and No. The human being says Yes and No and thus attains his happiness or unhappiness abstractly and bathetically. Ratio and oratio are a tremendous simplification of life. . . . (p.7) (Translation mine).

Through semiotic means, grounded in anticipatory processes (attainment of happiness, for instance), individuals aggregate physical and cognitive capabilities in their effort. Indeed, group efforts make possible accomplishments that the individual could not obtain.

Obviously, this perspective is much more comprehensive than the foundation of semiotics on the confusing notion of the sign. In what I described, there is no sign to identify, rather a process of understanding, of reciprocal "reading" and "interpreting." The decisive aspect is the process; the representation is the unfolding of the process defining cognitive states. This view has the added advantage of explaining, though indirectly, the major cause why semiotics as the discipline of signs continues to remain more a promise than the "universal science" that Morris (1938) chose to qualify it. A discipline dependent upon a concept (on which no agreement is possible) is much less productive than a discipline associated with activities: What do semioticians do?

KNOWLEDGE IS A CONSTRUCT

Wittgenstein's views on knowledge led to the understanding of language as a tool for knowledge acquisition. We have access to a large body of shared knowledge on the evolution of humankind, in particular on the role of various forms of interaction among individuals and within communities. The entire history of science and technology is part of this body of shared knowledge. Also documented is the interaction between the human being and the rest of the world. This knowledge is available for persons seeking an understanding of semiotics in connection to practical activities—where the sign lives. This is not different from the situation of mathematics. Let us recall only that geometry originates in activities related to sharing space, and eventually to laying claim to portions of the surroundings, to ownership and exchange, to production and market processes. There are no triangles in the world, as there are no numbers in the world, or lines. To measure a surface, i.e., to introduce a scale, is related to practical tasks. Such tasks become more creative as improved means for qualifying the characteristics of the area are conceived and deployed. To measure is to facilitate the substitution of the real (the measured entity) with the measurement, i.e., representation of what is measured. To travel, to orient oneself, to navigate are all "children of geometry," extended from the immediacy of one's place to its representation. This is where semiotics shows up. The experiences of watching stars and of observing repetitive patterns in the environment translate into constructs, which are integrated in patterns of activity. Rosen took note of "shepherds (who) idly trace out a scorpion in the stars..." (the subject of interest being "relations among components"). He also brought up the issue of observation: "Early man . . . could see the rotation of the Earth every evening just by watching the sky" (Rosen, 1985, p. 201). In the spirit of Hausdorff's definition of the semiotic animal, Rosen's suggestion is that inference from observations to comprehension is not automatic: An early observer "could not understand what he was seeing," as "we have been unable to understand what every organism is telling us" (p. 201). The "language" in which phenomena (astronomic or biological) "talk" to the human being is that of semiotics; the human being constructs its "vocabulary" and "grammar." This applies to our entire knowledge, from the most concrete to the most abstract.

Mathematics, in its more comprehensive condition as an expression of abstract knowledge, is a view of the world as it changes. It is expressed in descriptions such as points, lines, and intersections; in formal entities, such as circle, square, volume, etc. It is expressed numerically, e.g., in proportions, which means analytically, through observations of how things change or remain the same over time. It can as well be expressed synthetically, that is, how we would like to change what is given into something else that we can describe as a goal (using numbers, drawings, diagrams, etc.).

TO UNDERSTAND THE WORLD

Informed by mathematics, we gain an intuitive understanding of how humans, in making themselves, also make their comprehension of the world part of their own reality. The perspective from which we observe reality is itself definitory for what we "see" and "hear," for our perceptions, and for our reasoning. This should help in realizing that the foundation of semiotics is, in the final analysis, a matter of the angle from which we examine its relevance. The hypothesis we shall address is that the definition upon the ill-defined notion of the sign is the major reason why semiotics remains more a promise than an effective theory. The failure of semiotics is semiotic: the representation of its object of inquiry through the entity called *sign* is relatively deceptive. It is as though someone were to establish mathematics around the notion of the number, or the notion of an integral, or the notion of sets. Indeed, there have been mathematicians who try to do just that; but in our days, those attempts are at best documented in the fact that there is number theory (with exceptional accomplishments), integral calculus, and set theory (actually more than one). But none defines mathematics and its goals. They illustrate various mathematical perspectives and document the multi-facetedness of human abstract thinking.

If we focus on the sign, we can at most define a subset of semiotics: sign theory, around classical definitions (as those of Saussure, Peirce, Hjelmslev, for example). But semiotics as such is more than these; and it is something else. Interaction being the definitory characteristic of the living, and semiotics its underlying condition, we could identify as subfields of interest the variety of forms of interaction, or even the variety of semiotic means through which interactions take place. Alternatively, to make interactions the subject of semiotics (as Sadowski attempted 2010) will also not do because interactions are means towards a goal. *Goals* define activities. Activities integrate actions. Actions are associated with representations.

What is semiotics?" not unlike "What is mathematics?" or for that matter "What is chemistry, biology, or philosophy?" are abbreviated inquiries. In order to define something, we actually differentiate. Semiotics is not mathematics. It does not advance a view of the world, but it provides mathematics with some of what it needs to arrive at a view of the world—with a language. Mathematicians do not operate on pieces of land, or on stones (which mathematics might describe in terms of their characteristics), or on brains, on cells, etc. They produce and operate on *representations*, on semiotic entities conjured by the need to replace the real with a description. The *goal* of the mathematicians' activity, involving thinking, intuition, sensory and motoric characteristics, emotions, etc., is abstraction. Their *activity* focuses on very concrete semiotic entities that define a specific language: topology, algebra, category theory, etc.

Among many others, Nietzsche (1975, p. 3)) observed that "Our writing tools are also working, forming our thoughts." As we program the world, we reprogram ourselves: Taylor's assembly line "reprogrammed" the worker; so do word and image processing programs; so do political programs, and the programs assumed by organizations and publications.

TO REPRESENT IS TO PRESENT AGAIN

To represent is one of the fundamental forms of human activity. *To express* is another such form. The fact that there might be a connection between how something (e.g., pain) is expressed (through a scream) and what it expresses is a late realization in a domain eventually defined as cognition. The relation between what (surprise, for example, can also lead to a scream) is expressed and how expression (wide-open eyes) becomes representation is yet another cognitive step. Furthermore, there is a relation between what is represented (e.g., fear) and the means of representation, which can vary from moving away from the cause of the fear to descriptions in words, images, etc. Moreover, to represent is to present one's self—as a living entity interacting with other living entities (individuals, as well as whatever else a person or person interact with)—as an identity subject to generalizations and abstractions. There are signs (usually called *symbols*, cf. Cassirer (1923-1929)) in mathematics, chemistry, and physics; more symbols are to be found in genetics, computer science, and artificial intelligence. But in these knowledge domains, they are not present as semiotic entities—i.e., as relevant to our understanding of interaction—but rather as convenient representations (of mathematical, chemical, or physical aspects), as formal entities, as means for purposes other than the acquisition and dissemination of semiotic knowledge. They are condensed representations. The integral sign $\int$ stands for a limit of sums. It represents the operation (e.g., calculate an area, a volume). Let us recall Lewis Mumford's observations: No computer can make a new symbol out of its own resources" (1967, p. 29).

The abbreviated inquiries invoked earlier—What is semiotics? What is mathematics? What is chemistry?—are relevant because behind them are explicit questions: What, i.e., which specific form of human activity, do they stand for? What do they mediate? What semiotics, or mathematics, or chemistry stands for means: What are their specific pragmatic justifications? What can you do with them? Moreover, while mathematics does not depend upon other sciences, can the same be said about semiotics?

If we could aggregate all representations (Figure 2) we would still not capture reality in its infinite level of detail; nor could we capture dynamics. The living unfolds beyond our epistemological boundaries. We are part of it and therefore every representation will contain the observed and the observer. The infinite recursivity of our observations explains why the phase space of the living is variable: with each new observation, the state of the observer and of the observed change. This is unavoidable. Considering the sequence of observations, translated into representations, we can say that the narrative of observation is by necessity incomplete.

The representation of different parts of the human body in the primary somatosensory cortex is a very clear example of the role of semiotic processes. Those representations change as the individual's activity changes. They facilitate preparation for future activities; they predate decisions and activities. They are in anticipation of change. The semiotics of the process is pragmatically driven. Let's recall Wittgenstein's observation on language as part of an activity. The narrative of life integrates semiotic representation. Think about the fascination with text messaging and how the fingers involved are represented in the cortex. The fact that text messaging affects driving (and leads to accidents) is only the next sequence in the narrative of living language. Semiotics understood in this vein returns knowledge regarding how technology empowers us, as it reshapes our cognitive condition at the same time.

Figure 2. The subset of possible partial representations (musical score, drawing, video or film associated with melody, metaphor, visualization, etc.) complete the descriptions. Representations are always an open ended selection, from which representations of representations etc. can also be generated

ONCE MORE ABOUT KNOWLEDGE

Wittgenstein took note of the fact that language is deceptive. His view was that philosophical problems (and for that matter problems in general) arise from our language—the games we play—not in the world. His world—World War II is the broad context—is, for all practical purposes, not fundamentally different from ours (a continuous state of war, reflecting the competitive nature of capitalism). Migration (the millions seeking a new life away from war, misery, intolerance, terrorism, etc.), political instability, and climate change concerns are expressed in our language—i.e., in the deceptive semiotics of the media—at a scale different from that of reality. Words never corresponded to things; they only re-presented them, and even in this re-presentation they lie. "Semiotics is concerned with everything that can be taken as a sign. (…) Semiotics is in principle the discipline studying everything which can be used in order to lie. (Eco 2017, p. 68). However, the plans for building a dam, the design of a hammer or of a car carries knowledge that makes the real (dam, hammer, car, etc.) possible. This prompts the question of whether what we call *knowledge*—shared understanding of everything pertaining to life—is peculiar to semiotics as it is to the making of things. The idea that narration is what semiotics is about—i.e., the sequences of actions leading to conceiving what will become a dam, a hammer, a car, is no longer an abstract representation, but a concrete instantiation: we are what we do. The story is the outcome and its interpretations in use.

The reference is always the human being animated by the practical need to know in order to succeed, or at least to improve efficiency of effort under specific circumstances (context). Thus, "What is semiotics?" translates as "What defines and distinguishes human interactions from all other known forms of interaction?" Indeed, the interaction of chemical elements (i.e., chemical reaction) is different from that of two individuals. Obviously, some chemistry is involved; however, the interaction characteristic of the living is not reducible to chemistry. "Mind reading" is not *abracadabra*; it is not picking up some mysterious or real waves (electro or whatever); it is not second-guessing the biochemistry of neuronal processes. It is modeling in one's own mind what others are planning, what goals they set for themselves. In some way, this involves adaptive percept-action processes (Morris & Ward, 2005; Pinegger, Hiebel, Wriessnegger, & Müller-Putz, 2017).

Physical interaction at the atomic level is quite different from that at the molecular and macroscopic levels, and even more different at the scale of the universe. As exciting as it is in its variety and precision, the physical interaction of masses (as in Newton's laws of mechanics) does not explain aggregation, e.g., the behavior of crowds, or the "wisdom of crowds." In the end, "What is semiotics?" means not so much to define its concepts (sign, sign processes, meaning, expression, etc.) as it means to address the question of whether whatever semiotics is, does it correspond to all there is, or only to a well-defined aspect of reality. Neither mathematics, nor chemistry, nor any other knowledge domain encompass all there is. One specific knowledge domain is not reducible to others. If the same holds true for semiotics, the specific knowledge domain would have to correspond to a well-defined aspect of reality. It is obvious, but worth repeating, that semiotics (not unlike mathematics, chemistry, physics, etc.) is a human product, a construct subject to our own evaluation of its significance.

Before there was mathematics, or chemistry, or physics, there was an activity through which individuals did something (e.g., kept records using knots, used a lever, mixed substances with the aim of making new ones). In this activity, they constituted themselves as mathematicians, physicists, or chemists; and were recognized as such by others (even before there was a label for activities qualifying, in retrospect, as mathematics, physics, chemistry, etc.). In retrospect, we label such activities as semiotic: this is the narration of semiotics itself.

Returning to mathematics: Is the integrating view of the world it facilitates exclusively a human-generated representation of gnoseological intent and finality? Or can we identify a mathematics of plants or animals, of physical processes (such as lightning, earthquakes, the formation of snowflakes)? Does nature "make" mathematics? The fact that mathematics describes the "geometry" of plants, the movement of fish in water, and volcanic activity cannot be automatically translated as "plants are geometricians," or "fish are analysis experts," or "volcanoes are topologists." Rather, watching reality through the lenses of mathematics, we identify characteristics that can be described in a language (or several) that applies not to one specific flower or leaf, not to one specific fish or school of fish, not to one volcano, but to all activity, regardless where it takes place. The generality of mathematical descriptions, moreover mathematical abstraction, is what defines the outcome of the activity through which some individuals identify themselves as mathematicians (professional or amateur).

For the sake of clarity: Nature does not make mathematics, as it does not make semiotics. Anthropomorphism is convenient—"the language of plants," the "symbols of nature"—but confusing. Only with awareness of the activity is it epistemologically legitimized. There are no signs of nature, or semiotic processes of nature; there are human-constructed models for understanding nature. The same applies to machines: there is no semiotics in the functioning of a machine. It is made of parts assembled in such a way that it turns an input into a desired (or not) output. The human being projects semiotics into interaction with machines. Of course, there are signals, best expressed through values defining the physical process (e.g., electrons traveling along circuits). But to confuse signal—physical level—and sign—semiotic level—means to make semiotics irrelevant. Too many well-intended researchers operate in the space of ill-defined entities.

THE IDENTITY OF THE SEMIOTICIAN

No doubt, identity is a concept anchored in Saussure's semiology. But here we pose a different question: Is there some generality, or level of abstraction, that can define the identity of a semiotician? Again, Wittgenstein would answer in the negative. Or are we all, regardless of what we do, semioticians (or *sémiologues*?), given that interaction, characteristic of all the living, cannot be avoided. Moreover, given that we all indulge in representations and act upon representations, does this not qualify us as semioticians? Given that we all interpret everything—regardless of the adequacy of our interpretations—does this make us all semioticians? The entire domain of the living, not only that of human existence, is one of expression and interaction that seems to embody semiotics in action. Mental states are associated with neuronal activity. The physics and biochemistry, and the thermodynamics for this activity form one aspect. The other aspect is the understanding of each instance of the process, of the aggregate state to which it leads. However, there is a distinction between the activity and awareness of its taking place, of its consequences. Based on knowledge from different disciplines (biology, genetics, neuroscience, etc.), the following statement can be made: Semiotics at the genetic level, semiotics at the molecular level, and semiotics at the cell level, in association with information processes, are prerequisites for the viability of the living as such. Furthermore, it can be ascertained that bottom-up and top-down semiotic processes define life as semiosis, in parallel to its definition as information, i.e., energy related process (going back to the laws of thermodynamics) (Nadin, 2010). Awareness of semiotic processes is not characteristic of genes or molecules; neither is information awareness located where information processes take place. Awareness (of semiotics, or of information processes) corresponds to the meta-level, not to the object level.

What can we learn about semiotics—assuming that semiotics is a legitimate form of knowledge—by examining the world? First and foremost, that interaction, as a characteristic of the living, is extremely rich, and ubiquitous. Second, and not least important, life being change, interactions not only trigger change, but they themselves are subject to change. Observation yields evidence that some interactions seem more patterned than others (and accordingly predictable). Take the interaction between a newborn (human, animal) and parent. There is a definite pattern of nurturing and protection—although there are also cases of filial cannibalism (eating one's young, as do some fish, bank voles, house finches, polar bears). These patterns correspond to representations of the present and future, i.e., they are connected to anticipatory processes (underlying evolution). Or take sexual interactions (a long gamut, extended well beyond evolutionary advantage in the life of human beings); or interactions between the living and the dying. The epistemological condition of semiotics derives from the fact that life would continue even if there were no semioticians to ever observe it and report on what they "see" as they focus on interactions, or on the constructs we call sign processes. The existence of life, or the making of life, does not depend on adding semiotic ingredients to the combination of whatever might be necessary to make it. For that matter, it does not depend on adding mathematics or physics or chemistry to the formula. The awareness resulting from a semiotic perspective leads to the acknowledgment of such phenomena as living expression. Indeed, in the absence of representations, life would cease.

ENGINEERING INTERACTIONS

But things are not as simple as a cookbook for life. The mathematics for the cookbook (also known as *algorithm*) is important in defining quantities and sequences in time (first bring water to a boil, add ingredients in a certain order, simmer). The semiotics is relevant not so much for cooking for oneself, but in supporting preparation of the meal for others. This is what representations do as they are passed along in the organism. Cells "work" for each other; a cell's state depends on the states of the adjacent or remote cells. (This is what inspired Conway's "Game of Life" 1970). The organism is the expression of all that is needed in terms of means of interaction—semiotic and informational—to make possible an aggregated whole of a nature different from that of its components. It is on account of complexity that this aggregation takes place and lasts as long as what we call *life*.

Expressed differently, semiotics is relevant for "engineering" interactions: recipes are the "shorthand" of cooking. They carry explicit instructions and implicit rules, that is, assumptions of shared experiences. Semiotics embodies the sharing, but does not substitute for the experience. The informational level corresponds to "fueling" the process, providing the energy. Taken literally, even the simplest recipe is disappointing. There is always something expected from those who will try it out. No recipe is or can be complete (in the same manner in which the use of a screwdriver presents an open-ended list of possibilities). The possibility to discover on your own what cannot be encapsulated in words, numbers, procedures, or images opens up the process of self-discovery. In this sense, semiotics is relevant for dealing with the question of what the future will bring: you mixed egg yolk and oil, and instead of getting mayonnaise, the ingredients start to separate. What now? At the level of the living, life, not mayonnaise, is continuously made. At the end of the life cycle, the ingredients separate, the semiotics disappears, information degrades. Semiotics encodes in generating representations, and decodes in interpreting representations. These are distinct practical functions otherwise inconceivable. *Encode* means as much as semiotic operations performed on representations. *Decode* means the reverse, but without the guarantee that the encoded will be retrieved. Quite often, we find a different "encoded" reality: semiotic processes are non-deterministic.

WHY DO WE STILL IGNORE WINDELBAND?

The narrative of philosophy, or that of a science for that matter, can be compressed into a time series of names—authors who contributed ideas that made a difference. Another narrative could be that of names dropped, names forgotten or ignored, which can mean many things: ideas significant once upon a time are less (or not at all) meaningful; ideas associated with one name or several were taken over by others, further developed, the original contributors forgotten. Or, even, that we are still not able (or willing) to accept viewpoints not aligned with the paradigm in place (Kuhn, 1962), which is another way of saying not accepted by those in "power." (Science itself is, as Lakatos 1970 argued, a power game.) In this section I shall focus on a precise example, with the hope that semioticians will take note of a contribution pertinent to their work.

It comes as no surprise to anyone that interactions can be mathematically (or genetically) described. But mathematical descriptions (or genetic, as well) can only incompletely characterize them. More precisely: the mathematics of interactions is, after all, the description of assumed or proven laws of interaction. In this respect, law is a repetitive pattern. Physical phenomena are acceptably described in mathematical descriptions called *laws*. This is what Windelband (1894)—the name left out of the narrative I discuss—defined as the *nomothetic* (derived from *nomothé* in Plato's *Cratylus,* 360 BCE). The same cannot be said of living interactions, even if we acknowledge repetitive patterns. No living entity is identical with another. The living is infinitely diverse. Therefore, semiotics could qualify as the attempt to acknowledge diversity unfolding over time as the background for meaning, not for scientific truth. This is what Windelband defined as the *idiographic*. Remember the primitive man watching the sky and not knowing the "truth" he was seeing (Earth's rotation). Organisms, while not devoid of truth (corresponding to their materiality) are rather expressions of meaning. Representations can be meaningful or meaningless. They are perceived as one or the other in a given context.

With meaning as its focus, semiotics will not be in the position to say what is needed to make something—as chemistry and physics do, with the help of mathematics—but rather to identify what meaning it might have in the infinite sequence of interactions in which representations will be involved. This applies to making rudimentary tools, simple machines, computer programs, or artificial or synthetic entities. Semiotic knowledge is about *meaning as process*. And this implies that changing a machine is very different from changing the brain. Inadequate semiotics led to the metaphor of "hardwired" functions in the brain. There is no such thing. The brain adapts. Activities change our mind: we become what we think, what we do. We are our semiotics.

THE MEANING OF INTERACTIONS

The fact that signs—better yet, representations—are involved in interactions is an observation that needs no further argument. Being entities that stand for other entities, signs might be considered as agents of interaction. Evidently, with the notion of agency we introduce the expectation of signs as no longer "containers" of representation, but rather as intelligent entities interacting with each other, self-reproducing as the context requires. Consequently, one might be inclined to see interaction processes mirrored into sign processes (i.e., what Peirce named *semiosis*). But interactions are more than sign processes. Better yet: sign processes describe only the meaning of interactions, but not the energy processes undergirding them. This needs elaboration, since the question arises: What does "ONLY the meaning of interactions" mean? Is something missing?

A Rejected Distinction Revisited

To describe interactions pertinent to non-living matter (the physical) is way easier than to describe interactions in the living, or among living entities. For such descriptions we rely on the physics of phenomena—different at the nano-level in comparison to the scale of reality or to the cosmic scale. Quantum mechanics contributed details to our understanding of physical interactions (for instance, in bringing to light the entanglements of phenomena at the quantum level of matter). Focusing on signs caused semiotics to miss its broader claim to legitimacy: to provide not only descriptions of the meaning of interactions, but also knowledge regarding the meaning of the outcome of interactions, the future. When the outcome can be derived from scientific laws, we infer from the past to the future. Statistical distribution and associated probabilities describe the level of our understanding of all that is needed for physical entities to change. When the outcome is as unique as the living interaction itself, we first need to acknowledge that the living is driven by goals—which is not the case with the physical, where, at best, we recognize attractors: the "teleology" of dynamic systems. Therefore, we infer not only from the past, but also from the future, as projection of the goals, or understandings of goals pursued by others. Possibilities describe the level of our understanding of what is necessary for living entities to change, i.e., to adapt to change. This is the domain of anticipation, from which semiotics ultimately originates. In addition to my arguments, Nadin (1991) on this subject, see Emmeche, Kull, Stjernfelt (2002), Hoffmeyer (2008), Kull, Deacon, Emmeche, Hoffmeyer, Stjernfelt, 2009). Therefore, semiotics should be more than the repository of meaning associated with interaction components.

As information theory—based on the encompassing view that all there is, is subject to energy change—emerged (Shannon and Weaver 1949), it took away from semiotics even the appearance of legitimacy. Why bother with semiotics, with sign processes, in particular (and all that terminology pertinent to sign typology), when you can focus on energy? Energy is observable, measurable, easy to use in describing information processes understood as the prerequisite for communication. Information is more adequate than semiotics for conceiving new communication processes, which, incidentally, were also iterative processes. But there is also a plus side to what Shannon suggested: information theory made it so much more clear than any speculative approach that semiotics should focus on meaning and significance rather than on truth.

Over time, semiotics attracted not only praise, but also heavy criticism (our own will be formulated in a later section). In general, lack of empirical evidence for some interpretations remains an issue. The obscurity of the jargon turned semiotics into an elitist endeavor. Structuralist semiotics (still dominant) fully evades questions of semiotic synthesis and the interpretant process. Too often, semiotics settled on synchronic aspects, a-historic at best (only Marxist semioticians take historicity seriously, but at times to the detriment of understanding semiotic structures). Closer to our time, semiotics has been criticized for turning everything into a sign, such semioticians forgetting that if everything is a sign, nothing is a sign. In one of his famous letters to Lady Welby, Peirce writes:

It has never been in my power to study anything—mathematics, chemistry, comparative anatomy, psychology, phonetics, economics, the history of science, whist, men and women, wine, metrology—except as a study of semiotics (Peirce, 1953, p. 32).

The message here is that semiotics is inclusive, and that it should not be arbitrarily fragmented. Peirce does not bring up a semiotics of mathematics, chemistry, comparative anatomy, etc. because it is nonsensical to dilute the "study of semiotics" into partial semiotics. Nobody who understands logic would advance sub-disciplines such as "logic of feminism," "logic of genetics," "logic of politic," etc. Those who lobby for all kinds of sub-semiotics deny semiotics its comprehensive perspective.

Parallel to this recognition is the need to assess meaning in such a manner that it becomes relevant to human activity. So far, methods have been developed for the experimental sciences: those based on proof, i.e., the expectation of confirmation and generalization. But there is nothing similar in respect to meaning, not even the realization that generalization is not possible; or that semiotic knowledge is not subject to proof, rather to an inquiry of its singularity. The *nomothetic* comprises positivism; the *idiographic* is the foundation of the constructivist understanding of the world (Piaget, 1955, von Foerster, 1981).

The Falsifiable

Mathematicians would claim that their proofs are absolute. Indeed, they make the criterion of falsifiability (Popper 1934) one of their methods: Let's assume, *ad absurdum*, that parallels meet. If they do, then what? No scientific ascertainment can be proven to the same level of certainty as the mathematical, because it is a projection of the mind. By extension, this applies to computer science and its many related developments, in the sense that automated mathematics is still mathematics. (Mathematicians themselves realize that in the future, mathematical proofs will be based on computation.) Science lives from observation; it involves experiment and justifies itself through the outcome. If the experiment fails, the science subject to testing fails. That particular observation is not absolute in every respect. Let us name some conditions that affect the outcome of experiments: selection (what is observed, what is ignored); evaluation (degrees of error); expression (how we turn the observation, i.e., data, into knowledge).

Experiments are always reductions. To reproduce an experiment is to confirm the reduction, not exactly the claim of broader knowledge. The outcome might be disappointing in respect to the goal pursued: for example, the various drugs that have failed after being tested and approved. But the outcome might, as well, prove significant in respect to other goals. Drugs that are dangerous in some cases prove useful in treating different ailments: thalidomide for arthritic inflammations, mouth and throat sores in HIV patients; botox for treating constricted muscles.

Failed scientific proofs (Deutsch 2012) prompt many fundamental reassessments. Compare the scientific theory of action at distance before Newton and after Newton's foundations of physics; compare Newton's view to Einstein's; and compare Einstein's science to quantum entanglement. Compare the views of biology prior to the theory of evolution, or to the discovery of the genetic code. Given the epistemological condition of mathematics, new evidence is not presented in the jargon of mathematics. A new mathematical concept or theorem is evidence. Probably more than science, mathematics is art. It is idiographic, not nomothetic knowledge. As we know from Turing and Gödel, it cannot be derived through machine operations (Hilbert's challenge). If there is a cause for mathematics, it is the never-ending questioning of the world appropriated by the mind at the most concrete level: its representation. The outcome is abstraction. This is what informed Hausdorff as he described human nature. There is, of course, right and wrong in mathematics, as there is right and wrong in art. But neither a Beethoven symphony nor Fermat's conjecture (proven or not) is meant as a hypothesis to be experimentally confirmed. Each has an identity, i.e., a semiotic condition. Each establishes its own reality, and allows for further elaborations. Not to have heard Beethoven's symphonies or not to have understood Fermat's law does not cause bridges to collapse or airplanes to miss their destinations.

The Art of Mathematics and the Art of Healing

By its nature, semiotics is not a discipline of proofs. Not even Peirce, obsessed with establishing semiotics as a logic of vagueness (Nadin 1980, 1983) produced proofs. In physics, the same cause is associated with the same effect (in a given context). Take the example of thalidomide, first used as a sedative, which led to birth defects ("thalidomide babies") when pregnant women took it. Now consider the reverse: the medicine is used for alleviating painful skin conditions and several types of cancer. The semiotics behind symptomatology concerns the ambiguous nature of disease in the living. The ambiguity of disease is reflected in the ambiguity of representations associated with disease. Better doctors are still "artists," which is not the case with software programs that analyze test results. Diagnosis is semiotics, i.e., representation and interpretation of symptoms, that is both art and science. Machine diagnosis is information processing at work. Human diagnosis is the unity of information and meaning.

When mathematicians, or logicians, translate semiotic considerations into mathematical descriptions, they do not prove the semiotics, but the mathematics used. For example, Marty (1990) provided the proof that, based on Peirce's definition of the sign and his categories, there can indeed be only ten classes of complete signs. But this brilliant proof was a contribution to the mathematics of category theory. Goguen's brilliant algebraic semiotics (1999) is in the same situation. "In this setting (i.e., user interface considered as representation, our note), representations appear as mappings, or morphisms . . . which should preserve as much structure as possible."

My own attempts at proving that signs relationally defined as fuzzy automata (Nadin 1977), Figure 3) are more a contribution to automata theory than to semiotics. No semiotician ever cared about these attempts; none took such proofs to mean anything in examining signs in action or in understanding semiotics. For their art, which is the art of semiotic interpretation, the mathematical proof is of no relevance. The same holds true for the classes of signs. There are no such signs as *icons*, *symbols*, or *indexes*. These are types of representation. But to deal with the ten classes that Peirce advanced is cumbersome, to say the least. To deal with the 66 classes of signs corresponding to his triadic-trichotomic view is even more arcane.

Figure 3. Sign and fuzzy automata. In this case, a Nerode automaton for S = f (O,R,I,o,i)

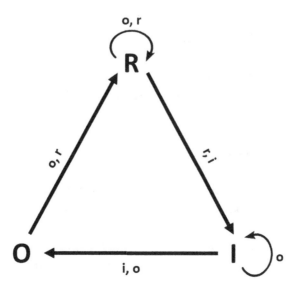

Acceptance

This extended preliminary discussion deals with how we might define a foundation of semiotics not around a formal concept—i.e., the sign. Since the concept is subject to so many different interpretations, none more justifiable than another, we need to avoid it. The goal is to make the reader aware of why even the most enthusiastic semioticians end up questioning the legitimacy of their pursuit. Before further elaborating on my own foundational statement for semiotics—this text is only an introduction to it—I shall proceed with a survey of the semiotic scene. This should produce arguments pertinent to the entire endeavor. I derive no pleasure from reporting on the brilliant failure of a discipline to which I remain faithful. Let's be clear: it is not because semioticians (of all stripe) come from different perspectives, and use different definitions, that semiotics does not emerge as a coherent approach. Rather, because it does not yet have a well-defined correlate in reality, in respect to which one could infer from its statements to their legitimacy and significance. Only because we can practice semiotics, or put on the hat that qualifies someone as semiotician (professor or not), does not justify semiotics as something more than quackery. Can semiotics have a defined correlate in reality? Can it transcend the speculative condition that made it into a discourse of convenience spiked with technical terminology? Jack Solomon (1988) argued that its own principles disqualify it from having universal validity.

Everyone in the more affluent part of the world knows that society can afford supporting the unemployed, or helping people without insurance, or providing for self-proclaimed artists. But this by-product of prosperity, and the general trend to support everything and anything, cannot justify semiotics more than the obsession with gold once justified alchemy, or the obsession with cheap oil justifies wars in our time. In order to earn its legitimacy, semiotics (i.e., semioticians) must define itself in relation to a compelling aspect of the living, something in whose absence life itself—at least in the form we experience it—would not be possible. If this sounds like a very high-order test of validity, those readers not willing to take it are free to remain insignificant, whether they call themselves semioticians or something else. With the demotion of Aristotelian inspired *vitalism*, life was declared to be like everything else. As our science evolved, the "knowledge chickens" came home to roost: We pay an epistemologically unbearable price for having adopted the machine as the general prototype of reality. The semiotic animal is not reducible to a machine (even though signs, in Peirce's definition, are reducible to fuzzy automata; cf. Figure 3).

GROUCHO MARX AS SEMIOTICIAN

The reader who still opens any of today's publications on semiotics—journals, proceedings, even books—often has cause to wonder: Is semiotics an exercise in futility? Authors of articles, conference papers, books, and other publications will probably present arguments such as:

- There is a peer-review process in place that legitimizes their efforts;
- The situation in semiotics is not different from that in any other knowledge domain;
- There are no evaluation criteria to help distinguish the "wheat" from the "chaff." In the democratic model of science (semiotics and other fields), "Anything goes."

Each argument deserves attention. But first an observation (which might not seem related to the subject): The quality of education and research in general seems to diminish as more money is spent for them. Stated differently: The gap between excellence—yes, excellence still exists—and mediocrity is widening. By contamination, mediocrity threatens to set a *very* low common denominator. Pretty soon, a Ph.D. will be as common (and insignificant) as membership in those clubs that Groucho Marx refused to join because they would have him. However, this is not the place to address the way in which expectations of higher efficiency (Nadin 1997), characteristic of our current state of civilization, translate into the politics and economics and education of mediocrity. A different aspect is worthy of discussion here: Some disciplines are focused on relevant aspects of science, humanities, and current technology. They define vectors of societal interest. It does not take too much effort to identify the life sciences as a field in the forefront of research and education; or, for better or worse, computer science, in its variety of directions. Nanotechnology is yet another such field. It originated in physics (which, in its classic form, became less relevant) only in order to ascertain its own reason for being well beyond anyone's expectations. Some readers might recall the time when scientists (Smalley (2001)) claimed that nanotechnology would not work, despite the scientific enthusiasm of the majority of scientists in the field[1]. In the meanwhile, nanotechnology has prompted spectacular developments that effected change in medicine and led to the conception of new materials and processes. Computer science met nanotechnology at the moment Moore's law, promising the doubling of computer performance every eighteen months, reached its physical limits.

Besides semiotics, many other disciplines (including traditional philosophy) live merely in the cultural discourse of the day, or in the past. More precisely, they live in a parasitic state, justifying themselves through arcane requirements, such as the famous American declaration, "We need to give students a liberal arts education" (a domain in which semiotics is often based). They do not even understand what *liberal arts* or *humanities* means today: using Twitter and the iPhone, or reading the Constitution? Being on social media or reading the "Great Books"? These are questions of a semiotic nature.

The Past Is Reified in Institutions

Semiotics as it is practiced, even by dedicated scholars, certainly does not qualify as groundbreaking, no matter how generous we want to be. Rather, it illustrates what happens to a discipline in which its practitioners, most of them in search of an academic identity—a placeholder of sorts—regurgitate good and bad from a past of promise and hopes never realized. What strikes the reader is the feeling that semiotics deals more with its own questions than with questions relevant to today's world. Even when some subjects of current interest come up—such as the self-defined niche of biosemiotics (cf. Uexküll 1934/2010, Barbieri 2007, Favareau, 2009)—they are more a pretext for revisiting obscure terminology or for resuscitating theories dead on arrival. Congresses, the major public event of a society formed around a discipline, are the occasion for defining the state of the art in a particular knowledge domain. The ten international congresses on semiotics held so far make up a revealing story of how the enthusiastic beginnings of modern semiotics slowly but surely morphed into a never-ending funeral. There is a dead body carried in that casket—semiotics—and there are endless speeches about its greatness. Like all institutions, the International Association is more concerned with its own perpetuation than with the growth and quality of the discipline it is supposed to represent.

The founding members of the IASS (Greimas, Jakobson, Kristeva, Beneviste, Sebeok) had in mind the promotion of semiotic research in a scientific esprit: "…promouvoir les recherches sémiotiques dans un esprit scientifique." (French dominated at that time.) This important function is specifically mentioned on the IASS website. Even in its so-called new form, the website, seen from the perspective of semiotics, is a rather telling example of how limited the contribution of semiotics is in providing new means and methods of communication and interaction. An inadequate website is not yet proof of the inadequacy of the current contributions to semiotics. It is a symptom, though. In the spirit of the dedication to a scientific agenda, Eco, Marcus, Pelc, Segre—to name a few—contributed to a better reputation of semiotic research. They, and a few others (e.g., Deledalle 1997/2001, Marty 1990, Bouissac 1977, Nöth 1985/1995), and the followers of the Stuttgart School succeeded in producing works worthy of respect. In the present, very few distinctive centers of semiotic research can be identified. One is located at the University of Toronto[2]; the other at the University of Tartu, Estonia[3]. They deserve recognition beyond the lines I dedicate to them in this study.

But a closer look at what continues to be produced under the guise of semiotics, all over the world, leads to the realization that the initial optimism of the "founders" was either groundless, or did not reflect the potential of the many self-proclaimed semioticians. On behalf of the first congress (Milan 1974), Umberto Eco (1975) wrote (in the Preface to the *Proceedings*) about a "fundamental" and an "archeological" task. The first would be the justification for the existence of semiotics; the second, to derive from its past a unified methodology and, if possible, a unified objective. Very little has been clarified regarding the initial existential questions: What justifies the existence of semiotics? What are its objectives? What is its methodology? The only significant aspect is that, despite their irrelevance, events such as congresses (and publication of the associated *Proceedings*) continue to take place! In keeping with the mercantile spirit of the time, the International Association for Semiotic Studies even came up with a scheme for a congress franchise.

Obviously, the statements made above require substantiation. Some of those persons alluded to might suspect the settling of some score (there is nothing to settle since there is no score to keep). Others might suspect a generational conflict, or even an attempt to idealize the past (the romantic notion of "heroic beginnings"). Obviously, such possible interpretations cannot be avoided. Nevertheless, the issue brought up—lack of significance—and the motivation—the reason for addressing it as a subject worthy of attention—are quite distinct. Therefore, I shall proceed in three directions:

1. A short presentation of today's major themes in the humanities, the sciences, and technology;
2. A short historic account of developments in semiotics;
3. A methodological perspective.

The intention is not to cast aspersion upon work produced in the field in recent, and less than recent, years, but rather, to show that this is probably the time of the most interesting (i.e., rewarding) subjects for semiotics. This is the time of new opportunity for semiotics to make its case as a viable discipline and to confirm its necessity. I do not write here delayed reviews of the many articles I indirectly refer to; neither do I write letters of evaluation for one or another author. To watch some presentations under the heading "Semiotics" on YouTube, or similar media, is embarrassing. But mediocrity in this case is not so much congenial to the subject as it is an expression of mediocrity as the new standard of acceptance on social media, and intellectual endeavor in general. To stimulate a discussion on the sad state of semiotics today is, to a great extent justified by the realization that defining semiotics in a manner counter-productive to its development explains its shortcomings. Why is semiotics, with very few exceptions, in a lamentable condition today? This is a valid interrogation, similar to one articulated regarding physics after the obsession with nuclear energy. Or, for that matter, why medicine, practiced as a reactive endeavor, is failing society. Concerning the "Why?" of the position I take: The attempt to redefine the foundation of semiotics is intended as an invitation to everyone dedicated to the subject, not to its occasional visitors. I do not promise miraculous solutions. This study is an expression of the love and passion I have for semiotics, and of the conviction that it can deliver more than fancy phrasings. The fact that it comes from an "outsider" (i.e., a semiotician who remains unaffiliated) should not be seen as an attack against the semiotic establishment. I've no ax to grind (and no time to do so), and aspire to no glory and to no office (national or international).

The Broader Perspective

The Human Genome Project (HGP)—an impressive undertaking that made powerful sequencing tools available—is seen by some as a huge success, and by no few others as a miserable failure—an example of "spin science," as it was recently labeled (Chu, Grundy, Bero, 2017). To indulge in a discussion of the argument could easily fill pages of books. What does not, however, require the same attention to detail and does not lead to shallow judgments ("Did it or did it not live up to the promises made?") is the realization that there is no such thing as a unique semiotics underlying genetics. The four letters of the genetic code are involved in an open-ended narration, different from person to person. Wittgenstein, again, would have identified instances of genetic expression and warned us that, "there is no shared constituent to be discovered" (cf. the Schlick dictation cited above). How many semioticians involved themselves in the project? (I do not ask how many were invited—the answer is None!) How many, after the HGR, took it upon themselves to decipher information made available (AAAS (2001))? Besides the rhetoric of the question, there is the reality of the fact that semioticians prefer to discuss terminology, compare their preferred authors (Peirce, Saussure, Eco, Barthes, Lotman, etc.), discuss movies and feminism, interpret religious or other codes—but do not acquire the knowledge needed to competently discuss the meaning of DNA, the semiosis of RNA, the individual genetic code, and similar subjects.

The most captivating mathematics (a subject I place in the humanities), the most brilliant attempts to understand language, the most dedicated effort to understand the human condition—these are themes impossible to even conceive of without acknowledging their semiotic condition. Take again the attempt to prove Fermat's Theorem. Fundamentally, the approach extends deep into the notion of representation. The very elaborate mathematical apparatus, at a level of abstraction that mathematics never reached before, makes the whole enterprise semiotically very relevant. The entire discussion that accompanied the presentation of the proof, expressions of doubt, commentaries, and attempts to explain the proof are *par excellence* all subjects for semiotics. The subject is interpretation, the "bread and butter" of semiotics, its *raison d'être*. A question that begs the attention of semioticians is, "How far from the initial mathematical statement (Fermat's Theorem) can the proof take place?" That is, how far can the representation of representation of representation, and so on extend the semiotic process before it becomes incoherent or incomprehensible?

Fermat's short message in Latin ("Cubem autem in duos cubos, etc." Figure 4) on his copy of a translation of Diophantus' *Arithmetica* (3rd century CE) is a theorem represented in words, i.e., in a "natural" language. It is relatively easy to interpret. Later (1637), this theorem was "translated" into mathematical formulae. Fermat's Last Theorem states that no nontrivial integer solutions exist for the equation:

$a^n + b^n = c^n$ if n is an integer greater than 2.

Figure 4. Fermat's Theorem in Latin

OBSERVATIO DOMINI PETRI DE FERMAT.

CVbum autem in duos cubos, aut quadratoquadratum in duos quadratoquadratos & generaliter nullam in infinitum vltra quadratum poteſtatem in duos eiuſdem nominis fas eſt diuidere cuius rei demonſtrationem mirabilem ſane detexi. Hanc marginis exiguitas non caperet.

One did not need to know Latin but had to be familiar with mathematical symbols in order to understand the equation (and even use it for some examples). Computation changed the way mathematicians (but not only) think. Therefore, mathematicians say that in order to prove Fermat's Theorem, they would have to prove a conjecture (the Taniyama-Shimura Conjecture) that deals with elliptic curves. Understanding the conjecture implies highly specialized knowledge. Wiles (1995) submitted a brilliant piece of mathematics as proof, and further worked on details once some colleagues challenged his results. Chances are that no other discipline besides semiotics can assist in giving meaning to the effort. Now it's time to explain this assertion.

Semiotics is, of course, a knowledge domain different from mathematics. Within its knowledge domain, the mathematical question and the proof concern Peirce's *interpretant process*. Fermat's description in Latin was unequivocal; the translation into mathematical symbolism is also unambiguous. The mathematical proof, however, is so far removed from the simplicity of the Theorem that one can question the semiosis: from simple to exceedingly complicated. Under which circumstances is such a semiosis (i.e., epistemology) justified? This goes well beyond Fermat; it transcends mathematics. It becomes an issue of relevance because many semiotically based activities (such as genetics, visualizations, virtual reality, ALife, synthetic life) pertinent to the acquisition of knowledge in our age tend to evolve into complicated operations not always directly connected to what is represented. The HGP mentioned above is another example of the same. This is an issue of meta-knowledge. If knowledge acquisition, expression, and communication are indeed semiotically based, then this would be the moment to produce a semiotic foundation for meta-knowledge.

Would Peirce, given his very broad horizon, have missed the opportunity to approach the subject? I doubt it. By the way: as Einstein produced his ground-breaking theory, Cassirer found it appropriate to offer an interpretation informed by his semiotics (1921)[4]. In other words, there is proof that semiotics can do better than indulge in useless speculative language games, as it does in our time.

What I suggest is that specialization—such as in the mathematics required to produce the proof, or the mathematics that Einstein mastered, or the genetics needed for evaluating the HGP—is a necessary condition for the progress of science. But not sufficient! Specialists—and there are more and more of them—ought to relate their discoveries to other fields, to build bridges. For this they need semiotics as an integral part of their way of thinking, as a technique of expression, and as a communication guide.

We are experiencing various attempts to integrate computation, genetics, anthropology, philosophy, and more into understanding how language emerged and diversified. Never before has language—in its general sense, not only as the language we speak—been as central to research as it is today. Hausdorff, the mathematician who understood the semiotic nature of the human being, anticipated this; that is, he acted according to this understanding. And since semiotics has, more often than not, been understood as the semiotics of language (in this sense, Saussure succeeded with his semiology), it would be only natural to expect semioticians of all stripes to get involved in it. Genetics is, in fact, the study of DNA "expression," of a particular kind of language defining the narrative and the associated stories that make up the "texts" and "books" of life. Or, as I shall argue, the narrative and the associated stories defining the unfolding of life over time. "Sentences" of a genetic nature identify not only criminals in a court of law, but also genetic mechanisms related to our health. Would Saussure have missed the chance to collaborate with researchers who uncover the first "language genes"? Would Hjelmslev? No one expects semioticians to clarify the relation between brain activity and language. Brain imaging opened access to cerebral activity. But language is not necessarily housed in the brain, or only in the brain; it extends to the entire body, always engaged when we express ourselves through language. Natural language is the most ubiquitous medium of interaction. It is involved in knowledge acquisition, in its expression, communication, and validation. Semiotics, if founded not around the sign concept—quite counter-intuitive when it comes to language (to the sign in the alphabet, the word, the sentence)—but with the understanding of the interactions that languages make possible, would contribute more than descriptions, usually of no consequence to anyone, and *post facto* explanations. That is why I am trying to suggest a foundation in the narrative, the timeline of everything we do.

"Living Mirrors of the Universe"

The monkey that Nicolelis (2001) used in order to "download" the thinking that goes on when games are played does not qualify as an example for using language. The monkey initially acted upon the joystick in order to score. But once it noticed that the signals associated with its actions—for instance, with what it wanted to do—it chose the economy and speed of motoric expression. Are downloaded streams of data describing brain processes made up of signs? Since everything can be interpreted as a sign, to dismiss such data as being only representative of the physics and chemistry of the monkey's brain activity would be as preposterous as making reference to "prove" spirituality. What such streams of data are is relatively clear: representations of quantitative processes, of measurements. However, the monkey is pursuing a goal associated with the classic reward mechanism, the banana in this case. (Talk about stereotypes in science!) Therefore, intentionality—recall the bacterium swimming upstream and Margolis—cannot be ignored. Once we associate data and meaning (as Wheeler[5] suggested, cf. Nadin 2011), we have access to information. The semiotics is implicit in the observation that thinking and acting upon representations can be connected. In a different context (Nadin 2016), I proved that the entire body is the brain. This applies to the human being as it applies to the Nicolelis monkey. The sensorial and the cognitive are associated. Motoric expression is not an execution of commands, but rather an expression of the holistic nature of the process. Moreover, the monkey condition is not equivalent to what we call the human condition. Humans play entire games of chess (or any other game) in their minds, not by necessarily moving pieces on a chessboard. For them, the pawn does not have to be on the chessboard in order to be identified as constitutive of the game.

As speculative as the notion of the human condition is, we have finally arrived at the juncture where very good models of the human condition, understood in its dynamics, can be conceived, constructed, and tested. The underlying element here is actually what Hausdorff defined as the *zoon semiotikon*, and what Cassirer defined as *animal symbolicum*. Hausdorff, a distinguished mathematician, could have defined the human being as "mathematical animal," but to him the qualifier *semiotic* meant a more general, more encompassing level. Cassirer was a philosopher; to him, generating symbols seemed more relevant than generating new philosophies. Before Hausdorff, and before Cassirer, many other scholars in the humanities considered the qualifier "semiotic" as co-extensive of being human. (Some extended it to animals, as well.) Leibniz, with his *miroirs vivants de l'universe*, inspired Cassirer's definition of the symbol and his attempt to define the human condition in semiotic terms. Locke (1690) found a place for semiotics in a precise domain, i.e., the ways and means whereby the knowledge … is attained and communicated. His definition:

Nor is there anything to be relied upon in Physick, but an exact knowledge of medicinal physiology (founded on observation, not principles), semeiotics, method of curing, and tried (not excogitated, not commanding) medicines.

The active role of the Russian and Czech semioticians in explaining the role of language in the making of humankind, and Roland Barthes' subtle analysis of language and culture, are convincing arguments that would not have failed to be in the forefront of the semiotic research associated with the current attempts to define the human condition. For more on the subject, see Nadin (1986, p. 163) and (2014).

The subject ought to be understood as broadly as possible. This means that within the realm of the living, there is a whole gamut—from the mono-cell to *homo faber*—of representations to consider. Is there anything that qualifies as semiotically relevant across the various forms of the living? Interaction is probably the most obvious aspect. At a closer look, the making of the living consists of integrated interactions—from the level of the cell to that of organisms and among them (not to say their interaction with the world, living or not). At all these levels, representations are exchanged. Interactions transcend unidirectional processes—which are the expression of causality. Therefore, semiotic processes appear as a characteristic of the whole (organism) in an integrated world, but also as one among organisms (same or different). For reasons of illustrating this idea, I will make reference to the interactions involving the human microbiome (i.e., microbes or microorganisms that inhabit part of the human body). To understand all of this, semioticians are not invited to become biologists, rather to engage biological knowledge (acquired in specific experiments) in order to generalize the notion of semiotic process. That which lives is defined not only by the physics, chemistry, or energy of the process, but also by the various representations exchanged and the ability to interpret them. The living is the domain of meaning. There was interactivity in every previous stage of evolution. Interactivity involving the living implied interpretation—the outcome depended on it—but never at the scale at which society makes semiotic-based interactions its major form of activity. Society also hopes to have the guidance of science, in particular semiotics, in giving meaning to such semiotic processes. The availability of such guidance will help avoid costly consequences—such as those experienced in recent years: terrorism, technological errors, speculation, etc. Medicare fails when data substitutes for meaning. The aging of humankind is probably even more consequential in this sense. Success and failure depend upon interpretation. Machines are better at processing data, but not really better than humans at interpreting it. They can handle way more quantitative descriptions than can the people who build them. But quantity does not automatically lead to improved comprehension in a changing context. Machines are cursed to be blind and deaf to meaning.

The major themes in the sciences beg no less for the contribution of semiotics. Computation is, for all practical purposes, semiotics at work, at a syntactic level, in communication with what is called information processing. Artificial intelligence, in its many flavors, cannot be conceived without integrating semiotic concepts in its concrete implementations. Learning, deep or not, implies considerations that transcend the quantitative. To emulate (a player of chess or of Go, a composer, or a painter) implies descriptions at a level of detail that cries out for more forms of qualitative distinctions. Why one option, from among a huge number ("brute force" in action), is better than others can be "learned" through "training" (in this case, of deep neural networks). That probability considerations dominate is normal: no new question is formulated, the past informs the future. For authentic creative endeavors, probability will have to be complemented by possibility (the possible future); reaction will have to make way for anticipation. The new forms of computation—genetic, quantum, DNA, etc.—are all forms of processes with a semiotic component. More specifically: No information process (e.g., computer, sensor-based information harvesting, intelligent agents-based activities) is possible without representation. Representation is the definitory subject of semiotics (in awareness of it, or in total disregard of it). While electrons move through circuits, and while logic is emulated in hardware (circuits performing logical operations), operations on representations are the prerequisite for any information processing. Unfortunately, the variety of representations (for which Peirce delivered the types, i.e., indexical, iconic, symbolic) and their specific dynamics are superficially understood, if at all (Sowa, 2017). The focus should be on the living—a distinction which the academic world still resists—and on human activity in general. This would make possible the semiotic processes implicit in mechanisms of life. Major research directions—cells or membrane biochemistry, for example—show that we are getting better at understanding the object level and in describing the associated representational level. To realize the unity between the focus on data and the semiotic focus on meaning is a major scientific challenge. It will not happen by itself. Institutionalized science always resists new viewpoints.

For the sake of clarity: Representation is not reducible to the entity we call *sign*, regardless of how it is defined. Signs are media for representation, like letters in the alphabet are media for words, sentences, texts. The process we call *representation* cannot be reduced to one or several signs (Figure 2). Pursuing the parallel mentioned earlier (mathematics, chemistry, biology, etc. and semiotics), we arrive at the realization that the definition of semiotics based on the sign is at least as unsatisfactory as a definition of mathematics would be if it were based on numbers alone, or of chemistry based on elements, or of biology based on cells, or of linguistics based on the alphabet. Representation would have to be further defined as a process, uniting information (measurable) and meaning (result of interpretation). It is in this condition that representation proves to be significant for the understanding of the living, of mathematics (a specific form of human activity), of science, of the arts, of communication, and of interaction. Despite this peculiarity, semioticians are so removed from the major scientific and humanistic themes of the day that they don't even know that this is their greatest chance—ever! The entire stem cell debate could have taken a different path had competent semioticians contributed to an understanding of stem cell "semiosis" and the relation to the broader issues of creativity.

LANGUAGES OF INTERACTION

I will finish this compressed exposition by stating that technology is shaped by questions that, at first glance, impress as being semiotic in nature. Technological artifacts of all kind—from games to virtual reality labs in which new materials are conceived—rely on various types of semiotic entities, on representations in the first places, and their interpretation. They make sense, and can become a relevant subject of inquiry, only as new "languages of interaction." The global scale of life makes an integrative approach necessary, but not in the sense of the economy of profits at a scale never before experienced. Globality was discussed at one of the semiotic congresses, or in previous meetings (*Signs of the World*,

Lyon, 2004, where "interculturality and globalization" were the convenient slogans of the semiotics community). Nevertheless, semioticians rather take a passive role instead of advancing a critical position. In our extremely controversial time, there is a need for semiotics based on acknowledging diversity, while simultaneously providing means of expression, communication, and signification that pertain to the new scale of human activity. The social dimension of semiotics, specifically brought up by Saussure could be reached by working out evaluation criteria. Opportunistic semiotics (on the bandwagon of many causes) is impotent. Creative ideas for addressing the increased abuse of the public (fake news, fake movements, then never-ending surveillance of the individual, etc.) are yet to be affirmed. The GPS facility, accessible world wide, was the first major embodiment of semiotics in action. I do not, of course, expect semioticians to start writing articles on what kind of a sign a GPS indicator is, but rather to contribute semiotic concepts that will make the language of the system so much easier to understand and use. Monitoring, regardless of purpose, is a form of limiting privacy; assisting, when desired, is helpful. When and if semiotics partakes in the process, GPS data will seamlessly integrate in what we do—drive, visit new places, connect to others, for example. That is, when it becomes part of our language, semiotics could support a concrete accomplishment. Hopefully, semioticians will be able to understand this opportunity.

On this note, a simple observation: Brain imaging revealed that taxi drivers in some of the big cities (London was the first address researched), difficult to navigate, developed in the process measurable new faculties. Of course, these are semiotic in nature: Understanding of representations and the ability to match goals and means (a request such as "Get me to Piccadilly in the shortest time," involves quite a number of parameters). The emergence of GPS-based navigation might lead to the loss of those faculties. Semioticians should be aware of the fact that the world before maps and the world after maps became available are very different realities. This example is only illustrative of the formative power of our representations (Figure 5).

As technology further evolves, more and more automated systems guide our navigation—in libraries, on the worldwide web, in air travel, on high-speed trains, on highways and toll roads, utilization of drones, etc. Aaron Koblin (2008-2015) documented this process in visualizations of extreme semiotic significance. So did Albert-László Barabási. If Within (the name of Koblin's company) where Koblin develops virtual stories, had been the invention of semioticians, I could define today's state of semiotics as excellent. But it was not. Neither was the work of Barabási and his group inspired by semiotics, but by networks. And if the Worldwide Web, through which many publications (including a few of semiotic interest) are presented, had involved the least participation of semiotics, we would have had a Web that is

Figure 5. Air traffic paths and flight patterns

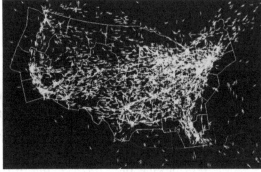 

not syntactically driven. The inventor of the Web (Tim Berners-Lee 1998), awarded with knighthood for his work, is still dreaming of a semantic stage (although what ontology engineers deliver seems to satisfy his criteria). Many work on this project, in particular the ontology engineers, who provided computers with machine-readable encyclopedias. (For me, personally, only a pragmatically driven Web makes real sense. But this is a different subject.) While the GPS actually changes the nature of our relation to space, and indirectly to time, its semiotics is a legitimate question waiting to be addressed because it involves a new semiotic condition for the human being. The military purpose of the orientation system is spectacularly transcended by rich semioses that, strangely enough, emerged without any input from semioticians. The autonomous car (the new obsession among technologists) is actually a semiotic integrated device: make sense of the world as you move from one location to another, and exchange data with similar devices. If today semiotics were to contribute to a semantic Web, we would avoid the many errors that have affected the growth of the Web into the monster it is now. But ontology engineers and semioticians don't work together. We find data on the Web, to the extent of overwhelming the user, but we do not really find information, and almost never meaning. If this is not a challenging semiotic project, then I don't know of any. At the drawing board for autonomous cars, ships, airplanes, etc., semioticians should have a say—if their semiotic competence were up to the task.

Some years ago, I acknowledged semiotics at work in the activity of Luc Steels, Stevan Harnad, and Juyan Weng, João Queiroz, and Angelo Loula, and especially in the work of Sieckenius de Souza's semiotic engineering work (Nadin 2017). And yes, in the AI domain, resuscitated by neural networks-based deep learning applications, there is a definite awareness of the semiotic component of intelligence. Tony Belpaeme (Professor of Cognitive Systems and Robotics) and Angelo Cangelosi (Professor in Artificial Intelligence and Cognition) come to mind in this vein. Few, if any semioticians made the effort to understand the semiotics of machine learning (ML) or the semiotics of neural network training. Therefore, they could not even serve as dialog partners to the mathematically focused community of deep learning researchers. But the work of such researchers is not presented at semiotics meetings and congresses or in the regular semiotics publications.

Obviously, this short account is not exhaustive, and it is less systematic than it would be in a different context. The intention is only to suggest that semiotics has a very fertile ground to cultivate, if semioticians care to work at it, or if professionals from other disciplines pay more attention to semiotics. It is not too late! In a different context (Nadin, 2005), I brought up *The Semiotic Engineering of Human-Computer Interaction*, a book written by a computer science professor (trained as a linguist), Clarisse Sieckenius de Souza (2005), who "spread" the semiotic word in the HCI community. We have here an example of an applied understanding of semiotics informed by the desire to advance issues of interaction—to make it into a foundation for new forms of engineering. It is modest proof, if anyone needed more proof, that so much can be done, provided that semiotics competence guides the effort. Aware of this characterization of her book (which semioticians managed to ignore), she recently wrote to me by e-mail:

Having studied semiotics does make a difference (…) I have the impression…that HCI professionals and students educated in North America tend to have a "What is in it for me?" approach. (…) As you know, the answer is, "a whole new world," but it will take a lot of critical thinking to get it.

She was not sure that making her thoughts public would help.

PRAGMATIC RELEVANCE

Semiotic awareness, such as expressed in biosemiotics or even semiotic engineering, has led to more than one attempt to define its knowledge domain and its specific methods. Still, so it seems, each start was relatively short-lived. The generically defined "ancient times" had such a start, with works such as Plato's *Sophistes* (360 BCE), Aristotle's *Poetica* (350 BCE), and the Stoics, mentioned in almost every account of history. It is worth mentioning that Sextus Empiricus (in *Adversus Mathematicos*, VIII) made the distinction between what is signified, what signifies, and the object. Early attempts to understand semiotics are focused on the verb *to introduce* something. The object and the signifier are material; the signified (*lekton*) is not, but it can only be right (adequate) or not (inadequate). Indian Buddhism and Brahmanism, the Christian infatuation with signs (St. Augustine's *De Doctrina Cristiana*, 397 CE, and St. Anselm's *Monologion*, 1075-1076), and Avicenna's explorations in medicine and theology remain documentary repositories of the many questions posed by two very simple questions: How can something in the world be "duplicated" in the mind? The duplication suggests that the question is not about signs (standing for some thing), but about re-presentation. Moreover, once we think about it (the reality duplicated in the mind), can we know it, or assume that what we know corresponds to reality? Or does knowing actually involve a practical activity with a desired outcome?

Edward O. Wilson (1984) came up with a provocative statement of significance not only to semiotics: "Scientists do not discover in order to know, they know in order to discover." The inversion of purpose (the causality) points to opportunity. Reading classical texts (such as those mentioned above)—and very few semioticians care to do that—reveals that the sign was only the trigger of the interactions it made possible, not associated with their meaning, and even less with their significance. From the beginning, the fascination was with semiotic knowledge: what we learn from observing interactions, and how these are subject to betterment. It is not the history that is important here, but rather the attempt to understand the need for semiotics—if a need indeed exists. The premise guiding this effort is *pragmatic relevance*: If semiotics does not make a difference, as mathematics, chemistry, and physics do, why bother with it? After the rather modest beginning of semiotic inquiry (within the broader questions of philosophy), interest in formulating semiotic interrogations diminished. However, the still controversial "Middle Ages" were yet another start. The works of Roscelin (representative of extreme nominalism); Guillaume de Champeaux (who maintained that universals exist independent of names), and Abélard (on logic) stand as examples for the enthusiasm of those seeking in semiotics answers to the many challenges of those times. Let's be clear: The fundamental opposition between nominalism and realism is a test case. If things are only names, semiotics would be in charge of the world. If, alternatively, the world, in its manifold materiality, were to look at names and call them a poor attempt at describing it, semiotics would be useless. Jean de Salisbury (*Metalogicon*) suggested that abstractions are not related to signs and take the role of names and naming. It is a fascinating journey to read Occam, William of Shyreswood, Lambert d'Auxerre, and Roger Bacon, first and foremost because their questions, extended to the domain of rationality, will inspire the third attempt at restarting semiotics in the classical age. To put it succinctly, it was not much more successful than the previous beginning. Hobbes (*Leviathan*, 1651) the *Logique de Port Royal*, (or *The Art of Thinking*, 1662) John Locke (the forms of reasoning and *The Division of Sciences*, 1690), and foremost, Leibniz (symbolic and mathematical thought, 1672-1696) are precursors of the modern rebirth associated with Saussure and Peirce and the already mentioned biosemiotics and semiotic engineering.

Important, even for those disinclined to seek guidance in works of the past, is the distinction between language associated with convention or law (*nomoi*)—such as programming languages—or with nature (*phusei*)—such as the genetic code. Nobody expects today's semioticians to become historians. But in the absence of a broader understanding of their concepts, semioticians will continue to explore, blindfolded, new continents (of thought and action). I do not doubt that Saussure and Peirce are valid references, but I suggest that Hermann Paul's (1880) *diachrony* is far more conducive to understanding the specific dynamics of languages. This is only one example. Nikolai Sergeyevitch Troubetzkoy (1939) might be another, as is Louis Hjelmslev. Even Uexküll deserves better.

Opportunism Testifies to Shallowness

The modern rebirth of semiotics eventually legitimized what others were doing within their respective disciplines: philologists, structuralists, scholars in literary theory, and morphology. Many fascinating ideas were advanced, and it seemed that a promising new age had begun. But the effort had one major weakness: it remained focused on the sign. Once the new territory of semiotics was defined—mainly by connecting it to Peirce's semiotics—many moved into it, while actually continuing to do what they had always done: interpretations of art and literature, with the help of scientific-sounding terminology. This is not unusual. The most recent example is the morphing of mathematicians and physicists into computer scientists. It took a while until the "new science" (if "new" can be justified in having Leibniz as the final reference) settled into its "language" and "methods." But in the case of semiotics, those who ran over the border and sought "political asylum" in the "free country" of semiotics actually remain faithful ("captive" would be a more accurate descriptor) to their old questions and methods. The new terminology was not revealing, but obstructive.

Semiotics at a rather superficial level became the stage for literary critics, art historians, confused structuralists, and even for some linguists, mathematicians, and sociologists; even some philosophers ventured on the stage. Before long, we had the semiotics of feminism, multiculturalism, human rights, sexuality, food, and even the semiotics of wine; we had gay, lesbian and transgender semiotics, environmental semiotics, and even climate change or sustainability semiotics. *But no semiotics!* Semiotics in this form became a critical discourse of convenience for everything opportunistic. Instead of a rigorous dedication to meaning, these semiotic exercises mimicked everything that the sciences had already provided. Philosophy, in its classical form (i.e., as a speculative endeavor), could have performed the same without the heavy terminology that alienated even those who were convinced that semiotics was a legitimate endeavor. While all the subjects—and there are way more than what is listed—are, of course, relevant within the broader context of culture and civilization, the qualifier *semiotic* at most justified the opportunistic take around the sign as identifier, but did not essentially contribute anything constructive. Jokes about semiotics ("Is it the half of *otics*?") replaced jokes about weather forecasters and statisticians.

LANGUAGE AND SEMIOTICS

While semiotics realized early on that language is the most complex sign system, the semiotic investigation into language was not really productive. I brought up Wittgenstein's views, especially the realization that philosophical problems are in language, not in the world, because more than the celebrated semioticians of language, he grounded language in human activity. To repeat: my main criticism of the semiotics of language concerns the abdication to the notion that semiotics is about the sign. That "language is the most complex sign system," as stated above, was helpful in enlisting language competence—of linguists, grammarians, anthropologists, etc.—but also limiting. Moreover, it confirmed a logocratic view—language as dominant—to the detriment of other forms of expression and communication. This ideology went unchallenged until Peirce, and later Cassirer, each in his own way realized that a variety of semiotic processes complement the semiotics of language. From a different perspective, Roland Barthes thematized the totalitarian nature of language within culture. Within his views, totalitarian regimes rely upon the authority of language in order to consolidate their power. Even the sciences (physics, mathematics, chemistry, etc.) can at times consolidate their "power" through the "languages" they cultivate, to the detriment of alternative understandings in their object domain. Computer science and genetics (the language of gene expression) fully illustrate this thought.

Attempts were made within semiotics to challenge the logocratic model. For instance, some scholars, in the tradition of Locke, tried to advance semiotic notions connected to human activity; others (inspired by Jakob von Uexküll, as author of theoretical biology (1934/2010)) reached beyond the human being into the larger domain of nature. But within semiotics itself, dominated by scholars who fled language studies, such attempts were at best tolerated, but never taken as a scientific challenge. If, finally, semiotics could in our days free itself from the obsession with sign-based language as object of its inquiry, it could make the progress everyone expected. A meaningful dialog among those who acknowledge images, sounds, smell, and tactility as relevant to interactions would certainly benefit semiotics. The fact that a musical score is much more than a string of notes (the syntax) is an almost trivial observation. That the score is, in the sense I suggest for a new foundation of semiotics, a narration invites a different understanding of semiotic processes. Indeed, from the narration to the stories it makes possible—the variety of interpretations, of performances, of meanings—the semiosis transcends that associated with signs. The dynamic dimension gives meaning to the semiotic approach. Similar reflections can be suggested for the narration embodied in images—regardless of whether they are realistic or abstract, photographs, typographic, video, mixed media—or for that matter associated with the sense of taste. A recipe is the "score" for the food which will be eventually prepared, cooked, eaten, enjoyed. Logocratic semiotics is simply incapable of effectively capturing the meaning on non-linguistic semiotic processes. Sign-based semiotics does not capture the meaning of the narration of the activity through which signs come to life.

Even though I have made some historical references, I'm not trying to rewrite the history of semiotics (in which very convincing work was already done). I am not even trying to associate moments in history with the currency of a particular subject. We are not short of histories as we are short of better semiotics. What I attempt here is to point to a development that explains the linguistic bent of even some of the best works produced at the end of the last century. The brilliant literary accomplishments of the French School, as well as the powerful arguments of the Vienna School, of the Russian-Prague formalists and the Soviet school, and even the German and American elaborations of the 1980s and 1990s are pretty much driven by the same implicit understanding that natural language is paradigmatic. A sign-focused semiotics further consolidated this position, instead of questioning it. We will not be able to escape the deadly embrace of this limited understanding unless and until semioticians establish a fresh perspective.

They should at least acknowledge that language is not always language. This is important because languages are structurally different, we miss the opportunity to take advantage of the characteristics of other cultures. (We have even generalized from the Indo-European languages to the new language of programming.) Moreover, we have generalized from Indo-European languages to images, sounds, and other expressive means, although their semiotic conditions are different. If the logocratic model is problematic in the first place, it becomes even more so when it generalizes on account of a particular language experience instead of integrating as many as possible (corresponding to the richness of human activities unfolding in various contexts). However, at the periphery—i.e., exactly that part of the world that was ignored by Western semiotics—semiotic awareness "outside the box" has developed quite convincingly and semiotics gained in significance. Of course, the periphery was "colonized;" English is the *lingua franca*, and semiotics was imported like so many Western-based intellectual endeavors. But recently, awareness of language and logic characteristics of practical experiences not reducible to those of Western civilization started to inform alternative understandings.

Just as an example: French, typical of Western language and logic, and Japanese, of a very different language and logic, are difficult to reconcile. (To elaborate extends beyond the scope of this text. See Nadin (1997, pp. 168-169, 214, 325)). And so is the phonetic writing of many western languages different from the synthetic Korean alphabet and from the Chinese. Within the space of examples promised, there is one example of the compression of writing (Figure 6).

But the semiotic process is even more evident in respect to artifacts of more recent date. For example, the word *thermos* in Korean (Figure 7).

Of course, the narration of the function of this industrial product is broader: how does the thermos work? In Chinese, the narration is more conspicuous: water is imprisoned and attached to a plate, and thus kept warm. The Korean narration is compressed, the English (dictionary definition): a container that keeps a drink or other fluid hot or cold by means of a double wall enclosing a vacuum.

Figure 6. From iconic representation to Chinese ideograms (cf. Dongguo (2008))

갑골문 　 금문 　 소전 　 예서 　 해서

Figure 7. Korean "shorthand for the word thermos

水 + 囚 + 皿 = 溫

물 수 + 가둘 수 + 그릇 명 = 따뜻할 온

Water + Imprison + Plate = Warmth

Along the line of the argument for the formative role that narration plays, I could have used the evolution of number representation: from the fingers and toes to the more compressed notation of Arabic numerals. Writing the word for each number exceeding a certain scale is, obviously, less efficient than the mathematical notation: one million seven hundred thirty-eight thousand five hundred six vs. 1,738,506. The compression can go even further.

Quite interesting semiotics is practiced today in China, eager to embrace all sciences; in Korea, the world center of digital interaction; in Japan, which capitalized on semiotics more than any other economy; and in India. The latter is the recipient of most of Western outsourcing, which is often semiotic work by the way: translations, word processing, scanning, record keeping, programming, etc. While the sign is not discarded, the focus of such a work is rather on broader semiotic entities (text, narrative, game, etc.). This suggests, indirectly, an interest in issues of representation, which are not affected by differences in languages and the associated differences in logic (from the 2-valued Aristotelian logic to the Oriental multi-valued logical systems).

If only Baumgarten's sketchy semiotics, which is part of his attempt to provide a foundation for aesthetics (*Aesthetica* 1750), were to be considered, semioticians would at least, instead of generalizing from the language-defined sign, seek a broader understanding of the sign as such, as Peirce attempted. Such an understanding will in the end have to translate into the most important dimension of the sciences: predictive power. Humankind is pretty advanced in the predictive aspects of the physical world. Nevertheless, we are still at a loss in regard to predictive aspects of living processes—medical diagnostic and treatment come first to mind. Let it be pointed out here that the logographic-driven semiotics focused on the sign could at best deliver explanations for semiotic processes concluded (e.g., characteristic of the physical reality). Analytical performance characterizes this attempt. But even in the best of cases, it could not serve as knowledge on whose basis future semiotic processes could be envisaged or, for that matter, designed, tested, and validated as means to support human activity. A semiotics running after, instead of leading to, desired semiotic processes cannot serve as a bridge among sciences, and even less as an innovative field of human activity.

These lines are only an indirect argument in favor of more semiotics of the visual or of multimedia, of learning from the differences in various languages, and of discovering the underlying shared elements of such languages. Whether we like it or not, language ceased being the dominant means of knowledge acquisition, just as it ceased being the exclusive means of knowledge dissemination (Again, Nadin (1997). Representations in expressions other than in language—computer models and simulations, for instance—are the rule, not the exception. Moreover, representation, in its broad sense, shapes human interaction to the extent that it renders the semiotics of natural language an exercise in speculative rhetoric.

The fact that means of representation are simultaneously constitutive of our own thinking and acting is not yet reflected in the semiotic elaborations of our time. Some researchers, unfortunately ignoring each other, rushed to establish a computational semiotics, and even cognitive semiotics, not realizing that the fashionable qualifiers "computational" and "cognitive" mean, after all, a semiotics of semiotics. What semiotics does not need is a new way of packaging the worn-out speculations resulting from the ceremonial of an old-fashioned dance around the sign—the elusive princess at a ball where everyone seems blessed with eternal oblivion.

Since computational semiotics was mentioned (cf. Stephan 1996, Rieger 1997, 2003, Gudwin & Queiroz 2005), it is appropriate to ask whether such a discipline is possible. Computers, in the form used in our days (i.e., Turing machines performing algorithmic operations), are syntactic engines. Without a semantic dimension, and furthermore without a pragmatic opening, they are limited to a language of two letters and a very constricting logic (Boolean). The semiotics of this language is very limited. However, the broad agreement that knowledge is expressed more and more in computational form could translate into a well-defined goal: express semiotic knowledge computationally. Of course, we are referring here to the meta-level. As such, the goal deserves attention because even though deterministic machines are inadequate for capturing nondeterministic processes, we can work towards conceiving new forms of processing that either mimic the living—such as neuronal networks or membrane computing—or even integrate the living (hybrid computation). Computational semiotics—making reference to Dmitri Pospelov and Eugene Pendergraft, to James Albus, to "language games"(behind which Wittgenstein is suspected), to Luis Rocha and Cliff Joslyn, and even to Leonid Perlovsky and his intelligent target tracker—is more than looking for justification for AI research, or for some computer-based terminology associated with signs. It would be encouraging to engage those interested in foundational aspects of semiotics in a computational effort. One possible result could be a semiotic engine conceived as a procedure for generating representations and for supporting interpretation processes.

THE SEMIOTIC METHOD

The possibility of a semiotic engine brings up the third and last aspect I listed above: What defines the semiotic method? Our concepts, whether semiotic or not, are a projection of our own reality: who we are, what we are made of, how we change, how we interact. The world in which we live embodies matter in an infinite variety of expression. Its dynamics results from energy-related processes, themselves of infinite variety. There is change, including our own; there is the rate of change, testifying to an acceleration related to improved performance, but not necessarily to better understanding of what and why we do what we do. There is also failure. The Internet exemplifies this suggestively: more possibilities, more liabilities. For the new freedom (access to data, communication, social media, etc.), we pay with a sense of vulnerability that undermines not only individual integrity, but even the foundation of democracy. "The Internet, our greatest tool of emancipation, has been transformed into the most dangerous facilitator of totalitarianism we have ever seen" (Assange, 2012, p. 1). The broader the scale of human endeavors, the bigger the scale at which we experience failure. For all practical purposes, a powerful earthquake and a massive tsunami are of a scale comparable to a nuclear power plant breakdown (and its many consequences). And there is the human being: *We are what we do* defines the living, including the human being. We are currently experiencing the computational condition of research and activity: a growing number of possibilities, immense risk. The computational extension of our reality opens new horizons; it also affects the nature of human existence, undermining the known in favor of possibilities that might erode the viability of the human species. Or, alternatively, increase it.

Among other things, humans observe nature (while being part of it) more through the deployment of computational means. And they attempt to change the world according to needs they have, desires they form, goals they express, capabilities they acquire. In this encompassing process of the human being's continuous self-making, humans are semiotic animals able to operate not only on what is available (from stones, tree branches, edible vegetation, to swiftly running rivers and combustible matter), but also on representations of what the world actually is. Computation is representation driven—and generates more representations. This ability is acquired, tested, and continuously changing. To operate on representations is to transcend the immediate, the present. Only the *zoon semiotikon* (and similarly the *animal symbolicum*) has an awareness of the future in the sense that they can affect the dynamics of existence.

Only through the intermediary of semiotic processes of representation do human beings free themselves from the immediate—but at the price of mortgaging their own future. Only awareness of meaning can inform a course of action that will bring opportunity and risk into some balance.

THE GAME OF LIFE

Although it has the power of a Turing machine, Conway's cellular automaton is not algorithmic: the "game" evolves only on account of its initial state (Figure 8). The initial patterns—living organisms on a checkerboard—is changed by using Conway's rules (for birth, death, survival). Here they are: every living cell adjacent to two or three neighboring living cells survives; if the number of adjacent cells with four neighbors dies (overpopulation), a single neighboring living cell dies from isolation. Each empty cell adjacent to *exactly* (i.e., no more and no less) three neighbors is a birth cell. What does this mathematical game have to do with semiotics? As an undecidable entity—i.e., it cannot be fully and consistently described—it is a representation of change and self-configuration. It is, of course, not a sign, but rather, as it is played, a narration resulting in visual patterns.

The life of the narrative is its interpretation—in stories. These are the outcome of the dynamics of the game of life. But we are again a bit ahead of ourselves. Let's step back to representation.

Representation is the prerequisite for natural or artificial reproduction (simulated in Conway's game). The sperm and the egg to be fertilized are embodied representations of the particular male and female. The stem cell, unfolding under complex anticipatory dynamics, is part of the process. Computer programs "translate" algorithms—describing a course of action for reaching a well-defined goal—into operations on representations. Computer viruses, probably more than other successful programs, illustrate artificial reproduction as it results from a dynamics associated with pre-defined operations. The reverse engineered Stuxnet—the virus deployed by secret services to control friend and foe (Iran, at that time)—is a good introduction to the subject. Many other stealth programs are at work at spying on those connected to the Internet: for commercial purposes, for political reasons, for criminal activities. (If you are on Facebook, for example, you are spied on: each click is recorded and meaning is extracted.) Adaptive characteristics of the living and adaptive mechanisms in the world of machines, as different as they are, correspond to two different modalities for generating representations appropriate to changing contexts of existence or functioning. In adapting, the living experiences information processes, corresponding to energy- and matter-related phenomena, and semiotic processes, corresponding to meaning, and embodied in the narrative of life and its many associated stories.

Figure 8. Visual patterns in the Game of Life

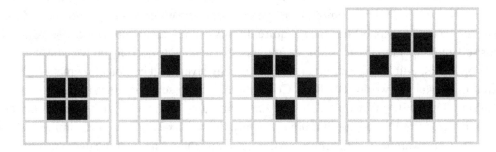

Space and time are constitutive representations characteristic of our epistemological focus (we want to know). Consequently, it is epistemologically suicidal not to realize that concepts, which are representations, help to both describe and constitute the world. We perceive the world empowered (when not blinded) by our thinking and supported (when not handicapped) by artificially extended perceptions. We "see" today much, much more than what we see (just think about the "invisible" micro-level of matter, or the phenomena in the universe in which we exist); we "hear" today much more than what our ears bring to us. But in the end, we never escape the epistemological circularity of our perspectives. This applies to mathematics as it does to semiotics. For people focused on a sign-centered semiotics, a sign definition is as adequate as we can make it adequate. But it is a construct, always subject to questioning, as Sadowski (2010) recently questioned Peirce's definition, or as I (Nadin 1983) questioned Saussure's definition (notwithstanding the relevance of his linguistic contributions (cf. Bouissac, 2010). Something else is at stake: not the adequacy of sign-based semiotic concepts, but the ability to support, to guide practical experiences. The first integrated VLSI (i.e., integrated circuits), celebrated as one of the major accomplishments in the technology of the last 50 years, was a project in applied physics. Today, we integrate billions of transistors in a chip, or achieve technological performance in myriad ways, Physics and awareness of the characteristics of the living fuse into a new perspective. Deeper and deeper neural networks, i.e., mathematical constructs mimicking the real neuron (infinitely diverse) afford learning patterns that imitate human training. But after all is said and done, the entire effort is focused on *representations*—of arithmetic, calculus, geometry, physics, etc. No doubt, the chip remains a magnificent outcome of mathematics, physics, chemistry, and technology, i.e., engineering. The artificial neuron is yet another example in the same league. But what is "condensed" on the chip or on the artificial neuron is knowledge—representations, not signs, expressed in digital form. Ultimately, this knowledge is a representation of all we know about arithmetic, calculus, geometry, etc., of what we know about graphics, color, form, shape, etc. The most recent (and probably soon-to-be improved upon) computational "winner" of the Go game (against the world champion) owns the past of all Go games, but could not come up with a new game. In video games, the victory of information processing (implementation of the binocular parallax) is associated with a semiotic accomplishment: the meaning of 3D in situations of search, hiding, and exploring realistic representations of landscapes, etc. Machines playing against machines is the new form of gladiatorial combat: no more blood and death, but a lot of resources. An AI program can reverse engineer the game engine used in making the game! Playing hide-and-go-seek involves our individual characteristics, our *ad hoc* knowledge pertinent to hiding and seeking. Playing an MMOG (massively multi-player online game) involves embodied knowledge. If this knowledge reflects the reductionist-deterministic view of the world, the game will be a good simulation of this perspective—but not a new perspective of our own being, of our condition as semiotic animals. This is a world of action-reaction. Playing with others, located around the world, via the medium of the game recovers anticipation. This is a victory for semiotics, even if semioticians have to date missed the meaning of such innovative applications.

MONSIEUR JOURDAIN DID NOT KNOW HE WAS SPEAKING PROSE

Monsieur Jourdain: And this, the way I speak. What name would be applied to...?

Figure 9. The Bourgeois Gentleman (Molière)

Philosophy Master: The way you speak?

Monsieur Jourdain: Yes.

Philosophy Master: Prose

Monsieur Jourdain: It's prose. Well, what do you know about that! These forty years now, I've been speaking in prose without knowing it.

But what are semioticians doing while the world changes drastically and a new human being emerges? The old soup of psychoanalytic extraction is warmed up again and again; literary criticism is disguised as semiotic analysis; structuralist considerations are rewritten in semiotic jargon; linguistic terminology is made to appear semiotic. To forever analyze popular culture (after Barthes and Eco exhausted the theme), film, music, new media, and video games might lead to texts published by editors as clueless as the writers, but not to the knowledge that society has the legitimate right to expect from semiotics. Books on the semiotics of games will never replace the experience of the game itself, or of conceiving the game. One alternative, among many possible, could be the opening of a "Story Lab" where semiotics can be practiced in generating new stories, corresponding to the fast dynamics of the present, instead of continuing the impotent discourse on narrativity (without understanding the difference between narration and story). No less exciting is the goal of providing semiotic methods for the human interactions of the future, not just attempts to explain what these interactions were. (Useless analytical exercises in semiotics have already perverted the field and damaged its reputation!)

Have I given the impression that conditions were ideal in the "good old days" of the semiotic revival of the early 1970s (or earlier)? I hope not. Have I incited a conflict between succeeding generations of semioticians? Probably, in the sense that I still hold to the notion (Peircean, by the way) that without an ethics of terminology, each of us will be talking about something (the sign, let's say) and understanding something else. The best example of this is the use of the word *sign*, and the tendency to substitute *sym-*

bol for *sign* (or *vice versa*). Those falling for YouTube elaborations on semiotics would be well advised to undergo some form of decontamination or some training in the ethics of terminology. For this ethics to emerge, we also need an encompassing semiotic culture: more people who read primary sources, not approximate derivations, and more people with *original* ideas who actually read what has already been written on a topic—and give credit where credit is due. Yes, there was more scholarship before, despite the absence of Google or Wikipedia—sources of generalized mediocrity, which some believe substitute for true research effort. Without the realization of the need for scholarship, well-intended newcomers will rediscover "continents" that were already explored, and consequently miss their chance to contribute fresh thoughts.

Mediocrity corresponds to a new semiotic condition of the human being. Within shorter cycles of change, and under the inescapable pressure of faster dynamics, there is no room left for depth. Humankind is shaping itself as a species of shallow enterprise, an existence focused on breadth not depth, contributing spectacularly to its own end (within a perspective of time that makes the end still far away). I know of "distinguished professors" (their official, but not earned, title) who cannot distinguish between semiology and semiotics, between meaning and significance, between data and information, and who cannot even pronounce "Peirce" and "Saussure" correctly. But they don't shy away from disseminating their ignorance to young people who trust them by virtue of their institutional identity. Authority built on language games ends up as academic charlatanism.

In various attempts at making up "specialized" semiotics—of music, law, sex, and so on and so on—mostly left in some state of indeterminacy, well-intentioned authors decided to use the concept of the sign in order to deal with particular objects of their interest. Obviously, someone can take a ruler to measure how long a carrot is, or how short a mouse's nose. Appropriateness of perspective, and thus of qualifiers for a certain action or tool, is a methodological prerequisite for any scientific endeavor. Philosophy is not measured in gallons; a work of art is not reducible to the number of knots in the canvas; music is not the map of sound frequencies. The sign, well- or ill-defined, can be the identifier of choice for pragmatic reasons: How well does the STOP sign perform its function? (Keep in mind: when the car is fully automated, i.e., the driverless, autonomous vehicle, the sign as such becomes obsolete.) How appropriate are the various components of a sign such as a logo in a corporate identity "language"? (But when the life of a corporation is no longer than the life of its only product, identity is consumed.) Why is a certain selection made (color, shape, rhythm) in the attempt to establish conventions for communication purposes, or within a culture? (Such choices will change as fast as anything does in our time.)

Semiotics is not reducible to signs, or to the formal relation among signs (what is called *syntax*). Those who do not realize this irreducibility might at times generalize in a manner not beneficial to semiotics. The best example is that of semioticians forcing their contrived terminology on hot domains of knowledge. *Biosemiotics* (Barbieri (2007)) is such a domain; and many self-delusional attempts have been made to find semiotics in biology, instead of first asking the question of how semiotics might be relevant to advancing biology. Biological processes consist of both informational and semiotic processes: they are narrations. This could be important to semioticians, but only after they find out what this means. However, more important than the syntax of life is life itself, a narration that encompasses semiotics and pragmatics. Its deviations in stories (disease, accident, birth and death, etc.) are far more conducive to knowledge than inventories of signs.

Kull and Velmezova (2014) honored by assessment:

The day when scholars and students of semiotics become the hottest commodity in the labor market and are traded like neurosurgeons, high-performance programmers, footballs players, movie stars, or animators, we will all know that semiotics finally made it. Currently, semiotics is of marginal interest, at most, in academia. Nobody hires semioticians. I am convinced that this can change. But for this change to come about, everyone involved in semiotics will have to think in a different way, to redefine their goals. Semioticians need the patience and dedication necessary for working on foundational aspects, starting with defining the specific domain knowledge and the appropriate methodology. And they need to define a research agenda for semiotics above and beyond the speculative.

As a dedicated scholar of the respectable Tartu School (usually identified with Lotman), Kalevi Kull might have decided to quote these words because he belongs to those who candidly wish that semiotics will do better than it now does.

ACKNOWLEDGMENT

Dr. Marcel Danesi saw value in my attempt at reassessing the foundations of semiotics and invited me to pursue the matter further. This encouraged me to revisit the main ideas formerly articulated as hypotheses. The initial text prompted Umberto Eco's endorsement and a lively discussion in Tartu with Kalevi Kull and Stuart Kauffman (Tartu Summer School of Semiotics, August 17-20, 2015). Dean Anne Balsamo, responding to suggestions from graduate students and PhD candidates at the ATEC Program (Art, Technology, Emerging Communication) at the University of Texas at Dallas, twisted my arm in requesting a class in semiotics. After 34 years of absence from teaching semiotics, I designed a curriculum that extends from the past of semiotics to ontology engineering, machine learning (including the new Deep Learning applications), biosemiotics. It is a test for my new foundation for semiotics. Of course, I am indebted to all those I interacted with on this captivating question.

Gratitude is expressed to Dr. Asma Naz for her contribution to many matters regarding this study. Joyce Hong is a patient assistant trying to keep up with a very demanding job in the antÉ Lab. Elvira Nadin provided invaluable research in support of my attempt to provide a broad context. I would like to express my gratitude also to the antÉ—Institute for Research in Anticipatory Systems, The University of Texas at Dallas, and to the Hanse Wissenschaftskolleg (Hanse Institute for Advanced Studies), Delmenhorst, Germany, for supporting the research that made this article possible.

REFERENCES

American Association for the Advancement of Science. (2001). Special Issue: The Human Genome. *Science, 291*(5507). Retrieved from http://science.sciencemag.org/content/291/5507

Aristotle. (350 BCE). *On interpretation, 1.16a 4-9* (E. M. Edgehill, E.M., Trans.). Retrieved from http://classics.mit.edu/Aristotle/interpretation.html

Assange, J. (2012). *Cypherpunks: Freedom and the future of the Internet*. New York: OR Books.

Barbieri, M. (Ed.). (2007). *Introduction to biosemiotics. The new biological synthesis*. Heidelberg, Germany: Springer. doi:10.1007/1-4020-4814-9

Baumgarten, A. G. (1750). *Aesthetica*. Frankfurt-an-der-Oder, Germany: Kley.

Berners-Lee, T. (1998). *Semantic web roadmap*. Retrieved on March 31, 2011 from http://www.w3.org/DesignIssues/Semantic.html

Bouissac, P. (1977). *Circus and culture: A semiotic approach (Advances in Semiotics)*. Bloomington, IN: Indiana University Press.

Bouissac, P. (2010). *Saussure. A guide for the perplexed*. New York: Continuum International Publishing.

Brentano, F. (1874). *Von empirische Standpunkt. In Psychologie* (Vols. 1–2). Leipzig: Von Duncker & Humblot.

Cassirer, E. (1921). *Zur Einstein'sche Relativitätstheorie. In Substance and function and Einstein's theory of relativity*. Chicago: Open Court Publishing Co.

Cassirer, E. (1955). *Philosophy of symbolic forms* (R. Manheim, Trans.). New Haven, CT: Yale University Press.

Chu, K., Grundy, Q., & Bero, L. (2017). "Spin" in published biomedical literature. A methodological system review. *PLoS Biology, 16*(9). Retrieved from http://journals.plos.org/plosbiology/article?id=10.1371/journal.pbio.2002173

Conway, J. H. (1970). See Gardner, M. (October 1970) Mathematical games—The fantastic combinations of John Conway's new solitaire game "life". *Scientific American, 223*, 120-123. Retrieved from http://rosettacode.org/wiki/Conway%27s_Game_of_Life

Deledalle, G. (1979). *Théorie et pratique du signe: Introduction à la sémiotique de Charles S. Peirce. Paris: Payot*.

Deutsch, D. (2012). *The Beginning of Infinity*. New York: Penguin Books.

Dewart, L. (2016). *Hume's Challenge and the Renewal of Modern Philosophy*. BookBaby Publishers.

Dingguo, S. (2008). *The Wisdom of Chinese Characters (English Edition)*. Beiging Language and Culture University.

Eco, U. (1975). Preface. In S. Chapman, U. Eco, & J.-M. Klinkenberg (Eds.), *A Semiotic Landscape/Panorama Sémiotique*. The Hague, The Netherlands: Mouton.

Eco, U. (1976). *A theory of wemiotics (Advances in Semiotics)*. Bloomington, IN: Indiana University Press. doi:10.1007/978-1-349-15849-2

Eco, U. (2017). *The philosophy of Umberto Eco* (S. Beardsworth & E. A. Randall, Eds.). Chicago: Open Court Publishing.

Emmeche, C., Kull, K., & Stjernfelt, F. (2002). *Reading Hoffmeyer, rethinking biology. Tartu Semiotics Library 3*. Tartu: Tartu University Press.

Favareau, D. (2009). *Essential readings in biosemiotics: Anthology and commentary (Biosemiotics, 3).* Dordrecht, The Netherlands: Springer. doi:10.1007/978-1-4020-9650-1

Gallese, V. (2001). The "Shared Manifold" hypothesis. From mirror neurons to empathy. *Journal of Consciousness Studies, 8,* 5–7, 33.

Gallese, V. (2009). Mirror neurons, embodied simulation, and the neural basis of social identification. *Psychoanalytic Dialogues, 19*(5), 519–536. doi:10.1080/10481880903231910

Goguen, J. (1999). An introduction to algebraic semiotics, with application to user interface design. In C. L. Nehaniv (Ed.), Lecture Notes in Computer Science: Vol. 1562. *Computation for metaphors, analogy, andaAgents* (pp. 242–291). Berlin: Springer. doi:10.1007/3-540-48834-0_15

Gudwin, R., & Queiroz, J. (2005). Towards an introduction to computational semiotics. *Proceedings of the 2005 IEEE International Conference on Integration of Knowledge Intensive Multi-Agent Systems,* 393-398. doi:10.1109/KIMAS.2005.1427113

Hoffmeyer, J. (Ed.). (2008). *A legacy for living systems. Gregory Bateson as precursor to biosemiotics.* Berlin: Springer.

Kauffman, S. (2011). *There are more uses for a screwdriver than you can calculate.* Retrieved from http://www.wbur.org/npr/135113346/there-are-more-uses-for-a-screwdriver-than-you-can-calculate

Koblin, A. (2008-2015). *Aaron Koblin Website: Information.* Retrieved on April 1, 2017 from http://www.aaronkoblin.com/info.html

Kuhn, T. S. (1962). *The structure of scientific revolutions.* Chicago: University of Chicago Press.

Kull, K., Deacon, T., Emmeche, C., Hoffmeyer, F., & Stjernfelt, F. (2009). Theses on biosemiotics: Prolegomena to a theoretical biology. *Biological Theory, 4*(2), 167–173. doi:10.1162/biot.2009.4.2.167

Kull, K., & Velmezova, E. (2014). What is the main challenge for contemporary semiotics? *Sign Systems Studies, 42*(4), 530–548. doi:10.12697/SSS.2014.42.4.06

Lakatos, I. (1970). Criticism and the methodology of scientific research programmes. *Proceedings of the Aristotelian Society, 69*(1968–1969), 149-186.

Locke, J. (1690). *An essay concerning human understanding.* Chapt. XXI, Book 4.

Lorusso, A. M. (2015). *Cultural semiotics.* London: Palgrave Macmillan. doi:10.1057/9781137546999

Lotman, J. M. (1990). *Universe of the mind: A semiotic theory of culture* (A. Shukman, Trans.). Bloomington, IN: Indiana University Press.

Margulis, L. (1995). Gaia is a tough bitch. In J. Brockman (Ed.), *The third culture: Beyond the scientific revolution.* New York: Simon and Schuster. Retrieved from https://www.edge.org/conversation/lynn_margulis-chapter-7-gaia-is-a-tough-bitchhttp

Marty, R. (1990). *L'Algèbre des signes: Essai de sémiotique scientifique d'après Charles Sanders Peirce (Foundations of Semiotics, 24).* Amsterdam: J. Benjamins Publishing Co. doi:10.1075/fos.24

Mitchell, C. J., De Houwer, J., & Lovibond, P. F. (2009). The propositional nature of human associative learning. *Behavioral and Brain Sciences*, *32*(2), 183–198. doi:10.1017/S0140525X09000855 PMID:19386174

Mongré, P. (1897). Sant' Ilario. Thoughts from Zarathustra's landscape. Leipzig: C.G. Nauman.

Morris, C. W. (1938). *Foundation of the theory of signs* (1st ed.). Chicago: University of Chicago Press.

Morris, R., & Ward, G. (Eds.). (2005). *The cognitive psychology of planning*. London: Psychology Press.

Mumford, L. (1967). *The myth of the machine: Technics and human development*. New York: Harcourt, Brace, Jovanovich.

Nadin, M. (1977). Sign and fuzzy automata. *Semiosis*, *1*, 5.

Nadin, M. (1980). The logic of vagueness and the category of synechism. *The Monist*, *63*(3), 351–363. doi:10.5840/monist198063326

Nadin, M. (1983). The logic of vagueness and the category of synechism. In E. Freeman (Ed.), *The relevance of Charles Peirce. La Salle* (pp. 154–166). The Monist Library of Philosophy.

Nadin, M. (1986). Pragmatics in the semiotic framework. In H. Stachowiak (Ed.), Pragmatik, II The rise of pragmatic thought in the 19th and 20th centuries (pp. 148–170). Academic Press.

Nadin, M. (1991). *Mind—anticipation and chaos*. Stuttgart, Germany: Belser.

Nadin, M. (1997). *The civilization of illiteracy*. Dresden, Germany: Dresden University Press.

Nadin, M. (2010). Anticipation and dynamics: Rosen's anticipation in the perspective of time. *International Journal of General Systems*, *39*(1), 3-33.

Nadin, M. (2011). Information and semiotic processes. The semiotics of computation. *Cybernetics & Human Knowing*, *18*(1-2), 153–175.

Nadin, M. (2012). Reassessing the foundations of semiotics: Preliminaries. *International Journal of Signs and Semiotic Systems*, *2*(1), 43–75. doi:10.4018/ijsss.2012010101

Nadin, M. (2014). Semiotics is fundamental science. In M. Jennex (Ed.), *Knowledge discovery, transfer, and management in the information age* (pp. 76–125). Hershey, PA: IGI Global. doi:10.4018/978-1-4666-4711-4.ch005

Nadin, M. (2016). *Anticipation and the brain, Anticipation and medicine*. Cham, Switzerland: Springer International Publishers.

Nadin, M. (2017). Semiotic engineering – An opportunity or an opportunity missed. In *Conversations around semiotic engineering*. Berlin: *Springer*.

Nicolelis, M. A. L., & Shuler, M. (2001). Thalamocortical and corticocortical interactions in the somatosensory system. In M. A. L. Nicolelis (Ed.), Progress in brain research: Vol. 130. *Advances in neural population coding* (pp. 89–110). Amsterdam: Elsevier Science. doi:10.1016/S0079-6123(01)30008-0

Nietzsche, F. (1975). *Kritische Gesamtausgabe Briefwechsel* (G. Colli & M. Montinari, Eds.). Berlin: Walter de Gruyter.

Nöth, W. (1985). *Handbuch der Semiotik. Stuttgart: Metzler. Published in English (1995), Handbook of semiotics (Advances in Semiotics)*. Bloomington, IN: University of Indiana Press.

Paul, H. (1880). Prinzipien der Sprachgeschichte. Halle: Max Niemeyer.

Peirce, C. S. (1953). *Letters to Lady Welby* (I. C. Lieb, Ed.). New Haven, CT: Whitlock.

Piaget, J. (1955). *The child's construction of reality*. London: Routledge and Kegan Paul.

Pinegger, A., Hiebel, H., Wriessnegger, S. C., & Müller-Putz, G. R. (2017). Composing only by thought: Novel application of the P300 brain-computer interface. *PLoS, 12*(9). Retrieved from https://www.ncbi. nlm.nih.gov/pubmed/28877175

Plato. (360 BCE). *Cratylus* (B. Jowett, Trans.). Retrieved on April 10, 2011 from, http://classics.mit. edu/Plato/cratylus.html

Popper, K. R. (1934). *Logik der Forschung. Vienna: Mohr Siebeck.*

Posner, R. (2004). Basic tasks of cultural semiotics. In G. Witham & J. Wallmannsberger (Eds.), *Signs of power – Power of signs. Essays in honor of Jeff Bernard* (pp. 56–89). Vienna: INST.

Rieger, B. B. (1997). Computational semiotics and fuzzy linguistics. On meaning constitution and soft categories. In A. Meystel (Ed.), *A learning perspective. International Conference on Intelligent Systems and Semiotics*. Washington, DC: NIST Special Publication.

Rieger, B. B. (2003). Semiotic cognitive information processing: Learning to understand discourse. A systemic model of meaning constitution. In R. Kühn et al. (Eds.), *Adaptivity and learning. An interdisciplinary debate* (pp. 347–403). Berlin: Springer. doi:10.1007/978-3-662-05594-6_24

Rosen, R. (1985). Organisms as causality systems which are not machines: An essay on the nature of complexity. In R. Rosen (Ed.), *Rosen: Theoretical biology and complexity* (pp. 165–203). Orlando, FL: Academic Press. doi:10.1016/B978-0-12-597280-2.50008-8

Sadowski, P. (2010). *Towards systems semiotics: Some remarks and (hopefully useful) definitions*. Retrieved from http://www.semioticon.com/semiotix/2010/03/towards-systems-semiotics-some-remarks-and-hopefully-useful-definitions/

Shannon, C. E., & Weaver, W. (1949). *The mathematical theory of communication*. Urbana, IL: University of Illinois Press.

Sieckenius de Souza, C. (2005). *The semiotic engineering of human-computer interaction*. Cambridge, MA: MIT Press.

Silverman, K. (1984). The subject of semiotics. Oxford University Press.

Smalley, R. E. (2001, September). Of chemistry, love and nanobots (nanofallacies). *Scientific American, 285*, 76–77. doi:10.1038/scientificamerican0901-76 PMID:11524973

Solomon, J. F. (1988). *The signs of our time. Semiotics: The hidden messages of environments, objects, and cultural images*. New York: St Martin's Press.

Sowa, J. (2017). *The virtual reality of the mind*. Retrieved from http://222.jfsowa.com/talks/vrmind.pdf

Stephan, P. (1996). Auf dem Weg zu Computational Semiotics. In D. Dotzler (Ed.), *Computer als Faszination* (p. 209). Frankfurt, Germany: CAF Verlag.

Trubetzkoy, N. S. (1939). Grundzüge der Phonologie. Traveaux du Cercle Linguistique du Prague, 7.

von Foerster, H. (1981). *Observing systems*. Seaside, CA: Intersystems Publications.

von Uxeküll, J. (1934). Streifzüge durch die Umwelt von Tieren und Menschen. Berlin: Julius Springer Verlag. doi:10.1007/978-3-642-98976-6

Wiles, A. (1995). Modular elliptic curves and Fermat's Last Theorem. *Annals of Mathematics, 141*(3), 443–551. doi:10.2307/2118559

Wilson, E. O. (1984). *Biophilia: The human bond with other species*. Cambridge, MA: Harvard University Press.

Windelband, W. (1894). Geschicthte und Naturwissenschaft, Präludien. Aufsätze und Reden zur Philosophie und ihrer Geschichte. Tubingen: J.C.B. Mohr.

Wittgenstein, L. (1953). *Philosophical investigations* (G. E. M. Anscombe, Trans.). London: Basil Blackwell.

Wittgenstein, L. (2003). *Diktat fur Schlick (Moritz Schlick, Dec. 1932). In The voices of Wittgenstein: The Vienna Circle: Ludwig Wittgenstein and Friedrich Waisman*. London: Routledge.

ENDNOTES

[1] "How soon will we see the nanometer-scale robots envisaged by K. Eric Drexler and other molecular nanotechologists? The simple answer is never."

[2] Semiotics and Communication Studies, Victoria College; the Toronto Semiotic Circle (founded in 1973) is still active. An International Summer Institute for Semiotic and Structural Studies takes place regularly.

[3] The Kaunas University of Technology hosts an International Semiotics Institute; there is also an online Semiotics Institute.

[4] "We make 'inner fictions or symbols' of outward objects, and these symbols are so constituted that the necessary consequences of the images are always images of the necessary consequences of the imaged objects."

[5] Famous as a physicist, John Archibald Wheeler insisted on the meaning of information (see Davies, 2004, pp. 8-10). Davies, C. W. P. (2004). John Archibald Wheeler and the clash of ideas. In J. D. Barrow, C. W. P. Davies, & C. L. Harper (Eds.), Science and ultimate reality (pp. 3-24). London: Cambridge University Press.

Chapter 7
Syntactic Semantics and the Proper Treatment of Computationalism

William J. Rapaport
The State University of New York at Buffalo, USA

ABSTRACT

Computationalism should not be the view that (human) cognition is computation; it should be the view that cognition (simpliciter) is computable. It follows that computationalism can be true even if (human) cognition is not the result of computations in the brain. If semiotic systems are systems that interpret signs, then both humans and computers are semiotic systems. Finally, minds can be considered as virtual machines implemented in certain semiotic systems, primarily the brain, but also AI computers.

INTRODUCTION

This essay treats three topics: computationalism, semiotic systems, and cognition (the mind), offering what I feel is the proper treatment of computationalism. From this, certain views about semiotic systems and minds follow (or, at least, are consistent): First, I argue that computationalism should not be understood as the view that (human) cognition *is* computation, but that it should be understood as the view that cognition (human or otherwise) is comput*able*. On this view, it follows that computationalism can be true even if (human) cognition is not the result of computations in the brain. Second, I argue that, if semiotic systems are systems that interpret signs, then both humans and computers are semiotic systems. Finally, I suggest that minds should be considered as virtual machines implemented in certain semiotic systems: primarily brains, but also AI computers. In the course of presenting and arguing for these positions, I respond to Fetzer's (2011) arguments to the contrary.[1]

DOI: 10.4018/978-1-5225-5622-0.ch007

THE PROPER TREATMENT OF COMPUTATIONALISM

Computationalism is often characterized as the thesis that cognition is computation. Its origins can be traced back at least to Thomas Hobbes:

For REASON, in this sense [i.e., as among the faculties of the mind], is nothing but reckoning—that is, adding and subtracting—of the consequences of general names agreed upon for the marking and signifying of our thoughts... (Hobbes, 1651, Part I, Ch. 5, p. 46).[2]

It is a view whose popularity, if not its origins, has been traced back to McCulloch & Pitts (1943), Hilary Putnam (1960 or 1961) and Jerry Fodor (1975) (see Horst, 2009, Piccinini, 2010). This is usually interpreted to mean that the mind, or the brain—whatever it is that exhibits cognition—computes, or *is* a computer. Consider these passages, more or less (but not entirely) randomly chosen:[3]

- A Plan is any hierarchical process in the organism that can control the order in which a sequence of operations is to be performed. A Plan is, for an organism, essentially the same as a program for a computer (Miller et al., 1960, p. 16).[4]
- [H]aving a propositional attitude is being in some *computational* relation to an internal representation. …Mental states are relations between organisms and internal representations, and causally interrelated mental states succeed one another according to computational principles which apply formally *to the representations* (Fodor, 1975, p. 198).
- [C]ognition ought to be viewed as computation. [This] rests on the fact that computation is the only worked-out view of *process* that is both compatible with a materialist view of how a process is realized and that attributes the behavior of the process to the operation of rules upon representations. In other words, what makes it possible to view computation and cognition as processes of fundamentally the same type is the fact that both are physically realized and both are governed by rules and representations (Pylyshyn, 1980, p. 111).
- [C]ognition *is* a type of computation (Pylyshyn, 1985, p. xiii.)
- The basic idea of the computer model of the mind is that the mind is the program and the brain the hardware of a computational system (Searle, 1990, p. 21).
- Computationalism is the hypothesis that cognition *is* the computation of functions. …The job for the computationalist is to determine…which specific functions explain specific cognitive phenomena (Dietrich, 1990, p. 135, my italics).
- [T]he Computational Theory of Mind…is…the best theory of cognition that we've got…. (Fodor, 2000, p. 1).
- Tokens of mental processes are 'computations;' that is, causal chains of (typically inferential) operations on mental representations (Fodor, 2008, pp. 5–6).
- The core idea of cognitive science is that our brains are a kind of computer…. Psychologists try to find out exactly what kinds of programs our brains use, and how our brains implement those programs (Gopnik, 2009, p. 43).
- [A] particular philosophical view that holds that the mind literally is a digital computer…, and that thought literally is a kind of computation…will be called the "Computational Theory of Mind"…. (Horst, 2009).

- Computationalism…is the view that the functional organization of the brain (or any other functionally equivalent system) is computational, or that neural states are computational states (Piccinini, 2010, p. 271).
- These remarkable capacities of computers—to manipulate strings of digits and to store and execute programs—suggest a bold hypothesis. Perhaps brains are computers, and perhaps minds are nothing but the programs running on neural computers (Piccinini, 2010, p, 277–278).
- Advances in computing raise the prospect that the mind itself is a computational system—a position known as *the computational theory of mind* (Rescorla, 2015).

That cognition is computation is an interesting claim, one well worth exploring, and it may even be true. But it is too strong: It is not the kind of claim that is usually made when one says that a certain behavior can be understood computationally (Rapaport, 1998). There is a related claim that, because it is weaker, is more likely to be true and—more importantly—is equally relevant to computational theories of cognition, because it preserves the crucial insight that cognition is capable of being explained in terms of the mathematical theory of computation.

Before stating what I think is the proper version of the thesis of computationalism, let me clarify two terms: (1) I will use 'cognition'[5] as a synonym for such terms as 'thinking,' 'intelligence' (as in 'AI,' not as in 'IQ'), 'mentality,' 'understanding,' 'intentionality,' etc. Cognition is whatever cognitive scientists study, including (in alphabetical order) believing (and, perhaps, knowing), consciousness, emotion, language, learning, memory, (perhaps) perception, planning, problem solving, reasoning, representation (including categories, concepts, and mental imagery), sensation, thought, etc. Knowing might *not* be part of cognition, insofar as it depends on the way the *world* is (knowing is often taken to be justified *true* belief) and thus would be independent of what goes on in the mind or brain; perception also depends on the way the world is (see below). (2) An "algorithm" for an executor E to achieve a goal G is, informally, a procedure (or "method") for E to achieve G, where (a) E is the agent—human or computer—that carries out the algorithm (or executes, or implements, or "follows" it), (b) the procedure is a set (usually, a sequence) of statements (or "steps," usually "rules" or instructions), and (c) G is the solution of a (particular kind of) problem, the answer to a (particular kind of) question, or the accomplishment of some (particular kind of) task. (See the Appendix for more details.)

Various of these features can be relaxed: One can imagine a procedure that has all these features of algorithms but that has no specific goal, e.g., "Compute 2+2; then read *Moby Dick*.," or one for which there is no executor, or one that yields output that is only *approximately* correct (sometimes called a 'heuristic'), etc. For alternative informal formulations of "algorithm," see the Appendix. Several different mathematical, hence precise, formulations of this still vague notion have been proposed, the most famous of which is Alan Turing's (1936) notion of (what is now called in his honor) a 'Turing machine.' Because all of these precise, mathematical formulations are logically equivalent, the claim that the informal notion of "algorithm" *is* a Turing machine is now known as "Turing's thesis" (or as "Church's thesis" or the "Church-Turing thesis," after Alonzo Church, whose "lambda calculus" was another one of the mathematical formulations).

Importantly, for present purposes, when someone says that a mathematical function or a certain phenomenon or behavior is "computable," they mean that there is an algorithm that outputs the values of that function when given its legal inputs[6] or that produces that phenomenon or behavior—i.e., that one could write a computer program that, when executed on a suitable computer, would enable that computer to perform (i.e., to output) the appropriate behavior. Hence: Computationalism, properly understood, should be the thesis that cognition is comput*able*, i.e., that there is an algorithm (more likely, a family of algorithms) that computes cognitive functions.

I take the basic research question of computational cognitive science to ask, "*How much* of cognition is computable?" And I take the working assumption (or expectation, or hope) of computational cognitive science to be that *all* cognition is computable. This formulation of the basic research question allows for the possibility that the hopes will be dashed—that some aspects of cognition might *not* be computable. In that event, the interesting question will be: *Which* aspects are not computable, and why?[7] Although several philosophers have offered "non-existence proofs" that cognition is *not* computable,[8] none of these are so mathematically convincing that they have squelched all opposition. And, in any case, it is obvious that *much* of cognition *is* computable (see Johnson-Laird, 1988, Edelman, 2008b, Forbus, 2010 for surveys). Philip N. Johnson-Laird (1988, pp. 26-27) has expressed it well:

The goal of cognitive science is to explain how the mind works. Part of the power of the discipline resides in the theory of computability. ...Some processes in the nervous system seem to be computations.... Others...are physical processes that can be modeled in computer programs. But there may be aspects of mental life that cannot be modeled in this way.... There may even be aspects of the mind that lie outside scientific explanation.

However, I suspect that *so* much of cognition will eventually be shown to be computable that the residue, if any, will be negligible and ignorable. This leads to the following "implementational implication:" If (or to the extent that) cognition *is* computable, then *anything* that implements cognitive computations would *be* (to that extent) cognitive. Informally, such an implementation would "really think." As Newell, Shaw, & Simon (1958, p. 153) put it (explicating Turing's notion of the "universal" Turing machine, or stored-program computer), "if we put any particular program in a computer, we have in fact a machine that behaves in the way prescribed by the program." The "particular program" they were referring to was one for "human problem solving," so a computer thus programmed would indeed solve problems, i.e., exhibit a kind of cognition. This implication is probably a more general point, not necessarily restricted to computationalism. Suppose, as some would have it, that cognition turns out to be fully understandable only in terms of differential equations (Forbus, 2010 hints at this but does not endorse it) or dynamic systems (van Gelder, 1995). Arguably, anything that implements cognitive differential equations or a cognitive dynamic system would be cognitive.

The more common view, that cognition is computa*tion*, is a "strong" view that the mind or brain *is* a computer. It claims that *how* the mind or brain does what it does is by computing. My view, that cognition is comput*able*, is a weaker view that *what* the mind or brain does can be *described* in computational terms, but that *how* it does it is a matter for neuroscience to determine.[9] Interestingly, some of the canonical statements of "strong" computationalism are ambiguous between the two versions. Consider some of Fodor's early statements in his *Language of Thought* (1975): "[H]aving a propositional attitude is being in some *computational* relation to an internal representation (p. 198, Fodor's emphasis).This could be interpreted as the weaker claim that the relation is comput*able*. The passage continues: "The intended claim is that the sequence of events that *causally* determines the mental state of an organism will be *describable as* a sequence of steps in a derivation.... (p. 198, my emphases). The use of 'causally' suggests the stronger—implementational—view, but the use of 'describable as' suggests the weaker view. There's more:

More exactly: Mental states are relations between organisms and internal representations, and caus-ally interrelated mental states succeed one another according to computational principles which apply formally to the representations (p. 198, my boldface, Fodor's italics).

If 'according to' means merely that they behave in accordance with those computational principles, then this is consistent with my—weaker—view, but if it means that they execute those principles, then it sounds like the stronger view. Given Fodor's other comments and the interpretations of other scholars, and in light of later statements such as the quote from 2008, above, I'm sure that Fodor always had the stronger view in mind. But the potentially ambiguous readings give a hint of the delicacy of interpretation.[10]

That cognition is comput*able* is a necessary—but not sufficient—condition for it to be computa*tion*. The crucial difference between cognition as being comput*able* rather than as *being* computa*tion* is that, on the weaker view, the implementational implication holds *even if humans don't implement cognition computationally.* In other words, it allows for the possibility that human cognition is comput*able* but is not comput*ed.* For instance, Gualtiero Piccinini (2005, 2007) has argued that "spike trains" (sequences of "action potential") in groups of neurons—which, presumably, implement human cognition—are not representable as strings of digits, hence not computational. But this does not imply that the *functions*[11] whose outputs they produce are not comput*able,* possibly by different mechanisms operating on differ-ent primitive elements in a different (perhaps non-biological) medium. And Makuuchi et al. (2009, p. 8362) say:

If the processing of PSG [phrase structure grammar] is fundamental to human language, the [sic] ques-tions about how the brain implements this faculty arise. The left pars opercularis (LPO), a posterior part of Broca's area, was found as a neural correlate of the processing of A^nB^n sequences in human studies by an artificial grammar learning paradigm comprised of visually presented syllables.... These 2 studies therefore strongly suggest that LPO is a candidate brain area for the processor of PSG (i.e., hierarchical structures).

This is consistent with comput*ability* without computa*tion.* However, Makuuchi et al. (2009, p. 8365) later say:

The present study clearly demonstrates that the syntactic computations involved in the processing of syntactically complex sentences is neuroanatomically separate from the non-syntactic VWM [verbal working memory], thus favoring the view that syntactic processes are independent of general VWM.

That is, brain locations where real computa*tion* is needed in language processing are anatomically distinct from brain locations where computation is *not* needed. This suggests that the brain *could* be computational, contra Piccinini. Similarly, David J. Lobina (2010), Lobina & García-Albea (2009) has argued that, although certain cognitive capabilities are recursive (another term that is sometimes used to mean "computable"), they might not be implemented in the brain in a recursive fashion. After all, algorithms that are most efficiently expressed recursively are sometimes compiled into more-efficiently executable, iterative (non-recursive) code.[12]

Often when we investigate some phenomenon (e.g., cognition, life, computation, flight), we begin by studying it as it occurs in nature, and then abstract away (or "ascend") from what might be called 'implementation details' (Rapaport 1999, 2005b) to arrive at a more abstract or general version of the phenomenon, from which we can "descend" to (re-)implement it in a different medium. When this occurs, the term that referred to the original (concrete) phenomenon changes its referent to the abstract phenomenon and then becomes applicable—perhaps metaphorically so—to the new (concrete) phenomenon. So, for instance, flight as it occurs in birds has been reimplemented in airplanes; 'flying' now refers to the more abstract concept that is multiply realized in birds and planes (cf. Ford & Hayes, 1998; Rapaport, 2000; Forbus, 2010, p. 2). And computation as done by humans in the late 19th through early 20th centuries[13] was—after Turing's analysis—reimplemented in machines; 'computation' now refers to the more abstract concept. Indeed, Turing's (1936) development of (what would now be called) a *computational* theory of *human* computation seems to me to be pretty clearly the first AI program! (See the Appendix, and below).

The same, I suggest, may (eventually) hold true for 'cognition' (Rapaport, 2000). (And, perhaps, for artificial "life.") As Turing (1950, p. 442, my emphasis) said,

The original question, 'Can machines think?' I believe to be too meaningless to deserve discussion. Nevertheless I believe that at the end of the century the use of words and general educated opinion will have altered so much that one will be able to speak of machines thinking without expecting to be contradicted.

"General educated opinion" changes when we abstract and generalize, and "the use of words" changes when we shift reference from a word's initial application to the more abstract or general phenomenon. Similarly, Derek Jones (2010) proposed the following "metaphysical thesis:" "Underlying biological mechanisms are irrelevant to the study of behavior/systems as such. The proper object of study is the abstract system considered as a multiply realized high-level object."

This issue is related to a dichotomy in cognitive science over its proper object of study: Do (or should) cognitive scientists study *human* cognition in particular, or (abstract) cognition in general? Computational psychologists lean towards the former; computational philosophers (and AI researchers) lean towards the latter (see note 9; cf. Levesque, 2012, who identifies the former with cognitive science and the latter with AI). We see this, for example, in the shift within computational linguistics from developing algorithms for understanding natural language using "human language concepts" to developing them using statistical methods: Progress was made when it was realized that "you don't have to do it like humans" (Lohr, 2010, quoting Alfred Spector on the research methodology of Frederick Jelinek; for further discussion of this point, see Forbus, 2010).

SYNTACTIC SEMANTICS

Three principles underlie computationalism properly treated. I call them "internalism," "syntacticism," and "recursive understanding." Together, these constitute a theory of "syntactic semantics."[14] In the present essay, because of space limitations, I will primarily summarize this theory and refer the reader to earlier publications for detailed argumentation and defense (Rapaport 1986, 1988, 1995, 1998, 1999, 2000, 2002, 2003a, 2005b, 2006, 2011).

Internalism

Internalism is the principle whereby cognitive agents have direct access only to internal representatives of external objects. This thesis is related to theses called "methodological solipsism" (Fodor, 1980, arguing against Putnam, 1975)[15] and "individualism" (see, e.g., Segal, 1989, arguing against Burge, 1986).[16] It is essentially Kant's point embodied in his distinction between noumenal things-in-themselves and their phenomenal appearances as filtered through our mental concepts and categories: We only have (direct) access to "phenomena," not noumena. Or, as expressed by a contemporary computational cognitive scientist, "My phenomenal world…[is] a neural fiction perpetrated by the senses" (Edelman, 2008b, p. 426). It is also related to many issues discussed by the Logical Positivists (see Coffa, 1991, p. 8, 67, 140, 176, 310ff, 364ff, & Ch. 9). Internalism is the inevitable consequence of the fact that "the output [of sensory transducers] is…the only contact the cognitive system ever has with the environment" (Pylyshyn 1985, p. 158).[17] As Ray Jackendoff (2002) put it, a cognitive agent understands the world by "pushing the world into the mind." Or, as David Hume (1777, part 1, p. 152) put it, "the existences which we consider, when we say, *this house* and *that tree*, are nothing but perceptions in the mind and fleeting copies or representations of other existences" (cf. Rapaport, 2000).

This can be seen clearly in two cases: (1) What I *see* is the result of a long process that begins outside my head with photons reflected off the physical object that I am looking at, and that ends with a qualitative mental image (what is sometimes called a "sense datum;" cf. Huemer, 2011) produced by neuron firings in my brain. These are internal representatives of the external object. Moreover, there is a time delay; what I am consciously aware of at time *t* is my mental representative of the external object as it was at some earlier time *t'*:

Computation necessarily takes time and, because visual perception requires complex computations, … there is an appreciable latency—on the order of 100msec—between the time of the retinal stimulus and the time of the elicited perception (Changizi et al. 2008, p. 460).

This is in *addition* to the time that it takes the reflected photons to reach my eye, thus *beginning* the computational process. I agree with Huemer (2011) that this implies that we do "not directly perceive *anything*…outside of" us (my emphasis). (2) Although my two eyes *look at* a single external object, they do so from different perspectives; consequently, they *see* different things. These two perceptions are combined by the brain's visual system into a single, three-dimensional perception, which is constructed and internal (cf. Julesz's 1971 notion of the "cyclopean eye"). Moreover, I can be aware of (i.e., perceive) the two images from my eyes simultaneously; I conclude from this that what I perceive is not what is "out there": There is only *one* thing "out there," but I perceive *two* things (cf. Hume 1739, Book I, part 4, §2, pp. 210-211; Ramachandran & Rogers-Ramachandran, 2009). Similarly the existence of saccades implies that "my subjective, phenomenal experience of a static scene" is internal ("irreal," a "simulation of reality") (Edelman, 2008b, pp. 410).

There is a related point from natural-language semantics: Not all words of a language have an external referent, notably words (like 'unicorn') for non-existents. In Rapaport 1981, I argued that it is best to treat *all* words uniformly as having *only* internal "referents." Just as my meaning for 'unicorn' will be some internal mental concept, so should my meaning for 'horse.' Consequently, both words and their meanings (including any external objects that serve as the referents of certain words) are represented internally in a *single* language of thought (LOT; see Fodor, 1975). "Methodological solipsism"—the (controversial) position that access to the external world is unnecessary (Fodor, 1980; cf. Rapaport, 2000)[18]—underlies representationalism:

If a system—creature, network router, robot, mind—cannot "reach out and touch" some situation in which it is interested, another strategy, deucedly clever, is available: it can instead exploit meaningful or representational structures in place of the situation itself, so as to allow it to behave appropriately with respect to that distal, currently inaccessible state of affairs (B.C. Smith, 2010).

For computers, the single, internal LOT might be an artificial neural network or some kind of knowledge-representation, reasoning, and acting system (such as SNePS; see Shapiro & Rapaport, 1987). For humans, the single, internal LOT is a biological neural network.[19]

It is this last fact that allows us to respond to most of the objections to internalism (see, e.g., Huemer 2011 for a useful compendium of them). For example, consider the objection that an internal mental representative (call it a "sense datum," "quale," whatever) of, say, one of Wilfrid Sellars's pink ice cubes is neither pink (because it does not reflect any light) nor cubic (because it is not a three-dimensional physical object). Suppose that we really are looking at an external, pink ice cube. Light reflects off the surface of the ice cube, enters my eye through the lens, and is initially processed by the rods and cones in my retina, which transduce the information[20] contained in the photons into electrical and chemical signals that travel along a sequence of nerves, primarily the optic nerve, to my visual cortex. Eventually, I see the ice cube (or: I have a mental image of it). Exactly how that experience of seeing (that mental image) is produced is, of course a version of the "hard" problem of consciousness (Chalmers, 1996). But we do know that certain neuron firings that are the end result of the ice cube's reflection of photons into my eyes are (or are correlated with) my visual experience of pink; others are (correlated with) my visual experience of cubicness. But now *imagine* a pink ice cube; presumably, the same or similar neurons are firing and are (correlated with) my mental image of pinkness and cubicness. In both cases, it is those neuron firings (or whatever it is that might be correlated with them) that constitute my internal representative. In *neither* case is there anything internal that *is* pink or cubic; in *both* cases, there *is* something that *represents* pinkness or cubicness (Shapiro, 1993, Rapaport, 2005b, Shapiro & Bona, 2010).

As I noted above, perception, like knowledge, might not be a strictly (i.e., internal) cognitive phenomenon, depending as it does on the external world.[21] When I see the ice cube, certain neuron firings are directly responsible for my visual experience, and I might think, "That's a pink ice cube." That thought is, presumably, also due to (or identical with) some (other) neuron firings. Finally, presumably, those two sets of neuron firings are somehow correlated or associated, either by the visual ones causing the conceptual ones or both of them being caused by the visual stimulation; in any case, they are somehow "bound" together.

My experience of the pink ice cube and my thought (or thoughts) that it is pink and cubic (or that there is a pink ice cube in front of me) occur purely in my brain. They are, if you will, purely solipsistic. (They are not merely *methodologically* solipsistic. Methodological solipsism is a research strategy: A third-person observer's theory of my cognitive processes that ignored the real ice cube and paid attention only to my neuron firings would be methodologically solipsistic.) Yet there are causal links between the neurological occurrences (my mental experiences) and an entity in the real world, namely, the ice cube.

What about a cognitively programmed computer or robot? Suppose that it has a vision system and that some sort of camera lens is facing a pink ice cube. Light reflects off the surface of the ice cube, enters the computer's vision system through the lens, and is processed by the vision system (say, in some descendent of the way that Marr 1982 described). Eventually, let's say, the computer constructs a representation of the pink ice cube in some knowledge-representation language (it may be a pictorial language). When the computer sees (or "sees") the ice cube, it might think (or "think"), "That's a pink ice cube." That thought might also be represented in the same knowledge-representation language (e.g., as is done in the knowledge-representation, reasoning, and acting system SNePS). Finally, those two representations are associated (Srihari & Rapaport, 1989, 1990).

The computer's "experience" of the pink ice cube and its thought (or "thoughts") that it is pink and cubic (or that there is a pink ice cube in front of it) occur purely in its knowledge base. They are purely solipsistic. Yet there are causal links between the computational representations and the ice cube in the real world. *There is no significant difference between the computer and the human.* Both can "ground" their "thoughts" of the pink ice cube in reality yet deal with their representations of both the phrase 'ice cube' and the ice cube in the same, purely syntactic, language of thought. Each can have a syntactic, yet semantic, relation between its internal representation of the linguistic expression and its internal representation of the object that it "means," and each can have *external* semantic relations between those internal representations and the real ice cube. However, neither can have direct perceptual access to the real ice cube to see if it matches their representation:

Kant was rightly impressed by the thought that if we ask whether we have a correct conception of the world, we cannot step entirely outside our actual conceptions and theories so as to compare them with a world that is not conceptualized at all, a bare "whatever there is" (Williams, 1998, p. 40).

Of course, both the computer (if equipped with effectors) and the human can grasp the real ice cube.[22] It might be objected that internalism underestimates the role of situatedness and embodiment of cognitive agents. Quite the contrary. First, any situated or embodied cognitive agent must internalize the information it receives from the environment that it is situated in, and it must process and respond to that information in the form in which it is received. Such information may be incomplete or noisy, hence not fully representative of the actual environment, but those are the cards that that agent has been dealt and that it must play with (cf. Shapiro & Rapaport, 1991). Second, the software- and hardware-embodied cognitive agents developed by colleagues in my research group operate (i.e., are situated) in real and virtual environments, yet are constructed on the "internal" principles adumbrated here, so they constitute a demonstration that situatedness and embodiment are not inconsistent with internal processing of symbols (Shapiro, 1998, 2006; Shapiro et al., 2001; Santore & Shapiro, 2003; Shapiro & Ismail, 2003; Goldfain, 2006; Anstey et al., 2009; Shapiro & Bona, 2010; and cf. Vera & Simon, 1993).

Finally, there is the issue of whether a cognitive computer (or a cognitively programmed computational agent) must have a sensory-motor interface to the real, or a virtual, environment in which it is situated. I am willing to limit my arguments to computers that do have such an interface to the real, external world. If one is willing to allow such an interface to a *virtual* world, however, and if that virtual world is completely independent from the computational agent, then the two situations (real vs. virtual) are parallel. If on the other hand, the virtual environment is completely internal to the computational agent (i.e., the agent believes falsely that it is really situated in that environment), then we have a situation about which there might be disagreement. However, I would maintain that such a (delusional!) agent is in a situation no different from the first kind of agent, because, in both cases, the agent must internalize its environment.

Syntacticism

It follows that words, their meanings, and semantic relations between them are all syntactic. Both 'syntax' and 'semantics' can mean different things. On one standard interpretation, 'syntax' is synonymous with 'grammar,' and 'semantics' is synonymous with 'meaning' or 'reference' (to the external world). But more general and inclusive conceptions can be found in Charles Morris (1938, pp. 6–7):

One may study the relations of signs to the objects to which the signs are applicable. ...the study of this [relation]...will be called semantics. [The study of]...the formal relation of signs to one another...will be named syntactics.

On the nature of syntax, consider this early definition from the *Oxford English Dictionary*: "Orderly or systematic arrangement of parts or elements; constitution (of a body); a connected order or system of things."[23] In a study of the history of the concept, Roland Posner (1992, p. 37) says that syntax "is that branch of Semiotics that studies the formal aspects of signs, the relations between signs, and the combinations of signs." On both of those senses, 'syntax' goes far beyond grammar. Throughout this essay, I use 'syntax' in that broader sense (except when quoting); when I intend the narrower meaning, I will use 'grammar' or 'grammatical syntax.'

On the nature of semantics, we might compare Alfred Tarski's (1944, p. 345) characterization: "*Semantics* is a discipline which, speaking loosely, *deals with certain relations between expressions of a language and the objects... 'referred to' by those expressions*" (original italics and original scare quotes around 'referred to,' suggesting that the relation need not be one of reference to *external* objects). But, surely, translation from one language to another is also an example of semantic interpretation, though of a slightly different kind: Rather than semantics considered as relations between linguistic expressions and objects in the world, in translation it is considered as relations between linguistic expressions in one language and linguistic expressions in another language. (We say that the French word 'chat' *means* "cat" in English.)

In fact, *all* relationships between two domains can be seen as interpretations of one of the domains in terms of the other—as a mapping from one domain to another. A mapping process is an algorithm that converts, or translates, one domain to another (possibly the same one). The input domain is the syntactic domain; the output domain is the semantic domain. (I have argued elsewhere that implementation, or realization, of an "abstraction" in some "concrete" medium is also such a mapping, hence a semantic interpretation *Rapaport, 1999, 2005b; cf. Dresner, 2010).

Generalizing only slightly, both syntax and semantics are concerned with relationships: syntax with relations *among* members of a *single* set (e.g., a set of signs, or marks,[24] or neurons, etc.), and semantics with relations *between two* sets (e.g., a set of signs, marks, neurons, etc., on the one hand, and a set of their meanings, on the other). More generally, semantics can be viewed as the study of relations between any two sets whatsoever, including, of course, two sets of signs (as in the case of translation) or even a set and itself; in both of those cases, semantics becomes syntax. Note that a special case of this is found in an ordinary, monolingual dictionary, where we have relations between linguistic expressions in, say, English and (other) linguistic expressions *also* in English. That is, we have relations *among* linguistic expressions in English. But this is syntax!

"Pushing" meanings into the same set as symbols for them allows semantics to be done syntactically: It turns *semantic* relations *between* two sets (a set of internal marks as discussed below and a set of, possibly external, meanings) into *syntactic* relations *among* the marks of a *single* (internal) LOT (Rapaport, 2000; 2003a, 2006, "Thesis 1" 2011). For example, both truth-table semantics and formal semantics are syntactic enterprises: Truth tables relate one set of marks (strings) representing propositions to another set of marks (e.g., letters 'T' and 'F') representing truth-values. Formal semantics relates one set of marks (strings) representing propositions to another set of marks representing (e.g.) set-theoretical objects (cf. B.C. Smith, 1982). The relations between sets of neuron firings representing signs and sets of neuron firings representing external meanings are also syntactic. *Consequently, symbol-manipulating computers can do semantics by doing syntax.* As Shimon Edelman (2008a, pp. 188-189) put it, "the meaning of an internal state (which may or may not be linked to an external state of affairs) for the system itself is most naturally defined in terms of that state's relations to its other states," i.e., syntactically.

This is the notion of semantics that underlies the Semantic Web, where "meaning" is given to (syntactic) information on the World Wide Web by associating such information (the "data" explicitly appearing in Web pages, usually expressed in a natural language) with more information ("metadata" that only appears in the HTML source code for the webpage, expressed in a knowledge-representation language such as RDF). But it is not "more of the same" kind of information; rather, the additional information takes the form of annotations of the first kind of information. Thus, relations are set up between information on, say, Web pages and annotations of that information that serve as semantic interpretations of the former. As John Hebeler has put it, "[T]he computer doesn't know anything more than it's a bunch of bits [i.e. (to rephrase his informal, spoken English), the computer only knows (i.e., doesn't know anything more than) that the Web is a bunch of bits].[25] So semantics merely adds extra information to help you with the *meaning* of the information" (quoted in Ray, 2010).

This is how data are interpreted by computer programs. Consider the following description from a standard textbook on data structures (my comments are interpolated as endnotes, not unlike the "semantic" metadata of the Semantic Web!):

[T]he concept of information in computer science is similar to the concepts of point, line, and plane in geometry—they are all undefined terms about which statements can be made but which cannot be explained in terms of more elementary concepts.[26] ...The basic unit of information is the bit, whose value asserts one of two mutually exclusive possibilities.[27] ...[I]nformation itself has no meaning. Any meaning can be assigned to a particular bit pattern as long as it is done consistently. It is the interpretation of a bit pattern that gives it meaning.[28] ...A method of interpreting a bit pattern is often called a data type. ...It is by means of declarations[29] [in a high-level language] that the programmer specifies how the contents of the computer memory are to be interpreted by the program. ...[W]e...view...data types as a method of interpreting the memory contents of a computer (Tenenbaum & Augenstein, 1981: p. 1, 6, 8).[30]

Recursive Understanding

Understanding is recursive: We understand one kind of thing in terms of another that is already understood; the base case is to understand something in terms of itself, which is syntactic understanding. There are two ways to understand something: One can understand something in terms of something else, or one can understand something directly (Rapaport, 1995). The first way, understanding one kind of thing in terms of another kind of thing, underlies metaphor, analogy, maps and grids (B. Smith, 2001), and simulation (see §8, below). It is also what underlies the relation between syntax and semantics. In the stereotypical case of semantics, we interpret, or give meanings to, *linguistic expressions*. Thus, we understand language in terms of the world. We can also interpret, or give meanings to, other kinds of (non-linguistic) *things*. For example, we can try to understand the nature of certain neuron firings in our brains: Is some particular pattern of neuron firings correlated with, say, thinking of a unicorn or thinking that apples are red? If so, then we have interpreted, hence understood, those neuron firings. And not only can we understand language in terms of the world, or understand parts of the world in terms of other parts of the world, but we can also understand *the world* in terms of *language about* the world: This is what we do when we learn something from reading about it.

Understanding of this first kind is recursive: We understand one thing by understanding another. It is "recursive," because we are understanding one thing in terms of another *that must already be understood*: We understand in terms of something understood. But all recursion requires a base case in order to avoid an infinite regress. Consequently, the second way of understanding—understanding something directly—is to understand a domain in terms of itself, to get used to it. This is the fundamental kind—the base case—of understanding.

In general, we understand one domain—call it a *syntactic* domain ('SYN_1')—*indirectly* by *interpreting it* in terms of a (different) domain: a *semantic* domain ('SEM_1'). This kind of understanding is "indirect," because we understand SYN_1 by looking *elsewhere*, namely, at SEM_1. But for this process of interpretation to result in real understanding, SEM_1 must be *antecedently understood*. How? In the same way: by considering it as a *syntactic* domain (rename it 'SYN_2') interpreted in terms of *yet another semantic* domain, which *also* must be antecedently understood. And so on. But, in order not to make this sequence of interpretive processes go on *ad infinitum*, there must be a base case: a domain that is understood directly, i.e., in terms of itself (i.e., not "antecedently"). Such direct understanding is syntactic understanding; i.e., it is understanding in terms of the relations among the marks of the system itself (Rapaport, 1986). Syntactic understanding may be related to what Piccinini (2008, p. 214) called "internal semantics"—the interpretation of an instruction in terms of "what its execution accomplishes *within* the computer." And it may be related to the kind of understanding described in Eco 1988, in which words and sentences are understood in terms of inferential (and other) relations that they have with "contextual" "encyclopedias" of other words and sentences—i.e., syntactically and holistically (cf. Rapaport, 2002).[31] (For more on the relation of syntax to semantics, see Rapaport 2017a.)

SYNTACTIC SEMANTICS VS. FETZER'S THESIS

Syntactic semantics implies that syntax suffices for semantic cognition, that (therefore) cognition is computable, and that (therefore) computers are capable of thinking. In a series of papers, James H. Fetzer has claimed that syntax does *not* suffice for semantic cognition, that cognition is *not* computable, and that computers are *not* capable of thinking. More precisely, Fetzer's thesis is that computers differ from cognitive agents in three ways—statically (or symbolically), dynamically (or algorithmically), and affectively (or emotionally)—and that simulation is not "the real thing." In the rest of this essay, I will try to show why I think that Fetzer is mistaken on all these points.

FETZER'S "STATIC" DIFFERENCE

In the forward to his 2001 collection, *Computers and Cognition*, as well as in his presentation at the 2010 North American Conference on Computing and Philosophy, Fetzer argued that "Computers are mark-manipulating systems, minds are not" on the following grounds of "static difference" (Fetzer 2001, p. xiii, my boldface):

Premise 1: Computers manipulate marks on the basis of their shapes, sizes, and relative locations.

Premise 2: [a] These shapes, sizes, and relative locations exert causal influence upon computers [b] but do not stand for anything for those systems.

Premise 3: Minds operate by utilizing signs that stand for other things in some respect or other for them as sign-using (or "semiotic") systems.

Conclusion 1: Computers are not semiotic (or sign-using) systems.

Conclusion 2: Computers are not the possessors of minds.

I disagree with all of the boldfaced phrases in Fetzer's static-difference argument. Before saying why, note that here—and in his arguments to follow—Fetzer consistently uses declaratives that appear to describe current-day computers: They *do* not do certain things, *are* not affected in certain ways, or *do* not have certain properties. But he really should be using modals that specify what he believes computers *cannot* do, be affected by, or have. Consider premise (2b), that the marks that computers manipulate "do not" stand for anything for those computers. Note that Fetzer's locution allows for the possibility that, although the marks *do* not stand for anything for the computer, they *could* do so. Insofar as they could, such machines *might* be capable of thinking. So Fetzer should have made the stronger claim that they "*could not* stand for anything." But then he'd be wrong, as I shall argue.[32]

Mark Manipulation

What is a "mark"? Fetzer 2011 does not define the term; he has used it many times before (e.g., Fetzer, 1994, 1998), also without definition, but occasionally with some clarification. It seems to be a term designed to be neutral with respect to such semiotic terms as 'sign,' 'symbol,' etc., perhaps even 'term' itself. It seems to have the essential characteristics of being the kind of entity that syntax is concerned with (i.e., marks are the things such that syntax is the study of *their* properties and relations) and of not having an intrinsic meaning. Many logicians use the terms 'sign' or 'symbol' in this way, despite the fact that many semioticians use 'sign' and 'symbol' in such a way that all signs and symbols are meaningful. So 'mark' is intended to be used much as 'sign' or 'symbol' might be used if the sign or symbol were stripped of any attendant meaning. Elsewhere (Rapaport, 1995), I have used 'mark' to mean "(perhaps) physical inscriptions or sounds, that have only some very minimal features such as having distinguished, relatively unchanging shapes capable of being recognized when encountered again;" they have "no intrinsic meaning. But such [marks] *get* meaning the more they are used—the more roles they play in providing meaning to *other*" marks. Fetzer's Static Premise 1 is true (i.e., I agree with it!): "computers manipulate marks on the basis of their shapes, sizes, and relative locations." But it is not the whole truth: They also manipulate marks on the basis of other, non-spatial relations of those marks to other marks, i.e., on the basis of the marks' syntax in the wide sense in which I am using that term.[33] Fetzer can safely add this to his theory. I also agree that this manipulation (or processing) is *not* (necessarily) independent of the meanings of these marks. But my agreement follows, not from Fetzer's notion of meaning, but from the principle of "syntacticism" (above): If *some* of the marks represent (or are) meanings of some of the *other* marks, then mark manipulation on the basis of size, shape, location, *and relations to other marks* includes manipulation on the basis of meaning!

However, this *is* independent of *external* reference. More precisely, it is independent of the actual, external referents of the marks—the objects that the marks stand for, if any (see §3.1)—in this way: Any relationship between a mark and an external meaning of that mark is represented by an internal (hence syntactic) relation between that mark and another mark that is the internal representative of the external referent. The latter mark is the output of a sensory transducer; in Kantian terms, it is an internal phenomenon that represents an external noumenon. This is the upshot of internalism. In this way, such marks *can* stand for something *for the computer*. Computers are, indeed, "string-manipulating" systems (Fetzer 1998: 374). But they are more than "mere" string-manipulating systems, for meaning can arise from (appropriate) combinations of (appropriate) strings.

COMPUTERS AND SEMIOTIC SYSTEMS

Fetzer's static-difference argument claims that computers are *not* semiotic systems. In an earlier argument to the same conclusion, Fetzer (1998), following Peirce, says that a semiotic system consists of something (S) being a sign of something (x) for somebody (z), where:

- Thing x "grounds" sign S
- Thing x stands in a relation of "interpretant (with respect to a context)" to sign-user z[34]
- And sign S stands in a "causation" relation with sign-user z.

This constitutes a "semiotic triangle" whose vertices are S, x, and z (Fetzer, 1998, p. 384, Fig. 1). This cries out for clarification:

1. What is the causation relation between sign-user z and sign S? Does one cause the other? Or is it merely that they are causally—i.e., physically—related?
2. What is the grounding relation between sign S and the thing x that S stands for (i.e., that S is a sign of)? If sign S is grounded by what it stands for (i.e., x), then is the relation of *being grounded by* the same as the relation of *standing for*?
3. And, if sign S stands for thing x *for* sign-user z, then perhaps this semiotic triangle should really be a semiotic "quadrilateral," with *four* vertices: sign S, user z, thing x, *and an interpretant I*, where the four sides of the quadrilateral are:
   a. User z "causes" sign S,
   b. Thing x "grounds" sign S,
   c. Interpretant I is "for" user z,
   d. And I stands for thing x.

There is also a "diagonal" in this quadrilateral: *I* "facilitates" or "mediates" *S*. (A better way to think of it, however, is that the two sides and the diagonal that all intersect at *I* represent the 4-place relation "interpretant *I* mediates sign *S* standing for object *x* for user *z*"; see Rapaport 1998, p. 410, Fig. 2). By contrast, according to Fetzer, a similar semiotic "triangle" for "input-output" systems, such as computers, *lacks* a relationship between what plays the role of sign *S* and what plays the role of thing *x*. (It is only a 2-sided "triangle"; see Fetzer, 1998, p. 384, Fig. 2. More precisely, for such input-output systems, Fetzer says that we have an input *i* (instead of a sign *S*), a computer *C* (instead of a sign-user *z*), and an output *o* (instead of a thing *x*, where there is still a causation relation between computer *C* and input *i*, and an interpretant relation between *C* and output *o*, but—significantly—no grounding relation between input *i* and output *o*.

Again, we may raise some concerns:

1. Computers are much more than mere input-output systems, because there is always an algorithm that mediates between the computer's input and its output.
2. The marks by means of which computers operate include more than merely the external input; there are usually *stored* marks representing what might be called "background knowledge" (or "prior knowledge")—perhaps "(internal) context" would not be an inappropriate characterization: Where are these in this two-sided triangle?
3. And what about the "stands for" relation? Surely, what the input marks stand for (and surely they stand for something) is not necessarily the *output*.
4. Finally, what does it mean for a sign *S* or an input mark *i* to stand for something *x* or *o*) yet not be "grounded" by *x*?

Fetzer's chief complaint about computers is not merely that they causally manipulate marks (premise 1) but that such causal manipulation is *all* that they can do. Hence, because such merely causal manipulation requires no mediation between input and output, computers are not semiotic systems. By contrast, semiosis does require such mediation; according to Peirce, it is a ternary relation:

I define a Sign as anything which is so determined by something else, called its Object, and so determines an effect upon a person, which effect I call its Interpretant, that the latter is thereby mediately determined by the former. My insertion of "upon a person" is a sop to Cerberus, because I despair of making my own broader conception understood (Peirce, 1908, pp. 80-81 [http://www.helsinki/fi/science/commens/ terms/sign.html], accessed 5 May 2011).

The fact that "upon a person" is a "sop to Cerberus" suggests that the effect need *not* be "upon a person;" it *could*, thus, be "upon a computer." That is, the mark-user need not be human (cf. Eco 1979, p. 15: "It is possible to interpret Peirce's definition in a non-anthropomorphic way…").[35]

Peirce and Computation

Given Fetzer's reliance on Peirce's version of semiotics (which I focus on rather than, say, on Saussure's version of semiotics, because Peirce is whom Fetzer focuses on), it is worth noting that Peirce had some sympathy for—and certainly an active interest in—computation, especially its syntactic aspect:

The secret of all reasoning machines is after all very simple. It is that whatever relation among the objects reasoned about is destined to be the hinge of a ratiocination, that same general relation must be capable of being introduced between certain parts of the machine. ...When we perform a reasoning in our unaided minds we do substantially the same thing, that is to say, we construct an image in our fancy under certain general conditions, and observe the result (Peirce, 1887, p. 168).[36]

There is even an anticipation of at least one of Turing's insights (cf. Appendix):[37]

[T]he capacity of a machine has absolute limitations; it has been contrived to do a certain thing, and it can do nothing else. For instance, the logical machines that have thus far been devised can deal with but a limited number of different letters. The unaided mind is also limited in this as in other respects; but the mind working with a pencil and plenty of paper has no such limitation. It presses on and on, and whatever limits can be assigned to its capacity to-day, may be over-stepped to-morrow (Peirce, 1887, p. 169, my italics).

Furthermore, Peirce had views on the relation of syntax to semantics that are, arguably, sympathetic to mind. As Brown (2002, p. 20) observes,[38]

Thus Peirce, in contrast to Searle, would not...allow any separation between syntax and semantics, in the following respect. He would claim that what Searle is terming "syntactic rules" partake of what Searle would consider semantic characteristics, and, generally, that such rules must so partake. However, if those rules were simple enough so that pure deduction, i.e., thinking of the first type of thirdness, was all that was required, then a machine could indeed duplicate such "routine operations" (Peirce, 1992d, p. 43). In this simple sense, for Peirce, a digital computer has "mind" or "understanding."

Thus, I find Fetzer's analysis vague and unconvincing at best. We need to bring some clarity to this. First, consider what Peirce actually says:

A sign, or representamen, is something which stands to somebody for something in some respect or capacity. It addresses somebody, that is, creates in the mind of that person an equivalent sign, or perhaps a more developed sign. That sign which it creates I call the interpretant of the first sign. The sign stands for something, its object. It stands for that object, not in all respects, but in reference to a sort of idea, which I have sometimes called the ground of the representamen. (Peirce (c. 1897), A Fragment, CP 2.228, accessed 5 May 2011 from[http://www.helsinki.fi/science/commens/terms/representamen.html].) [B]y "semiosis" I mean...an action, or influence, which is, or involves, a cooperation of three subjects, such as a sign, its object, and its interpretant, this tri-relative influence not being in any way resolvable into actions between pairs (Peirce 1907, EP 2.411; CP 5.484, accessed 5 May 2011 from [http:/www. helsinki.fi/science/commens/terms/semiosis.html]).

By 'representamen,' Peirce means "sign," but I think that we may also say 'mark.' So, the Peircean analysis consists of:

1.   A mark (or sign, or representamen):

a. A "mark" is, roughly, an uninterpreted sign—if that's not an oxymoron: If a sign consists of a mark (signifier) plus an interpretation (signified), then it must be marks that are interpreted; after all, signs (presumably) wear their interpretations on their sleeves, so to speak. Although many semioticians think that it is anathema to try to separate a sign into its two components, that is precisely what "formality" is about in such disciplines as "formal logic," "formal semantics," and so on: Formality is the separation of mark from meaning; it is the focus on the *form* or shape of marks (cf. B.C. Smith. 2010);

2. A mark user (or interpreter):

a. A sign has an Object and an Interpretant, the latter being that which the Sign produces in the Quasi-mind that is the Interpreter by determining the latter to a feeling, to an exertion, or to a Sign, which determination is the Interpretant (Peirce 1906, Prolegomena to an Apology for Pragmaticism, CP 4.536, accessed 5 May 2011 from [http://www.helsinki.fi/science/commens/terms/interpretant.html]);

3. An object that the mark stands for (or is a sign of);

4. An interpretant in the mark-user's mind, which is also a mark.

The interpretant is the mark-user's idea (or concept, or mental representative)[39]—but an idea or representative of what? Of the mark? Or of the object that the mark stands for? Peirce says that the interpretant's *immediate* cause is the mark and its *mediate* cause is the object.[40] Moreover, the interpretant is also a representamen, namely, a sign of the original sign.[41]

But there is also another item: Presumably, there is a causal process that produces the interpretant in the mark-user's mind. This might be an automatic or unconscious process, as when a mental image is produced as the end product of the process of visual perception. Or it might be a conscious process, as when the mark-user reads a word (a mark) and consciously figures out what the word might mean, resulting in an interpretant (as in "deliberate" contextual vocabulary acquisition; cf. Rapaport & Kibby, 2007, 2010).

One standard term for this process (i.e., for this relation between mark and interpretant) is 'interpretation.' The word 'interpretation,' however, is ambiguous. In one sense, it is a functional[42] *relation* from a mark to a thing that the mark "stands for" or "means." In another sense, it is the end product or *output* of such a functional relation. For present purposes, since we already have Peirce's term 'interpretant' for the *output*, let us restrict 'interpretation' to refer to the *process*. (What is the relationship between my *binary* relation of interpretation and Peirce's *ternary* relation of semiosis? (See the quote from Peirce 1907, above.)[43] "Interpretation" in my sense is only *part* of "semiosis" in Peirce's sense; it is a relation between only two of the three parameters of Peircean semiosis, namely, the sign and its interpretant).

With this by way of background, I will now present three arguments that computers *are* semiotic systems.[44] Recall that Peirce's view of semiotics does not require the mark-user to be human!

Computers Are Semiotic Systems I: Incardona's Argument

The first argument, due to Lorenzo Incardona (personal communication), consists of three premises and a conclusion.

1. Something is a semiotic system if and only if it carries out a process that mediates between a mark and an interpretant of that mark: The essential characteristic of a semiotic system is its ternary na-

ture; it applies (i) a mediating or interpretive process to (ii) a mark, resulting in (iii) an interpretant of that mark (and indirectly of the mark's object). A semiotic system interprets marks.[45] However, there must be a fourth component: After all, *whose* interpretant is it that belongs to a given mark? This is just another way of asking about "the" meaning of a word. What is "the" interpretant of 'gold'? Is it what *I* think gold is? What *experts* think it is (Putnam, 1975)? It is better to speak of "a" meaning "for" a word than "the" meaning "of" a word (Rapaport 2005a). Surely, there are many meanings for marks but, in semiosis, the only one that matters is that of (iv) the interpreter, i.e., the executor of the interpretive process (i, above) that mediates the mark (ii, above) and the (executor's) interpretant (iii, above).

2.   Algorithms describe processes that mediate between inputs and outputs. An algorithm is a *static text* (or an abstract, mathematical entity) that describes a *dynamic process*. That process can be thought of as the algorithm being executed. The output of an algorithm stands (by definition) in a functional relation to its input. It describes (or can describe) a meaningful relation between the input mark and the output mark. So, the output of an algorithm is an interpretant of its input. This is a causal relationship, but it is not *merely* a causal (or stimulus-response, "behavioral") correlation between input and output: In the case of computable functions (i.e., of a computable interpretation process), there is a mediating process, namely, an algorithm. Consequently, the input-output relation is grounded—or mediated—by that algorithm, i.e., the mechanism that converts the input into the output, i.e., the interpretation.[46]

3.   Clearly, computers are algorithm machines.[47] That's what computers do: They execute algorithms, converting inputs into outputs, i.e., interpreting the inputs as the outputs according to an algorithm. But a computer is (usually) *not a mere* input-output system. The (typical) algorithm for converting the input to the output (i.e., for *interpreting*) the input as the output) is not a mere table-lookup. First, there is internally stored data that can be used or consulted in the conversion process; this can be thought of as a "context of interpretation." Second, there are numerous operations that must be performed on the input together with the stored data in order to produce the output, so it is a dynamic process. Moreover, the stored data can be modified by the input or the operations so that the system can "learn" and thereby change its interpretation of a given input.

4.   Therefore, computers *are* semiotic systems.

Computers Are Semiotic Systems II: Goldfain's Argument From Mathematical Cognition

Does a calculator that computes greatest common divisors (GCDs) understand what it is doing? I think that almost everyone, including both Fetzer and I, would agree that it does not. But *could* a *computer* that computes GCDs understand what it is doing? I am certain that Fetzer would say 'no;' but I—along with Albert Goldfain (2008; Shapiro, Rapaport et al., 2007)—say 'yes': Yes, it could, as long as it had enough background, or contextual, or supporting information: A computer with a full-blown theory of mathematics at, say, the level of an algebra student learning GCDs, together with the ability to explain its reasoning when answering a question, could understand GCDs as well as the student. (Perhaps it could understand better than the student, if the student lacks the ability to fully explain his or her reasoning.)

Calculating becomes understanding if embedded in a larger framework linking the calculation to other concepts (not unlike the semantic contributions of Semantic Web annotations, which also serve as an embedding framework). So a computer *could* know, do, and understand mathematics, *if* suitably programmed with all the background information required for knowing, doing, and understanding mathematics. A computer or a human who can calculate GCDs by executing an algorithm does not (necessarily) thereby understand GCDs. But a computer or a human who has been taught (or programmed) all the relevant mathematical definitions can understand GCDs. (For further discussion, see Rapaport 1988, 1990, 2006, 2011.)

Goldfain (personal communication) has offered the following argument that marks can stand for something for a computer:

1. The natural numbers that a cognitive agent refers to are denoted by a sequence of marks that are unique to the agent. These marks exemplify (or implement; Rapaport, 1999, 2005b) a finite, initial segment of the natural-number structure (i.e., 0, 1, 2, 3, …, up to some number n).

2. Such a finite, initial segment can be generated by a computational cognitive agent (e.g., a computer, suitably programmed) via perception and action in the world during an act of counting (e.g., using the Lisp programming language's "gensym" function; for details see Goldfain, 2008). Thus, there would be a history of how these marks came to signify something for the agent (e.g., the computer).

3. These marks (e.g., b4532, b182, b9000, …) have no meaning for another cognitive agent (e.g., a human user of the computer) who lacks access to their ordering.

4. Such private marks (called 'numerons' by cognitive scientists studying mathematical cognition) are associable with publicly meaningful marks (called 'numerlogs').[48] For example, 'b4532' might denote the same number as the Hindu-Arabic numeral '1,' 'b182' might denote the same number as '2,' etc.

5. A computational cognitive agent (e.g., a computer) can do mathematics solely on the basis of its numerons. (See Goldfain, 2008 for a detailed demonstration of this; cf. Shapiro, Rapaport et al. 2007.)

6. Therefore, *these marks stand for something for the computer* (i.e., for the agent). Moreover, we can check the mathematics, because of premise 4.

7. Thus, such a computer would be a semiotic system.

Computers Are Semiotic Systems III: Argument From Embedding in the World

Besides being able to understand mathematics in this way, computers can also understand other, conventional activities. Fetzer (2008) gives the following example of something that a semiotic system can do but that, he claims, a computer cannot:

A red light at an intersection…stands for applying the brakes and coming to a complete halt, only proceeding when the light turns green, for those who know "the rules of the road" [My emphasis].

As Fetzer conceives it, the crucial difference between a semiotic system and a computer is that the former, but not the latter, can use a mark as something that stands for something (else) *for itself* (cf.

Fetzer, 1998). In that case, we need to ask whether such a red light can stand for "applying the brakes," etc. *for a computer*.

It could, *if* it has those rules *stored* (memorized) in a knowledge base (its mind). But merely storing, and even being able to access and reason about, this information is not enough: IBM's *Jeopardy*-winning Watson computer can do that, but no one would claim that such merely stored information is understood, i.e., stands for anything for Watson itself. At least one thing more is needed for a red light to stand for something for the computer itself: The computer must *use* those rules to drive a vehicle. But there *are* such computers (or computerized vehicles), namely, those that have successfully participated in the DARPA Grand Challenge autonomous vehicle competitions. (For a relevant discussion, see Parisien & Thagard, 2008.) I conclude that (such) computers can be semiotic systems.

Some might argue that such embedding in a world is not computable. There are two reasons to think that it is. First, all of the autonomous vehicles must internally store the external input in order to compute with it. Thus, the syntactic-semantic principle of internalism explains how the embedding is computable. Second, the vehicles' behaviors are the output of computable (indeed, computational) processes.[49]

FETZER'S "DYNAMIC" DIFFERENCE

Fetzer also argues that "computers are governed by algorithms, but minds are not," on the following grounds (Fetzer 2001, p. xv; my boldface and italics):

Premise 1: Computers are governed by programs, which are causal models of algorithms.

Premise 2: Algorithms are effective decision procedures for arriving at definite solutions to problems in a finite number of steps.

Premise 3: Most human thought processes, including dreams, daydreams, and ordinary thinking, are not procedures for arriving at solutions to problems in a finite number of steps.

Conclusion 1: Most human thought processes are not governed by programs as causal models of algorithms.

Conclusion 2: Minds are not computers.

Once again, I disagree with the boldfaced claims. The italicized claims are ones that are subtly misleading; below, I will explain why. I prefer not to speak as Fetzer does. First, let me point to a "red herring": Fetzer (2008) said that "if thinking is computing *and computing is thinking* and if computing is algorithmic, then thinking is algorithmic, but it isn't" (my emphasis). The second conjunct is false; fortunately (for Fetzer), it is also irrelevant: A computer executing a non-cognitive program (e.g., an operating system) *is* computing but is *not* thinking. (Of course, this depends on whether you hold that an operating system is a *non*-cognitive, computer program. Once upon a time, it was *humans* who operated computers. Arguably, operating systems, insofar as they do a task once reserved for humans, *are* doing a cognitive task. Alternatively, what the humans who operated computers were doing was a task requiring no "intelligence" at all.[50] Cf. my remark about Turing machines as AI programs, above; on performing intelligent tasks without intelligence, see Dennett, 2009.)

Algorithms

Premise 2 is consistent with the way I characterized 'algorithm' above, so I am willing to accept it. Yet I think that algorithms, so understood, are the wrong entity for this discussion. Instead, we need to relax some of the constraints and to embrace a more general notion of "procedure" (Shapiro, 2001):

1.   In particular, procedures (as I am using the term here) are like algorithms, but they do not necessarily halt of their own accord; they may continue executing until purposely (or accidentally) halted by an outside circumstance (cf. Knuth's 1973 notion of a (non-finite) "computational method"; see the Appendix). For instance, an automatic teller machine or an online airline reservation system should not halt unless turned off by an administrator (cf. Wegner, 1997).
2.   Also, procedures (as I am using the term) need not yield "correct" output (goal *G* need not be completely or perfectly accomplished). Consider a computer programmed for playing chess. Even IBM's celebrated, chess champion Deep Blue will not win or draw every game, even though chess is a game of perfect (finite) information (like Tic Tac Toe) in which, mathematically, though not practically, it is knowable whether any given player, if playing perfectly, will win, lose, or draw, because the "game tree" (the tree of all possible moves and replies) can *in principle* be completely written down and examined. But because of the practical limitations (the tree would take too much space and time to actually write down; cf. Zobrist, 2000, p. 367), the computer chess program will occasionally give the "wrong" output. But it is not behaving any differently from the algorithmic way it would behave if it gave the correct answer. Sometimes, such programs are said to be based on "heuristics" instead of "algorithms," where a *heuristic for problem P* is an *algorithm* for some other problem *P'*, where the solution to *P'* is "near enough" to a solution to *P* (Rapaport, 1998; Simon's 1956 notion of "satisficing"). Another kind of procedure that does not necessarily yield "correct" output is a "trial and error" (or Putnam-Gold) machine that continually generates "guesses" as to the correct output and is allowed to "change its mind," with its "final answer" being its *last* output, not its *first* output (Kugel, 2002).
3.   Finally, procedures (as I am using the term) include "interactive" programs, which are best modeled, not as Turing machines, but as Turing's "oracle" machines (a generalization of the notion of Turing machine; see Wegner, 1997, Soare, 2009).

In order for computational cognitive science's working hypothesis to be correct, the computation of cognition will have to be done by "algorithms" that are procedures in one or more of these senses.

Are Dreams Algorithms?

Fetzer (1998) argues that dreams are not algorithms and that ordinary, stream-of-consciousness thinking is not "algorithmic." I am willing to agree, up to a point, with Fetzer's Dynamic Premise 3: *Some human thought processes may indeed not be algorithms (or even procedures more generally). But that is not the real issue. The real issue is this: Could there be algorithms (or procedures) that produce dreams, stream-of-consciousness thinking, or other mental states or procedures, including those that might not *themselves* be algorithms (or procedures)?

The difference between an entity *being* computable and being *produced by* a computable process (i.e., being the *output* of an algorithm) can be clarified by considering two ways in which images can be considered computable entities. An image could be *implemented by* an array of pixels; this is the normal way in which images are stored in—and processed by—computers. Such an image is a computable, discrete data structure reducible to arrays of 0s and 1s. Alternatively, an image could be *produced by* a computational process that drives something like a flatbed plotter (Phillips, 2011) or produces a painting (as with Harold Cohen's AARON; see McCorduck, 1990). Such an image is *not* a discrete entity—it is, in fact, continuous; it is *not* reducible to arrays of 0s and 1s. Similarly, dreams need not themselves *be* algorithms in order to be producible *by* algorithms. (The same could, perhaps, be said of pains and other qualia: They might not themselves *be* algorithmically describable states, but they might be the *outputs* of algorithmic(ally describable) processes.)

What are dreams? In fact, no one knows, though there are many rival theories. Without some scientific agreement on what dreams are, it is difficult to see how one might say that they are—or are not—algorithmic or producible algorithmically. But suppose, as at least one standard view has it, that dreams are our interpretations of random neuron firings during sleep (perhaps occurring during the transfer of memories from short- to long-term memory),[51] interpreted as if they were due to external causes. Suppose also that *non*-dream neuron firings are computable. There are many reasons to think that they are; after all, the working assumption of computational cognitive science may have been challenged, but has not yet been refuted (*pace* Piccinini, 2005, 2007—remember, I am supposing that neuron firings are comput*able*, not necessarily computa*tional*; thus, Piccinini's arguments that neuron firings are *not* computational are irrelevant—and *pace* those cited in note 8). In that case, the neuron firings that constitute dreams would also be computable.

What about stream-of-consciousness thinking? That might be computable, too, by means of spreading activation in a semantic network, apparently randomly associating one thought to another. In fact, computational cognitive scientists have proposed computational theories of both dreaming and stream-of-consciousness thinking! (See Mueller, 1990, Edelman, 2008b, Mann, 2010.) It is not a matter of whether these scientists are right and Fetzer is wrong, or the other way around. Rather, the burden of proof is on Fetzer to say why he thinks that these proposed computational theories fail irreparably.

The important point is that whether a mental state or process is computable is at least an empirical question. Anti-computationalists must be wary of committing what many AI researchers think of as the Hubert Dreyfus fallacy: One philosopher's idea of a non-computable task may be just another computer scientist's research project. Put another way, what no one has *yet* written a computer program for is not thereby necessarily non-computable.

Are Minds Computers?

Fetzer's Dynamic Conclusion 2 is another claim that must be handled carefully: Maybe minds are computers; maybe they aren't. The more common formulation is that minds are *programs* and that *brains* are computers (cf. Searle, 1990, Piccinini, 2010). But I think that there is a better way to express the relationship than either of these slogans: A mind is a virtual machine, computationally implemented in some medium.

Roughly, a virtual machine is a computational implementation of a real machine; the virtual machine is executed as a process running on another (real or virtual) machine. For instance, there is (or was, at the time of writing) an "app" for Android smartphones that implements (i.e., simulates, perhaps emulates) a Nintendo Game Boy. Game Boy videogames can be downloaded and played on this "virtual" Game Boy "machine." (For a more complex example, of a virtual machine running on another virtual machine, see Rapaport, 2005b.) Thus, the human mind is a virtual machine computationally implemented in the nervous system, and a robot mind would be a virtual machine computationally implemented in a computer. Such minds consist of states and processes produced by the behavior of the brain or computer that implements them. (For discussion of virtual machines and this point of view, see Rapaport, 2005b, Hofstadter, 2007, Edelman, 2008b, Pollock, 2008.)

FETZER'S "AFFECTIVE" DIFFERENCE

Fetzer also argues that "Mental thought transitions are affected by emotions, attitudes, and histories, but computers are not," on the following grounds (Fetzer, 2008, 2010; my boldface and italics):

Premise 1: *Computers are governed by programs, which are causal models of algorithms.*

Premise 2: *Algorithms are effective decisions, which are not affected by emotions, attitudes, or histories.*

Premise 3: *Mental thought transitions are affected by values of variables that do not affect computers.*

Conclusion 1: *The processes controlling mental thought transitions are fundamentally different than those that control computer procedures.*

Conclusion 2: *Minds are not computers.*

Once again, I disagree with the boldfaced claims and find the italicized ones subtly misleading, some for reasons already mentioned. Before proceeding, it will be useful to rehearse (and critique) the definitions of some of Fetzer's technical terms, especially because he uses some of them in slightly non-standard ways. On Fetzer's view:

- The *intension* of expression $E =_{\text{def}}$ the conditions that need to be satisfied for something to be an *E*. (This term is not usually limited to noun phrases in the way that Fetzer seems to limit it ("to be an *E*"), but this is a minor point.)
- The *extension* of expression $E =_{\text{def}}$ the class of all things that satisfy *E*'s intension. (A more standard definition would avoid defining 'extension' in terms of 'intension;' rather, the extension of an expression would be the class of all (existing) things to which the expression *E* is applied by (native) speakers of the language.)
- The *denotation* of expression *E* for agent $A =_{\text{def}}$ the subset of *E*'s extension that *A* comes into contact with. (This notion may be useful, but 'denotation' is more often a mere synonym of 'extension' or 'referent.')

- The *connotation* of expression E for agent $A =_{def} A$'s attitudes and emotions in response to A's interaction with E's denotation for A. (Again, a useful idea, but not the usual use of the term, which is more often a way of characterizing the other concepts that are closely related to E (perhaps in some agent A's mind, or just the properties associated with (things called) E. Both 'denotation' and 'connotation' in their modern uses were introduced by Mill 1843, the former being synonymous with 'extension' and the latter referring to implied properties of the item denoted.)

Fetzer then identifies the "meaning" of E for A as E's denotation and connotation for A. Contra Fetzer's Affective Premises 2 and 3, programs *can* be based on idiosyncratic emotions, attitudes, and histories: Karen Ehrlich and I (along with other students and colleagues) have developed and implemented a computational theory of contextual vocabulary acquisition (Ehrlich, 1995; Rapaport & Ehrlich, 2000; Rapaport, 2003b, 2005a; Rapaport & Kibby, 2007, 2010). Our system learns (or "acquires") a meaning for an unfamiliar or unknown word from the word's textual context integrated with the reader's prior beliefs. These prior beliefs (more usually called 'prior knowledge' in the reading-education literature), in turn, can—and often do—include idiosyncratic "denotations" and "connotations" (in Fetzer's senses), emotions, attitudes, and histories. In fact, contrary to what some reading educators assert (cf. Ames, 1966, Dulin, 1970), a meaning for a word can*not* be determined *solely* from its textual context. The reader's prior beliefs are essential (Rapaport, 2003b, Rapaport & Kibby, 2010). And, clearly, the meaning that one reader figures out or attributes to a word will differ from that of another reader to the extent that their prior beliefs differ.

Furthermore, several cognitive scientists have developed computational theories of affect and emotion, showing that emotions, attitudes, and histories *can* affect computers that model them (Simon, 1967, Sloman & Croucher, 1981, Wright et al. 1996, Sloman, 2004, 2009, Picard, 1997, and Thagard, 2006), among others). Once again, the burden of proof is on Fetzer.

SIMULATION

I close with a discussion of "the matter of simulation." Fetzer argues that "Digital machines can nevertheless simulate thought processes and other diverse forms of human behavior," on the following grounds (Fetzer, 2001, p. xvii, 2008, 2010; my emphasis):

Premise 1: Computer programmers and those who design the systems that they control can increase their performance capabilities, making them better and better simulations.

Premise 2: Their performance capabilities may be closer and closer approximations to the performance capabilities of human beings without turning them into thinking things.

Premise 3: Indeed, the static, dynamic, and affective differences that distinguish computer performance from human performance preclude those systems from being thinking things.

Conclusion: Although the performance capabilities of digital machines can become better and better approximations of human behavior, they are still not thinking things.

As before, I disagree with the boldfaced claims. But, again, we must clarify Fetzer's somewhat non-standard use of terms. Computer scientists occasionally distinguish between a "simulation" and an "emulation," though the terminology is not fixed. In the *Encyclopedia of Computer Science*, Paul F. Roth (1983) says that *x simulates y* means that *x* is a model of some real or imagined system *y*, and we experiment with *x* in order to understand *y*. (Compare our earlier discussion of understanding one thing in terms of another) Typically, *x* might be a computer program, and *y* might be some real-world situation. In an extreme case, *x* simulates *y* if and only if *x* and *y* have the same input-output behavior.

And in another article in that *Encyclopedia*, Stanley Habib (1983) says that *x emulates y* means that either: (a) computer *x* "interprets and executes" computer *y*'s "instruction set" by implementing *y*'s operation codes in *x*'s hardware—i.e., hardware *y* is implemented as a virtual machine on *x*—or (b) software feature *x* "simulates"(!) "hardware feature" *y*, doing what *y* does exactly (so to speak) as *y* does it. Roughly, *x* emulates *y* if and only if *x* and *y* not only have the same input-output behavior, but also use the same algorithms and data structures. This suggests that there is a continuum or spectrum, with "pure" simulation at one end (input-output–equivalent behavior) and "pure" emulation at the other end (behavior that is equivalent with respect to input-output, all algorithms in full detail, and all data structures). So, perhaps there is no real distinction between simulation and emulation except for the degree of faithfulness to what is being simulated or emulated.

In contrast, Fetzer uses a much simplified version of this for his terminology (Fetzer, 1990, 2001, 2011):

- System *x simulates* system $y =_{def} x$ and *y* have the same input-output behavior.
- System *x replicates* system $y =_{def} x$ simulates *y* by the same or similar processes.
- System *x emulates* system $y =_{def} x$ replicates *y*, and *x* and *y* "are composed of the same kind of stuff."

At least the latter two are non-standard definitions and raise many questions. For instance, how many processes must be "similar," and how "similar" must they be, before we can say that one system replicates another? Or consider (1) a Turing machine (a single-purpose, or dedicated, computer) that computes GCDs and (2) a universal Turing machine (a multiple-purpose, or stored-program computer) that can be programmed to compute GCDs using exactly the same program as is encoded in the machine table of the former Turing machine. Does the latter emulate the former? Does the former emulate the latter? Do two copies of the former emulate each other? Nevertheless, Fetzer's terminology makes some useful distinctions, and, here, I will use these terms as Fetzer does.

The English word 'simulation' (however defined) has a sense (a "connotation"?) of "imitation" or "unreal": A simulation of a hurricane is not a real hurricane. A simulation of digestion is not real digestion. And I agree that a computer that simulates (in Fetzer's sense) some process *P* is not *necessarily* "really" doing *P*. But what, exactly, is the difference? A computer simulation (or even a replication) of the daily operations of a bank is not thereby the daily operations of a (real) bank. But I *can* do my banking online; simulations *can* be used as if they were real, as long as the (syntactic) simulations have causal impact on me (Goldfain, personal communication, 2011). And although computer simulations of hurricanes don't get real people wet (they are, after all, not emulations in Fetzer's sense), they *could* get *simulated* people *simulatedly* wet (Shapiro & Rapaport, 1991, Rapaport, 2005b):

[A] simulated hurricane would feel very real, and indeed can prove fatal, to a person who happens to reside in the same simulation (Edelman, 2011, 2n.3, my italics.)

The "person," of course, would have to be a *simulated* person. And it well might be that the way that the simulated hurricane would "feel" to that simulated person would differ from the way that a real hurricane feels to a real person. (On the meaningfulness of that comparison, see Strawson 2010, pp. 218–219.)

Paul Harris [2000] found that even two-year-olds will tell you that if an imaginary teddy [bear] is drinking imaginary tea, then if he spills it the imaginary floor will require imaginary mopping-up (Gopnik, 2009, p. 29.)

This is a matter of the "scope" of the simulation: Are people within the scope of the hurricane simulation or not? If they are not, then the simulation won't get them wet. If they are—i.e., if they are simulated, too—then it will. It should also be noted that sometimes things like *simulated* hurricanes *can* do something analogous to getting *real* people wet: Children "can have real emotional reactions to entirely imaginary scenarios" (Gopnik, 2009, p. 31), as, of course, can adults, as anyone who has wept at the movies can testify (cf. Schneider 2009).

But there are cases where a simulation *is* the real thing. For example:

- A scale model of a scale model of the Statue of Liberty *is* a scale model of the Statue of Liberty.
- A Xerox copy of a document *is* that document, at least for purposes of reading it and even for some legal purposes.
- A PDF version of a document *is* that document.

More specifically, a computer that simulates an "informational process" *is* thereby actually doing that informational process, because a computer simulation of information *is* information:

- A computer simulation of a picture *is* a picture, hence the success of digital "photography."
- A computer simulation of language *is* language. Indeed, as William A. Woods (2010, p. 605, my emphasis) said:

[L]anguage is fundamentally computational. Computational linguistics has a more intimate relationship with computers than many other disciplines that use computers as a tool. When a computational biologist simulates population dynamics, no animals die.[52] When a meteorologist simulates the weather, nothing gets wet.[53] But computers really do parse sentences. Natural language question-answering systems really do answer questions.[54] The actions of language are in the same space as computational activities, or alternatively, these particular computational activities are in the same space as language and communication.

- A computer simulation of mathematics *is* mathematics. As Edelman (2008b, p. 81) put it (though not necessarily using the terms as Fetzer does):

A simulation of a computation and the computation itself are equivalent: try to simulate the addition of 2 and 3, and the result will be just as good as if you "actually" carried out the addition—that is the nature of numbers.

- A computer simulation of reasoning *is* reasoning. This is the foundation of automated theorem proving.

And, in general, a computer simulation of cognition *is* cognition. To continue the just-cited quote from Edelman (2008b, p. 81):

Therefore, if the mind is a computational entity, a simulation of the relevant computations would constitute its fully functional replica.

CONCLUSION

I conclude that Fetzer is mistaken on all counts: Computers *are* semiotic systems and *can* possess minds, mental processes *are* governed by algorithms (or, at least, "procedures"), and algorithms *can* be affected by emotions, attitudes, and individual histories. Moreover, computers that implement cognitive algorithms really do exhibit those cognitive behaviors—they really do think. And syntactic semantics explains how this is possible.

ACKNOWLEDGMENT

This essay is based on my oral reply ("Salvaging Computationalism: How Cognition *Could Be* Computing") to two unpublished presentations by Fetzer: (1) "Can Computationalism Be Salvaged?," 2009 International Association for Computing and Philosophy (Chicago), and (2) "Limits to Simulations of Thought and Action," session on "The Limits of Simulations of Human Actions" at the 2010 North American Computing and Philosophy conference (Carnegie-Mellon University). I am grateful to Fetzer for allowing me to quote from his slides and from an unpublished manuscript; to our co-presenters at both meetings—Selmer Bringsjord and James H. Moor—for their commentaries; to our unique audiences at those meetings; and to Randall R. Dipert, Albert Goldfain, Lorenzo Incardona, Kenneth W. Regan, Daniel R. Schlegel, Stuart C. Shapiro, and members of the SNePS Research Group for comments or assistance on previous drafts.

REFERENCES

Ames, W. S. (1966). The development of a classification scheme of contextual aids. *Reading Research Quarterly*, 2(1), 57–82. doi:10.2307/747039

Andersen, P. B. (1992). Computer semiotics. *Scandinavian Journal of Information Systems*, 4(1), 1–30.

Anstey, J., Seyed, A. P., Bay-Cheng, S., Pape, D., Shapiro, S. C., Bona, J., & Hibit, S. (2009). The agent takes the stage. *International Journal of Arts and Technology*, 2(4), 277–296. doi:10.1504/IJART.2009.029236

Barsalou, L. W. (1999). Perceptual symbol systems. *Behavioral and Brain Sciences*, 22, 577–660. PMID:11301525

Brighton, H. (2002). Compositional syntax from cultural transmission. *Artificial Life, 8*(1), 25–54. doi:10.1162/106454602753694756 PMID:12020420

Brown, S. R. (2002). Peirce, Searle, and the Chinese room argument. *Cybernetics & Human Knowing, 9*(1), 23–38.

Burge, T. (1986). Individualism and psychology. *The Philosophical Review, 95*(1), 3–45. doi:10.2307/2185131

Cariani, P. (2001). Symbols and dynamics in the brain. *Bio Systems, 60*(1-3), 59–83. doi:10.1016/S0303-2647(01)00108-3 PMID:11325504

Chalmers, D. J. (1996). *The conscious mind: In search of a fundamental theory.* New York: Oxford University Press.

Church, A. (1936). An unsolvable problem of elementary number theory. *American Journal of Mathematics, 58*(2), 345–363. doi:10.2307/2371045

Cleland, C. E. (1993). Is the Church-Turing thesis true? *Minds and Machines, 3*(3), 283-312.

Coffa, J. A. (1991). *The semantic tradition from Kant to Carnap: To the Vienna Station.* Cambridge, UK: Cambridge University Press. doi:10.1017/CBO9781139172240

Crane, T. (1990). The language of thought: No syntax without semantics. *Mind & Language, 5*(3), 187–212. doi:10.1111/j.1468-0017.1990.tb00159.x

Crossley, J. N., & Henry, A. S. (1990). Thus spake al-Khwārizmī: A translation of the text of Cambridge University Library Ms. Ii.vi.5. *Historia Mathematica, 17*(2), 103–131. doi:10.1016/0315-0860(90)90048-I

Dennett, D. (2009). Darwin's 'Strange Inversion of Reasoning.' *Proceedings of the National Academy of Sciences, 106*(Suppl 1), 10061-10065.

Dietrich, E. (1990). Computationalism. *Social Epistemology, 4*(2), 135–154. doi:10.1080/02691729008578566

Dipert, R. R. (1984). Peirce, Frege, the logic of telations, and Church's Theorem. *History and Philosophy of Logic, 5*(1), 49–66. doi:10.1080/01445348408837062

Dresner, E. (2010). Measurement-theoretic representation and computation-theoretic Realization. *The Journal of Philosophy, 107*(6), 275–292. doi:10.5840/jphil2010107622

Dreyfus, H. L. (1965). *Alchemy and artificial intelligence.* Report P-3244. RAND Corp. Accessed 2 November 2011 from http://tinyurl.com/4hczr59

Dreyfus, H. L. (1972). *What computers still can't do: A critique of artificial reason.* New York: Harper & Row.

Dreyfus, H. L. (1979). *What computers still can't do: A critique of artificial reason* (revised ed.). New York: Harper & Row.

Dreyfus, H. L. (1992). *What computers still can't do: A critique of artificial reason.* Cambridge, MA: MIT Press.

Dulin, K. L. (1970). Using context clues in word recognition and comprehension. *The Reading Teacher*, *23*(5), 440–445, 469.

Eco, U. (1979). *A theory of semiotics*. Bloomington, IN: Indiana University Press.

Eco, U. (1988). On truth: A fiction. In U. Eco, M. Santambrogio, & P. Violi (Eds.), *Meaning and mental representations* (pp. 41–59). Bloomington, IN: Indiana University Press.

Edelman, S. (2008a). On the nature of minds, or: Truth and consequences. *Journal of Experimental & Theoretical Artificial Intelligence*, *20*(3), 181–196. doi:10.1080/09528130802319086

Edelman, S. (2008b). *Computing the Mind*. New York: Oxford University Press.

Edelman, S. (2011). Regarding reality: Some consequences of two incapacities. *Frontiers in Theoretical and Philosophical Psychology, 2*, 1–8. doi:10.3389/fpsyg.2011.00044

Ehrlich, K. (1995). *Automatic vocabulary expansion through narrative context*. Technical Report 95-09. Buffalo, NY: SUNY Buffalo Department of Computer Science.

Fetzer, J. H. (1990). *Artificial intelligence: Its scope and limits*. Dordrecht, The Netherlands: Kluwer Academic Publishers. doi:10.1007/978-94-009-1900-6

Fetzer, J. H. (1994). Mental algorithms: Are minds computational systems? *Pragmatics & Cognition*, *2*(1), 1–29. doi:10.1075/pc.2.1.01fet

Fetzer, J. H. (1998). People are not computers, (most) thought processes are not computational procedures. *Journal of Experimental & Theoretical Artificial Intelligence*, *10*(4), 371–391. doi:10.1080/095281398146653

Fetzer, J. H. (2001). *Computers and cognition: Why minds are not machines*. Dordrecht, The Netherlands: Kluwer Academic Publishers. doi:10.1007/978-94-010-0973-7

Fetzer, J. H. (2008, July 11). *Computing vs. cognition: Three dimensional differences*. Unpublished.

Fetzer, J. H. (2010). *Limits to simulations of thought and action*. Talk given at the 2010 North American Conference on Computing and Philosophy (NA-CAP), Carnegie-Mellon University.

Fetzer, J. H. (2011). Minds and machines: Limits to simulations of thought and action. *International Journal of Signs and Semiotic Systems*, *1*(1), 39–48. doi:10.4018/ijsss.2011010103

Fodor, J. A. (1975). *The language of thought*. New York: Crowell.

Fodor, J. A. (1980). Methodological solipsism considered as a research strategy in cognitive psychology. *Behavioral and Brain Sciences*, *3*(01), 63–109. doi:10.1017/S0140525X00001771

Fodor, J. A. (2000). *The mind doesn't work that way: The scope and limits of computational psychology*. Cambridge, MA: MIT Press.

Fodor, J. A. (2008). *LOT 2: The language of thought revisited*. Oxford, UK: Clarendon. doi:10.1093/ac prof:oso/9780199548774.001.0001

Forbus, K. D. (2010). AI and Cognitive Science: The Past and Next 30 Years. *Topics in Cognitive Science*, *2*(3), 345–356. doi:10.1111/j.1756-8765.2010.01083.x PMID:25163864

Ford, K., & Hayes, P. (1998). On computational wings. Scientific American Presents, 9(4), 78-83.

Franzén, T. (2005). *Gödel's theorem: An incomplete guide to its use and abuse*. Wellesley, MA: A.K. Peters. doi:10.1201/b10700

Gandy, R. (1988). The confluence of ideas in 1936. In R. Herken (Ed.), The universal Turing Machine: A half-century survey (2nd ed.; pp. 51-102). Vienna: Springer-Verlag.

Gelman, R., & Gallistel, C. R. (1986). *The child's understanding of number*. Cambridge, MA: Harvard University Press.

Goldfain, A. (2006). Embodied enumeration: Appealing to activities for mathematical explanation. In M. Beetz, K. Rajan, M. Thielscher, & R. Bogdan Rusu (Eds.), *Cognitive robotics: Papers from the AAAI Workshop (CogRob2006)* (pp. 69-76). Technical Report WS-06-03. Menlo Park, CA: AAAI Press.

Goldfain, A. (2008). *A computational theory of early mathematical cognition* (PhD dissertation). Buffalo, NY: SUNY Buffalo Department of Computer Science & Engineering. Retrieved from http://www.cse.buffalo.edu/sneps/Bibliography/GoldfainDissFinal.pdf

Gopnik, A. (2009). *The philosophical baby: What children's minds tell us about truth, love, and the meaning of life*. New York: Farrar, Straus and Giroux.

Habib, S. (1983). Emulation. In A. Ralston & E. D. Reilly Jr., (Eds.), *Encyclopedia of Computer Science and Engineering* (2nd ed.; pp. 602–603). New York: Van Nostrand Reinhold.

Harnad, S. (1990). The symbol grounding problem. *Physica D. Nonlinear Phenomena, 42*(1-3), 335–346. doi:10.1016/0167-2789(90)90087-6

Harris, P. (2000). *The work of the imagination*. Malden, MA: Blackwell.

Hobbes, T. (1651). *Leviathan*. Indianapolis, IN: Bobbs-Merrill Library of Liberal Arts.

Hofstadter, D. (2007). *I am a strange loop*. New York: Basic Books.

Horst, S. (2009). The computational theory of mind. In E. N. Zalta (Ed.), Stanford Encyclopedia of Philosophy. Accessed 6 May 2011 from http://plato.stanford.edu/archives/win2009/entries/computational-mind/

Huemer, M. (2011). Sense-data. In E. N. Zalta (Ed.), Stanford Encyclopedia of Philosophy. Accessed 6 May 2011 from http://plato.stanford.edu/archives/spr2011/entries/sense-data/

Hume, D. (1739). A treatise of human nature (L. A. Selby-Bigge, Ed.). London: Oxford University Press.

Hume, D. (1777). An enquiry concerning human understanding (L. A. Selby-Bigge, Ed.). London: Oxford University Press. doi:10.1093/oseo/instance.00046350

Incardona, L. (2012). *Semiotica e web semantico: Basi teoriche e metodologiche per la semiotica computazionale* [Semiotics and semantic web: Theoretical and methodological foundations for computational semiotics] (PhD dissertation). University of Bologna.

Jackendoff, R. (2002). *Foundations of language: Brain, meaning, grammar, evolution*. Oxford, UK: Oxford University Press. doi:10.1093/acprof:oso/9780198270126.001.0001

Johnson-Laird, P. N. (1983). *Mental models: Towards a cognitive science of language, inference, and consciousness*. Cambridge, MA: Harvard University Press.

Johnson-Laird, P. N. (1988). *The computer and the mind: An introduction to cognitive science*. Cambridge, MA: Harvard University Press.

Jones, D. (2010). *Animal liberation*. Paper presented at the 2010 North American Conference on Computing and Philosophy (NA-CAP), Carnegie-Mellon University.

Julesz, B. (1971). *Foundations of Cyclopean perception*. Chicago: University of Chicago Press.

Ketner, K. L. (1988). Peirce and Turing: Comparisons and conjectures. *Semiotica, 68*(1/2), 33–61.

Kirby, S. (2000). Syntax without natural selection: How compositionality emerges from vocabulary in a population of learners. In C. Knight (Ed.), *The evolutionary emergence of language: Social function and the origins of linguistic form* (pp. 303–323). Cambridge, UK: Cambridge University Press. doi:10.1017/CBO9780511606441.019

Kleene, S. C. (1952). *Introduction to metamathematics*. Princeton, NJ: D. Van Nostrand.

Kleene, S. C. (1967). *Mathematical logic*. New York: Wiley.

Knuth, D. E. (1972). Ancient Babylonian algorithms. *Communications of the ACM, 15*(7), 671–677. doi:10.1145/361454.361514

Knuth, D. E. (1973). Basic concepts: Algorithms. In *The Art of Computer Programming* (2nd ed.; pp. xiv-9). Reading, MA: Addison-Wesley.

Kugel, P. (2002). Computing machines can't be intelligent (…and Turing said so). *Minds and Machines, 12*(4), 563–579. doi:10.1023/A:1021150928258

Levesque, H. (2012). *Thinking as Computation: A First Course*. Cambridge, MA: MIT Press. doi:10.7551/mitpress/9780262016995.001.0001

Lobina, D. J. (2010). Recursion and the competence/performance distinction in AGL tasks. *Language and Cognitive Processes*. Accessed 4 November 2011 from http://tinyurl.com/Lobina2010

Lobina, D. J., & García-Albea, J. E. (2009). Recursion and cognitive science: Data structures and mechanisms. In N. Taatgen & H. van Rijn (Eds.), *Proceedings of the 31ˢᵗ Annual Conference of the Cognitive Science Society* (pp. 1347-1352). Academic Press.

Lohr, S. (2010, September 24). Frederick Jelinek, who gave machines the key to human speech, dies at 77. *New York Times*, p. B10.

Lucas, J. R. (1961). Minds, machines and Gödel. *Philosophy (London, England), 36*(137), 112–127. doi:10.1017/S0031819100057983

Makuuchi, M., Bahlmann, J., Anwander, A., & Friederici, A. D. (2009). Segregating the core computational faculty of human language from working memory. *Proceedings of the National Academy of Sciences, 106*(20), 8362-8367. doi:10.1073/pnas.0810928106

Mann, G. A. (2010). A machine that daydreams. *IFIP Advances in Information and Communication Technology, 333*, 21–35. doi:10.1007/978-3-642-15214-6_3

Markov, A. A. (1954). Theory of algorithms. Tr. Mat. Inst. Steklov, 42.

Marr, D. (1982). *Vision: A computational investigation into the human representation and processing of visual information*. New York: W.H. Freeman.

McCorduck, P. (1990). *Aaron's code: Meta-Art, artificial intelligence, and the work of Harold Cohen.* New York: W.H. Freeman.

McCulloch, W. S., & Pitts, W. H. (1943). A logical calculus of the ideas immanent in nervous activity. *The Bulletin of Mathematical Biophysics, 5*(4), 115–133. doi:10.1007/BF02478259

McGee, K. (2011). Review of K. Tanaka-Ishii's *Semiotics of programming. Artificial Intelligence, 175,* 930–931. doi:10.1016/j.artint.2010.11.023

Mill, J. S. (1843). *A system of logic.* Accessed 4 November 2011 from http://tinyurl.com/MillSystemLogic

Miller, G. A., Galanter, E., & Pribram, K. H. (1960). *Plans and the structure of behavior.* New York: Henry Holt. doi:10.1037/10039-000

Morris, C. (1938). *Foundations of the theory of signs.* Chicago: University of Chicago Press.

Mueller, E. T. (1990). *Daydreaming in humans and machines: A computer model of the stream of thought.* Norwood, NJ: Ablex.

Nadin, M. (2007). Semiotic machine. *Public Journal of Semiotics, 1*(1), 85-114. Accessed 5 May 2011 from http://www.nadin.ws/archives/760

Nadin, M. (2010). Remembering Peter Bøgh Andersen: Is the computer a semiotic machine? A discussion never finished. *Semiotix: New Series,* (1). Retrieved from http://www.semioticon.com/semiotix/2010/03/

Newell, A., Shaw, J. C., & Simon, H. A. (1958). Elements of a theory of human problem solving. *Psychological Review, 65*(3), 151–166. doi:10.1037/h0048495

Newell, A., & Simon, H. A. (1976). Computer science as empirical inquiry: Symbols and search. *Communications of the ACM, 19*(3), 113–126. doi:10.1145/360018.360022

Nöth, W. (2003). Semiotic machines. *S.E.E.D. Journal (Semiotics, Evolution, Energy, and Development), 3*(3), 81-99. Retrieved from http://www.library.utoronto.ca/see/SEED/Vol3-3/Winfried.pdf

Parisien, C., & Thagard, P. (2008). Robosemantics: How Stanley represents the world. *Minds and Machines, 18*(2), 169–178. doi:10.1007/s11023-008-9098-2

Pattee, H. H. (1969). How does a molecule become a message? *Developmental Biology,* (Supplement 3), 1–16.

Pattee, H. H. (1982). Cell psychology: An evolutionary approach to the symbol-matter problem. *Cognition and Brain Theory, 5*(4), 325–341.

Peirce, C. S. (1887). Logical machines. *The American Journal of Psychology, 1,* 165–170.

Peirce, C. S. (1908). Semiotic and significs: The correspondence between Charles S. Peirce and Victoria Lady Welby. Bloomington, IN: Indiana University Press.

Peirce, C. S. (1992). Some consequences of four incapacities. In N. Houser & C. Kloesel (Eds.), *The essential Peirce: Selected philosophical writings* (Vol. 1, pp. 186–199). Bloomington, IN: Indiana University Press.

Peirce, C. S. (1998). The essential Peirce: Selected philosophical writings (Vol. 2). Bloomington, IN: Indiana University Press.

Peirce, C. S. (1931-1958). Collected papers of Charles Sanders Peirce (vols. 1-8). Cambridge, MA: Harvard University Press.

Penrose, R. (1989). *The Emperor's new mind: Concerning computers, minds and the laws of physics.* New York: Oxford University Press.

Perlovsky, L. I. (2007). Symbols: Integrated cognition and language. In R. Gudwin & J. Queiroz (Eds.), *Semiotics and intelligent systems development* (pp. 121–151). Hershey, PA: Idea Group. doi:10.4018/978-1-59904-063-9.ch005

Phillips, A. L. (2011). The algorists. *American Scientist, 99*(2), 126. doi:10.1511/2011.89.126

Picard, R. (2010). *Affective computing.* Accessed 6 August 2010 from http://affect.media.mit.edu/

Picard, R. W. (1997). *Affective computing.* Cambridge, MA: MIT Press. doi:10.1037/e526112012-054

Piccinini, G. (2005). *Symbols, strings, and spikes: The empirical refutation of computationalism.* Retrieved from http://tinyurl.com/Piccinini2005

Piccinini, G. (2007). Computational explanation and mechanistic explanation of mind. In M. Marraffa, M. De Caro, & F. Ferretti (Eds.), *Cartographies of the mind: Philosophy and psychology in intersection* (pp. 23–36). Dordrecht, The Netherlands: Springer. doi:10.1007/1-4020-5444-0_2

Piccinini, G. (2008). Computation without representation. *Philosophical Studies, 137*(2), 205–241. doi:10.1007/s11098-005-5385-4

Piccinini, G. (2010). The mind as neural software? Understanding functionalism, computationalism, and computational functionalism. *Philosophy and Phenomenological Research, 81*(2), 269–311. doi:10.1111/j.1933-1592.2010.00356.x

Piccinini, G., & Scarantino, A. (2011). Information processing, computation, and cognition. *Journal of Biological Physics, 37*(1), 1–38. doi:10.1007/s10867-010-9195-3 PMID:22210958

Pollock, J. L. (2008). What am I? Virtual machines and the mind/body problem. *Philosophy and Phenomenological Research, 76*(2), 237–309. doi:10.1111/j.1933-1592.2007.00133.x

Posner, R. (1992). Origins and development of contemporary syntactics. *Languages of Design, 1*, 37–50.

Putnam, H. (1960). Minds and machines. In S. Hook (Ed.), *Dimensions of mind: A symposium* (pp. 148–179). New York: New York University Press.

Putnam, H. (1961). *Brains and behavior.* Presented at the American Association for the Advancement of Science, Section L (History and Philosophy of Science).

Putnam, H. (1975). The meaning of 'meaning.' In K. Gunderson (Ed.), Language, mind, and knowledge (pp. 131-193). Minneapolis, MN: University of Minnesota Press.

Putnam, H. (1988). *Representation and reality*. Cambridge, MA: MIT Press.

Pylyshyn, Z. W. (1980). Computation and cognition: Issues in the foundations of cognitive science. *Behavioral and Brain Sciences*, *3*(01), 111–169. doi:10.1017/S0140525X00002053

Pylyshyn, Z. W. (1985). *Computation and cognition: Toward a foundation for cognitive science* (2nd ed.). Cambridge, MA: MIT Press.

Ramachandran, V., & Rogers-Ramachandran, D. (2009). Two eyes, two views. *Scientific American Mind*, *20*(5), 22–24. doi:10.1038/scientificamericanmind0909-22

Rapaport, W. J. (1979). (1978. Meinongian theories and a Russellian paradox. *Noûs 12*, 153-180. *Noûs (Detroit, Mich.)*, *13*, 125. doi:10.2307/2214805

Rapaport, W. J. (1979). An adverbial Meinongian theory. *Analysis*, *39*(2), 75–81. doi:10.1093/analys/39.2.75

Rapaport, W. J. (1981). How to make the world fit our language: An essay in Meinongian semantics. *Grazer Philosophische Studien*, *14*, 1–21. doi:10.5840/gps1981141

Rapaport, W. J. (1985/1986). Non-existent objects and epistemological ontology. Grazer Philosophische Studien, 25/26, 61-95.

Rapaport, W. J. (1986). Searle's experiments with thought. *Philosophy of Science*, *53*(2), 271–279. doi:10.1086/289312

Rapaport, W. J. (1988). Syntactic semantics: Foundations of computational natural-language understanding. In J. H. Fetzer (Ed.), Aspects of artificial intelligence (pp. 81-131). Dordrecht, The Netherlands: Kluwer Academic Publishers. Retrieved from http://tinyurl.com/SynSemErrata1

Rapaport, W. J. (1990). *Computer processes and virtual persons: Comments on Cole's 'Artificial Intelligence and Personal Identity.'* Technical Report 90-13. Buffalo, NY: SUNY Buffalo Department of Computer Science. Retrieved from http://tinyurl.com/Rapaport1990

Rapaport, W. J. (1995). Understanding understanding: Syntactic semantics and computational cognition. In J. E. Tomberlin (Ed.), AI, connectionism, and philosophical psychology (pp. 49-88). Atascadero, CA: Ridgeview.

Rapaport, W. J. (1998). How minds can be computational systems. *Journal of Experimental & Theoretical Artificial Intelligence*, *10*(4), 403–419. doi:10.1080/095281398146671

Rapaport, W. J. (1999). Implementation is semantic interpretation. *The Monist*, *82*(1), 109–130. doi:10.5840/monist19998212

Rapaport, W. J. (2000). How to pass a Turing Test: Syntactic semantics, natural-language understanding, and first-person cognition. Journal of Logic, Language, and Information, 9(4), 467-490.

Rapaport, W. J. (2002). Holism, conceptual-role semantics, and syntactic semantics. *Minds and Machines*, *12*(1), 3–59. doi:10.1023/A:1013765011735

Rapaport, W. J. (2003a). What did you mean by that? Misunderstanding, negotiation, and syntactic semantics. *Minds and Machines*, *13*(3), 397–427. doi:10.1023/A:1024145126190

Rapaport, W. J. (2003b). What is the 'context' for contextual vocabulary acquisition? In P. P. Slezak (Ed.), *Proceedings of the 4th International Conference on Cognitive Science/7th Australasian Society for Cognitive Science Conference (ICCS/ASCS-2003; Sydney, Australia)* (Vol. 2, pp. 547-552). Sydney: University of New South Wales.

Rapaport, W. J. (2005a). In defense of contextual vocabulary acquisition: How to do things with words in context. In A. Dey, B. Kokinov, D. Leake, & R. Turner (Eds.), *Modeling and using context: 5th International and Interdisciplinary Conference, CONTEXT 05, Paris, France, July 2005, Proceedings* (pp. 396-409). Berlin: Springer-Verlag.

Rapaport, W. J. (2005b). Implementation is semantic interpretation: Further thoughts. *Journal of Experimental & Theoretical Artificial Intelligence, 17*(4), 385–417. doi:10.1080/09528130500283998

Rapaport, W. J. (2006). How Helen Keller used syntactic semantics to escape from a Chinese room. *Minds and Machines, 16*(4), 381–436. doi:10.1007/s11023-007-9054-6

Rapaport, W. J. (2011). Yes, she was! Reply to Ford's 'Helen Keller was never in a Chinese room.'. *Minds and Machines, 21*(1), 3–17. doi:10.1007/s11023-010-9213-z

Rapaport, W. J. (2015). On the relation of computing to the world. In T. M. Powers (Ed.), *Philosophy and computing: Essays in epistemology, philosophy of mind, logic, and ethics.* Springer. Retrieved from http://www.cse.buffalo.edu/~rapaport/Papers/covey.pdf

Rapaport, W. J. (2017a). Semantics as syntax. *American Philosophical Association Newsletter on Philosophy and Computers.* Retrieved from http://www.cse.buffalo.edu/~rapaport/Papers/synsemapa.pdf

Rapaport, W. J. (2017b). *Philosophy of computer science.* Retrieved from https://www.cse.buffalo.edu//~rapaport/Papers/phics.pdf

Rapaport, W. J., & Ehrlich, K. (2000). A computational theory of vocabulary acquisition. In Ł. M. Iwańska & S. C. Shapiro (Eds.), *Natural language processing and knowledge representation: Language for knowledge and knowledge for language* (pp. 347–375). Menlo Park, CA: AAAI Press/MIT Press. Retrieved from http://tinyurl.com/RapaportEhrlichErrata

Rapaport, W. J., & Kibby, M. W. (2007). Contextual vocabulary acquisition as computational philosophy and as philosophical computation. *Journal of Experimental & Theoretical Artificial Intelligence, 19*(1), 1–17. doi:10.1080/09528130601116162

Rapaport, W. J., & Kibby, M. W. (2010). *Contextual vocabulary acquisition: From algorithm to curriculum.* Retrieved from http://tinyurl.com/RapaportKibby2010

Ray, K. (2010). *Web 3.0.* Accessed 1 March 2011 from http://kateray.net/film

Rescorla, M. (2017). The computational theory of mind. In *The Stanford Encyclopedia of Philosophy.* Retrieved from https://plato.stanford.edu/archives/spr2017/entries/computational-mind/

Rosen, R. (1987). On the scope of syntactics in mathematics and science: The machine metaphor. In J. L. Casti & A. Karlqvist (Eds.), *Real brains, artificial minds* (pp. 1–23). New York: Elsevier Science.

Rosser, B. (1939). An informal exposition of proofs of Gödel's theorem. *The Journal of Symbolic Logic*, *4*(2), 53–60. doi:10.2307/2269059

Roth, P. F. (1983). Simulation. In A. Ralston & E. D. Reilly Jr., (Eds.), *Encyclopedia of Computer Science and Engineering* (2nd ed.; pp. 1327–1341). New York: Van Nostrand Reinhold.

Santore, J. F., & Shapiro, S. C. (2003). Crystal Cassie: Use of a 3-D gaming environment for a cognitive agent. In R. Sun (Ed.), *Papers of the IJCAI 2003 Workshop on Cognitive Modeling of Agents and Multi-Agent Interactions (IJCAII; Acapulco, Mexico, August 9, 2003)* (pp. 84-91). Academic Press.

Schagrin, M. L., Rapaport, W. J., & Dipert, R. R. (1985). *Logic: A computer approach*. New York: McGraw-Hill.

Schneider, S. (2009). The paradox of fiction. *Internet Encyclopedia of Philosophy*. Accessed 6 May 2011 from http://www.iep.utm.edu/fict-par/

Searle, J. R. (1980). Minds, brains, and programs. *Behavioral and Brain Sciences*, *3*(03), 417–457. doi:10.1017/S0140525X00005756

Searle, J. R. (1990). Is the brain a digital computer? *Proceedings and Addresses of the American Philosophical Association*, *64*(3), 21–37. doi:10.2307/3130074

Segal, G. (1989). Seeing what is not there. *The Philosophical Review*, *98*(2), 189–214. doi:10.2307/2185282

Shapiro, S. C. (1992). Artificial Intelligence. In S. C. Shapiro (Ed.), *Encyclopedia of Artificial Intelligence* (2nd ed.; pp. 54–57). New York: John Wiley & Sons.

Shapiro, S. C. (1993). Belief spaces as sets of propositions. *Journal of Experimental & Theoretical Artificial Intelligence*, *5*(2-3), 225–235. doi:10.1080/09528139308953771

Shapiro, S. C. (1998). Embodied Cassie. In *Cognitive Robotics: Papers from the 1998 AAAI Fall Symposium* (pp. 136-143). Technical Report FS-98-02. Menlo Park, CA: AAAI Press.

Shapiro, S. C. (2001). *Computer science: The study of procedures*. Accessed 5 May 2011 from http://www.cse.buffalo.edu/~shapiro/Papers/whatiscs.pdf

Shapiro, S. C. (2006). Natural-language-competent robots. *IEEE Intelligent Systems*, *21*(4), 76–77.

Shapiro, S. C., Amir, E., Grosskreutz, H., Randell, D., & Soutchanski, M. (2001). *Commonsense and embodied agents*. A panel discussion. At Common Sense 2001: The 5th International Symposium on Logical Formalizations of Commonsense Reasoning, Courant Institute of Mathematical Sciences, New York University. Accessed 20 October 2011 from http://www.cse.buffalo.edu/~shapiro/Papers/commonsense-panel.pdf

Shapiro, S. C., & Bona, J. P. (2010). The GLAIR cognitive architecture. *International Journal of Machine Consciousness*, *2*(02), 307–332. doi:10.1142/S1793843010000515

Shapiro, S. C., & Ismail, H. O. (2003). Anchoring in a grounded layered architecture with integrated reasoning. *Robotics and Autonomous Systems*, *43*(2-3), 97–108. doi:10.1016/S0921-8890(02)00352-4

Shapiro, S. C., & Rapaport, W. J. (1987). SNePS Considered as a fully intensional propositional semantic network. In N. Cercone & G. McCalla (Eds.), *The knowledge frontier: Essays in the representation of knowledge* (pp. 262–315). New York: Springer-Verlag. doi:10.1007/978-1-4612-4792-0_11

Shapiro, S. C., & Rapaport, W. J. (1991). Models and minds: Knowledge representation for natural-language competence. In R. Cummins & J. Pollock (Eds.), Philosophy and AI: Essays at the interface (pp. 215-259). Cambridge, MA: MIT Press.

Shapiro, S. C., Rapaport, W. J., Kandefer, M., Johnson, F. L., & Goldfain, A. (2007). Metacognition in SNePS. *AI Magazine, 28*(1), 17–31.

Simon, H. A. (1956). Rational choice and the structure of the environment. *Psychological Review, 63*(2), 129–138. doi:10.1037/h0042769 PMID:13310708

Simon, H. A. (1967). Motivational and emotional controls of cognition. *Psychological Review, 74*(1), 29–39. doi:10.1037/h0024127 PMID:5341441

Sloman, A. (2004). *What are emotion theories about?* Workshop on Architectures for Modeling Emotion, AAAI Spring Symposium, Stanford University.

Sloman, A. (2009). *The cognition and affect project.* Accessed 6 May 2011 from http://www.cs.bham.ac.uk/research/projects/cogaff/cogaff.html

Sloman, A., & Croucher, M. (1981). Why robots will have emotions. *Proceedings of the International Joint Conference on Artificial Intelligence.* Retrieved from http://www.cs.bham.ac.uk/research/projects/cogaff/81-95.html#36

Smith, B. C. (1982). Linguistic and computational semantics. In *Proceedings of the 20th Annual Meeting of the Association for Computational Linguistics* (pp. 9-15). Morristown, NJ: Association for Computational Linguistics. doi:10.3115/981251.981254

Smith, B. C. (1985). Limits of correctness in computers. ACM SIGCAS Computers and Society, 14-15(1-4), 18-26.

Smith, B. C. (2001). True grid. In D. Montello (Ed.), *Spatial Information Theory: Foundations of Geographic Information Science, Proceedings of COSIT 2001, Morro Bay, CA, September 2001* (pp. 14-27). Berlin: Springer.

Smith, B. C. (2010). Introduction. In B. C. Smith (Ed.), *Age of Significance.* Cambridge, MA: MIT Press. Retrieved from http://www.ageofsignificance.org/aos/en/aos-v1c0.html

Soare, R. I. (2009). Turing oracle machines, online computing, and three displacements in computability theory. *Annals of Pure and Applied Logic, 160*(3), 368–399. doi:10.1016/j.apal.2009.01.008

Srihari, R. K., & Rapaport, W. J. (1989). Extracting visual information from text: Using captions to label human faces in newspaper photographs. In *Proceedings of the 11th Annual Conference of the Cognitive Science Society* (pp. 364-371). Hillsdale, NJ: Lawrence Erlbaum Associates.

Srihari, R. K., & Rapaport, W. J. (1990). Combining linguistic and pictorial information: Using captions to interpret newspaper photographs. In D. Kumar (Ed.), Current Trends in SNePS—Semantic Network Processing System (pp. 85-96). Berlin: Springer-Verlag.

Strawson, G. (2010). *Mental reality* (2nd ed.). Cambridge, MA: MIT Press.

Tarski, A. (1944). The semantic conception of truth and the foundations of semantics. *Philosophy and Phenomenological Research, 4*(3), 341–376. doi:10.2307/2102968

Tenenbaum, A. M., & Augenstein, M. J. (1981). *Data structures using Pascal.* Englewood Cliffs, NJ: Prentice-Hall.

Thagard, P. (2006). *Hot thought: Mechanisms and applications of emotional cognition.* Cambridge, MA: MIT Press.

Toussaint, G. (1993). A new look at Euclid's second proposition. *The Mathematical Intelligencer, 15*(3), 12–23. doi:10.1007/BF03024252

Turing, A. M. (1936). On computable numbers, with an application to the *Entscheidungsproblem. Proceedings of the London Mathematical Society, Ser. 2, 42*, 230–265.

Turing, A. M. (1950). Computing machinery and intelligence. *Mind, 59*(236), 433–460. doi:10.1093/mind/LIX.236.433

Van Gelder, T. (1995). What might cognition be, if not computation? *The Journal of Philosophy, 92*(7), 345–381. doi:10.2307/2941061

Vera, A. H., & Simon, H. A. (1993). Situated action: A symbolic interpretation. *Cognitive Science, 17*(1), 7–48. doi:10.1207/s15516709cog1701_2

Wegner, P. (1997). Why interaction is more powerful than algorithms. *Communications of the ACM, 40*(5), 80–91. doi:10.1145/253769.253801

Weizenbaum, J. (1976). *Computer power and human reason: From judgment to calculation.* New York: W. H. Freeman.

Widdowson, H. G. (2004). *Text, Context, Pretext.* Malden, MA: Blackwell. doi:10.1002/9780470758427

Williams, B. (1998). The end of explanation? *The New York Review of Books, 45*(18), 40-44.

Winograd, T., & Flores, F. (1987). *Understanding computers and cognition: A new foundation for design.* Reading, MA: Addison-Wesley.

Woods, W. A. (2010). The right tools: Reflections on computation and language. *Computational Linguistics, 36*(4), 601–630. doi:10.1162/coli_a_00018

Wright, I., Sloman, A., & Beaudoin, L. (1996). Towards a design-based analysis of emotional episodes. *Philosophy, Psychiatry, & Psychology, 3*(2), 101–126. doi:10.1353/ppp.1996.0022

Wu, H. H. (2011). *Understanding numbers in Elementary school mathematics.* Providence, RI: American Mathematical Society. doi:10.1090/mbk/079

Zobrist, A. L. (2000). Computer games: Traditional. In A. Ralston, E. D. Reilly, & D. Hemmendinger (Eds.), Encyclopedia of Computer Science (4th ed.; pp. 364-368). New York: Grove's Dictionaries.

ENDNOTES

[1]   Although I take full responsibility for this essay, I am also using it as an opportunity to publicize two arguments for the claim that computers are semiotic systems. These arguments were originally formulated by my former students Albert Goldfain and Lorenzo Incardona. I am grateful to both of them for many discussions on the topics discussed here.

[2]   Throughout this essay, all italics in quotes are in the original, unless otherwise noted.

[3]   The authors quoted here include most of the principal philosophers who have written on computationalism. There are, of course, many researchers in both symbolic and connectionist AI who, without necessarily taking explicit philosophical positions on computationalism, are computationalists of one stripe or another. These include, according to one anonymous referee, Barsalou, Grossberg, Kosslyn, Kozma, Perlovsky, and Rocha, all of whom do computational modeling of (human) cognition. One must be cautious, however; Perlovsky, in particular, misunderstands at least one important logical claim: Contrary to what he consistently says in several of his writings (e.g., Perlovsky, 2007, p. 123), Gödel *never* "proved inconsistency of logic." Rather, Gödel proved, roughly, that *if* a formal system consisting of first-order logic together with Peano's axioms for the natural numbers was consistent, then it was incomplete. Thus, such a system of "arithmetic" (which contains, but is not the same as, "logic") would only be inconsistent if it were complete. The usual interpretation of Gödel is that arithmetic *is* (in all likelihood) consistent; hence, it is incomplete. See note 49 for further discussion. See Franzén 2005 to be disabused of misinterpretations of Gödel.

[4]   See the rest of their Ch. 1, "Images and Plans" for some caveats. E.g., their endnote 12 (p. 16) says: "comparing the sequence of operations executed by an organism and by a properly programmed computer is quite different from comparing computers with brains, or electrical relays with synapses."

[5]   Except when quoting or in bibliographic references, I use single quotes to form names of expressions, and I use double quotes as "scare quotes" or to mark quoted passages.

[6]   By 'legal,' I mean values in the domain of the function. So a partial function (i.e., one that is undefined on some set of values) would be such that those values on which it is undefined are "illegal." The notion of "legal" inputs makes sense in a mathematical context, perhaps less so in a biological one (Goldfain, personal communication, 2011). Biologically, presumably, all values are "legal," except that some of them are filtered out by our senses or produce unpredictable behavior.

[7]   Or, as Randall R. Dipert pointed out to me (personal communication, 2011), we might be able to understand only those aspects of cognition that *are* computable.

[8]   Notably, J.R. Lucas (1961), the ever-pessimistic Hubert Dreyfus (1965, 1972, 1979, 1992), John Searle (1980), Roger Penrose (1989), Terry Winograd & Fernando Flores (1987), to some extent Joseph Weizenbaum (1976, esp. Chs. 7,8), and even Putnam (1988) and Fodor (2000) themselves!

[9]   The "strong" and "weak" views are, perhaps, close to, though a bit different from, what Stuart C. Shapiro and I have called "computational psychology" and "computational philosophy," respectively; see Shapiro 1992, Rapaport 2003a.

[10]   Cf. also these passages from Newell, Shaw, & Simon 1958:

The theory [of human problem solving] postulates…[a] number of *primitive
information processes*, which operate on the information in the memories. Each
primitive process is a perfectly definite operation for which known physical
mechanisms exist. (The mechanisms are not necessarily known to exist in the

human brain, however—we are only concerned that the processes be described without ambiguity) (p. 151, their emphasis).

The parenthetical phrase can be given the "computable" reading. But I hear the stronger, "computational" reading in the next passage:

[O]   ur theory of problem solving…shows specifically and in detail how the
       processes that occur in human problem solving can be compounded out of
       elementary information processes, and hence how they can be carried out by
       mechanism (p. 152).

I admit that the 'can be' weakens the "computational" reading to the "computable" reading.

[11]   Here, 'function' is to be taken in the mathematical sense explicated in note 42, below, rather than in the sense of an activity or purpose. On the computational theory, the brain performs certain "purposeful functions" by computing certain "mathematical functions." Anti-computationalists say that the brain does not perform its "purposeful functions" by computing mathematical ones, but in some other way. Yet, I claim, those purposeful functions might be accomplished in a computational way. A computer programmed to compute those mathematical functions would thereby perform those purposeful functions.

[12]   Here, however, there might be a conflation of two senses of 'recursive': (a) as synonymous with 'computable' and (b) as synonymous with "inductive." See Soare 2009.

[13]   See an ad for a (human) "computer" in *The New York Times* (2 May 1892) [http://tinyurl.com/NYT-ComputerWanted].

[14]   An anonymous referee commented, "These views are popular, but it does not make them true. They are wrong, (1) there is evidence that syntax can evolve as a secondary feature in language (see Brighton [2002], Kirby [2000], others), (2) reclusiveness [*sic*] never was demonstrated to be useful, it follows from hierarchy, which is much more useful, (3) semantics refers to objects and events in the world, there can be no semantics inside a semiotic system, even if many people misunderstood semiotics this way." As will become clearer 2, by 'syntacticism' here I do not mean *grammatical* syntax, as the referee seems to think I do; hence, point (1), though of independent interest, is irrelevant. I am at a loss as to what to say about point (2), other than that I am not referring to *recursiveness* in language, but to a recursive feature of *understanding*, as will be seen. Finally, point (3) is simply false; semantics can "refer" to *mental* concepts as well as *worldly* objects and events. And at least one semiotician has remarked to me (personal communication) that many semioticians "misunderstand" semiotics in exactly that way: that talk about "objects and events in the world" is simplistic, if not blasphemous!

[15]   "[S]o long as we are thinking of mental processes as purely computational, the bearing of environmental information upon such processes is exhausted by the formal character of whatever the oracles ["analogs to the senses"] write on the tape [of a Turing machine]. In particular, it doesn't matter to such processes whether what the oracles write is *true*…. [T]he formality condition…is tantamount to a sort of methodological solipsism." (Fodor 1980: 65.)

[16]   "According to individualism about the mind, the mental natures of all a person's or animal's mental states (and events) are such that there is no necessary or deep individuative relation between the individual's being in states of those kinds and the nature of the individual's physical or social environments" (Burge 1986: 3–4).

[17]   An anonymous referee commented about this quote, "again, wrong, despite many famous people could be misinterpreted this way [*sic*]. Interaction between an organism and the world is a *process*.

In this process there are (1) intuition[s] about the world, (2) predictions and (3) confirmations or disconfirmations of these predictions. The same process is the process of science." I agree that organism-world interaction is a process. Pylyshyn's point, however, is that the *output* of that process is the *only* input to our cognitive system. Hence, intuitions, predictions, and (dis-)confirmations are all internal.

[18] Strict adherence to methodological solipsism would seem to require that LOT have syntax but no semantics. Fodor (2008: 16) suggests that it needs a purely referential semantics. I have proposed instead a Meinongian semantics for LOT, on the grounds that "non-existent" objects are best construed as internal mental entities; see Rapaport 1978, 1979, 1981, and, especially, 1985/1986.

[19] The terms (or "marks") of this LOT (e.g., nodes of a semantic network; terms and predicates of a language; or their biological analogues, etc.) need not all be alike, either in "shape" or function. E.g., natural languages use a wide variety of letters, numerals, etc.; neurons include afferent, internal, and efferent ones (and the former do much of the internalization or "pushing"). (Albert Goldfain, personal communication.) Moreover, some of the internal marks might well include Barsalou's (1999) "perceptual symbols."

[20] When I use 'information' (except when quoting), I am using it in a sense that is neutral among several different meanings it can have; in particular, I do not necessarily intend the Shannon theory of information. For further discussion, see Piccinini & Scarantino 2011.

[21] Perception is input; so, if one wants to rule it out as an example of a cognitive phenomenon, then perhaps *outputs* of cognitive processes (e.g., motor activity, or actions more generally) should also be ruled out. This would be one way to respond to Cleland 1993; see n. 46.

[22] One anonymous referee objected that "Despite the limitations on [human] sensation, there is an analog aspect and nonarbitrariness about the [human] cognitive representations that is not shared with a computer." But a computer's representations of visual input are surely just as non-arbitrary (or, conversely, both are equally arbitrary; it's just that we're more familiar with our own) and just as "analog" as ours are; this is one of the main points of Marr's (1982) computational vision theory.

[23] [http://www.oed.com/view/Entry/196559]

[24] My use of 'mark' derives from Fetzer 2011 and will be clarified below.

[25] For this reason, Barry Smith and Werner Ceusters (personal communication) prefer to call it the '*Syntactic* Web'!

[26] I.., they are pure syntax—WJR.

[27] Although this is suggestive of semantic interpretation, it is really just syntax: A bit is not something (a sign) that is then interpreted as something else (e.g., 0 or 1). Rather, a bit *is* 0 or else it *is* 1 (or it *is* high voltage or else it *is* low voltage; or it *is* magnetized or else it *is* not magnetized; and so on).—WJR

[28] This certainly *sounds* like ordinary semantics, but the very next passage clearly indicates a syntactic approach.—WJR

[29] "Declarations" are *syntactic* parts of a computer program.

[30] For more quotes along these lines, see my Web page "Tenenbaum & Augenstein on Data, Information, & Semantics,"
[http://www.cse.buffalo.edu/~rapaport/563S05/data.html].

[31] Shapiro (personal communication, 2011) suggests that, when we stop the recursion, we may only *think* that we understand, as in the case of the sentence, "During the Renaissance, Bernini cast a bronze of a mastiff eating truffles" (Johnson-Laird, personal communication, 2003; cf. Johnson-

Laird 1983: 225, Widdowson 2004). The claim is that many people can understand this sentence without being able to precisely define any of the principal words, as long as they have even a vague idea that, e.g., the Renaissance was some period in history, 'Bernini' is someone's name, "casting a bronze" has something to do with sculpture, bronze is some kind of (perhaps yellowish) metal, a mastiff is some kind of animal (maybe a dog), and truffles are something edible (maybe a kid of mushroom, maybe a kind of chocolate candy). Still, such understanding is syntactic—it is understanding of one's internal concepts.

[32] A referee commented, "it seems that the author assumes that language and cognition is the same thing.—this should be clarified. Syntax is a set of relations among signs. But how signs are related to external objects and situations—not a word so far." However, in these sections, I do not make this assumption. Moreover, a large part of the entire essay concerns how semantics in the sense of the relation of signs to external objects and situations can be handled by syntax.

[33] Crane 1990 also points out that shape alone is not sufficient for syntax (as he interprets Fodor 1980 as holding). But Crane uses 'syntax' in the narrow sense of "grammar"; he is correct that a sentence printed in all capital letters has a different shape from—but the same (grammatical) syntax as—the same sentence printed normally. But, as I use the term, although shape does not *suffice* for syntax, it is surely part of it.

[34] The notion of "interpretant" will be clarified below.

[35] I owe the observations in this paragraph to Incardona, personal communication, 2010.

[36] I am indebted to Incardona for directing me to both this article by Peirce and an interpretation by Kenneth Laine Ketner (1988, see esp. pp. 34, 46, 49). I have two problems with Ketner's interpretation, however: (1) Ketner (p. 49) cites Peirce's distinction between "corollarial" reasoning (reading off a conclusion of an argument from a diagram of the premises) and "theorematic" reasoning (in which the reasoner must creatively *add* something to the premises). But neither Peirce nor Ketner offers any arguments that theorematic reasoning cannot be reduced to corollarial reasoning. If corollarial reasoning can be interpreted as referring to ordinary arguments with all premises explicitly stated, and theorematic reasoning can be interpreted as referring to arguments (viz., enthymemes) where a missing premise has to be supplied by the reasoner as "background knowledge," then I would argue that the latter can be reduced to the former (see Rapaport & Kibby 2010). However, there are other interpretations (see Dipert 1984).

(2) Ketner seems to misunderstand the Church-Turing Thesis (pp. 51–52). He presents it as stating that computers can only do what they have been programmed to do (or "calculated to do," to use Peirce's phrase). But what it really asserts is that the *informal* notion of "algorithm" can be *identified* with the *formal* notions of Church's lambda calculus or Turing's machines. Ketner also seems to misunderstand the import of Turing's proof of the existence of non-computable functions: He thinks that this shows "that mathematical method is not universally deterministic" (p. 57). But what it really shows is that Hilbert's decision problem (viz., for any mathematical proposition P, is there an algorithm that decides whether P is a theorem?) must be answered in the negative. Unfortunately, in both cases, Ketner cites as an authority a logic text (Schagrin et al. 1985: 304–305) co-authored by me! Granted, my co-authors and I use Church's Thesis to justify a claim that "Computers can perform only what algorithms describe"—which *sounds like* "computers can only do what they have been programmed to do"—but our intent was to point out that, *if* a computer can perform a task, *then* that task can be described by an algorithm. Ketner's sound-alike claim is usually taken to mean that computers can't do anything other than what they were explicitly programmed to do

(e.g., they can't show creativity or initiative). But computers that can learn can certainly do things that they weren't explicitly programmed to do; the Church-Turing thesis claims that anything that they *can* do—including those things that they learned how to do—must be computable in the technical sense of computable in the lambda calculus or by a Turing machine.

[37] Dipert (1984: 59) suggests that Peirce might also have anticipated the Church-Turing Thesis.

[38] Again, thanks to Incardona, whose dissertation explores these themes.

[39] One anonymous referee commented, "[T]he mark's user is always the same as the interpretant… so the user is the same as the interpretant." This is a misreading, either of me or of Peirce. I am not using 'interpretant' to be synonymous with 'mark user.' Rather, I am following Peirce's use, roughly to the effect that an interpretant is a sign (or sign-like entity) *in* the user's mind (see quotes above).

[40] "That determination of which the immediate cause, or determinant, is the Sign, and of which the mediate cause is the Object may be termed the Interpretant…." (Peirce 1909, "Some Amazing Mazes, Fourth Curiosity," CP 6.347,
[http://www.helsinki.fi/science/commens/terms/interpretant.html].)

[41] "I call a representamen which is determined by another representamen, an interpretant of the latter." (Peirce 1903, *Harvard Lectures on Pragmatism*, CP 5.138, [http://www.helsinki.fi/science/commens/terms/interpretant.html].)

[42] Mathematically a binary *relation* defined on the Cartesian product of two sets A and B (i.e., defined on $A \times B$) is a set of ordered pairs ("input-output" pairs), where the first member comes from A and the second from B. A *function* is a binary relation f on $A \times B$ ("from A to B") such that any given member a of A is related by f to (at most) a unique b in B. I.e., if $(a, b_1) \in f$ and $(a, b_2) \in f$ (where $a \in A$ and $b_1, b_2 \in B$), then $b_1 = b_2$. Conversely, if $b_1 \neq b_2$, then f cannot map (or interpret) a as *both* b_1 *and* b_2 simultaneously. I.e., no *two* outputs have the same input. If the outputs are thought of as the meanings or "interpretations" (i.e., *interpretants*) of the inputs, then an interpretation *function* cannot allow for ambiguity.

[43] This question was raised by an anonymous referee.

[44] For further discussion of the relation of computers to semiotic systems, see Andersen 1992, who argues that "computer systems" are (merely) "sign-vehicles whose main function is to be perceived and interpreted by some group of users"; Nöth 2003, who argues that "none of the criteria of semiosis is completely absent from the world of machines"; and Nadin (2007, 2010), who seems to agree that "computers are semiotic machines." Other sources are cited in McGee 2011.

[45] In Rapaport 1999, 2005b, I argue that implementation as semantic interpretation is also a ternary relation: Something is (1) a (concrete or abstract) implementation (i.e., a semantic "interpretant") of (2) an abstraction in (3) a (concrete or abstract) medium.

[46] One anonymous referee said: "I disagree here: The output of a process (algorithm) need not be caused by a certain input. It could have been caused by something else." This is technically true: The output of a constant function is not causally or algorithmically related to any input. But in all other cases an algorithm's output *is* a function of its input. The referee continues: "For instance, me going to eat (output) is not necessarily an interpretant of my being hungry. It could be because I have an appointment with a friend to go out eating. Likewise, a given input can yield different outputs. When hungry, I may decide to eat, but also may decide to continue working." I agree about hunger and eating, but the relation between being hungry and going out to eat is not algorithmic; this is not sign interpretation. The referee continues: "what I mean to say is that both input and output—and

past experiences—are so tightly linked that the link/process should be part of the interpretant." If 'interpretant' means 'mark user,' then I agree that the "process should be part of" the user (or the computer), but this makes no sense on the Peircean characterization of 'interpretant.' Finally, the referee says, "the context (input) and process are crucial to meaning formation." I quite agree that all *three* of these are crucial, but it is important to note that context and input are *two* distinct things. A context is not input to an algorithm; rather, it is the environment in which the algorithm is executed. Elaboration of this would take us too far afield; the interested reader should consult Smith 1985, Cleland 1993, Soare 2009 for further discussion of the relationship between context and input.

[47]   Related to the notion of a semiotic system is that of a *symbol* system, as described by Newell & Simon 1976:116:

A physical symbol system consists of a set of entities, called symbols, which are physical patterns that can occur as components of another type of entity called an expression (or symbol structure). Thus, a symbol structure is composed of a number of instances (or tokens) of symbols related in some physical way (such as one token being next to another). At any instant of time the system will contain a collection of these symbol structures. Besides these structures, the system also contains a collection of processes that operate on expressions to produce other expressions: processes of creation, modification, reproduction and destruction. A physical symbol system is a machine that produces through time an evolving collection of symbol structures. Such a system exists in a world of objects wider than just these symbolic expressions themselves. ...An expression designates an object if, given the expression, the system can either affect the object itself or behave in ways dependent on the object. ...The system can interpret an expression if the expression designates a process and if, given the expression, the system can carry out the process.

In short, a physical symbol system is a computer.

[48]   "...in order to be able to refer separately to the general category of possible count tags and the subset of such tags which constitute the traditional count words[, w]e call the former *numerons*; the latter *numerlogs*. Numerons are any distinct and arbitrary tags that a mind (human or nonhuman) uses in enumerating a set of objects. Numerlogs are the count words of a language." (Gelman & Gallistel 1986: 76–77.)

[49]   A referee suggested that the views of Pattee, 1969, 1982; Rosen, 1987; and Cariani, 2001 may also be relevant, though I have my doubts. In any case, detailed discussion of their views is beyond the scope of the present essay, so a few words will have to suffice. As I emphasized earlier, I am not seeking to understand the (human) *brain* (as they are); I am seeking to understand *mind*, and not only the *human* mind, but mind more generally and abstractly, such that both cognitive computers and humans can be said to have minds. This requires an emphasis on symbols rather than matter, but must be accompanied by a study of implementation (or realization)—the ways in which symbols can be "grounded"—and its converse (as Pattee notes). For my thoughts on Harnad's (1990) theory of symbol grounding, in the context of syntactic semantics, see Rapaport 1995. For my thoughts on the relation between implementation and semantics, see Rapaport 1999, 2005b. Briefly, as I read Pattee 1969, he supports what I call syntacticism, and his 1982 notion of semantic closure seems to describe a symbolic (i.e., syntactic) system whose semantic interpretation is the system itself—but that would make it fully syntactic. Rosen offers two reasons why syntax is not sufficient for semantics: The first is a variation on the Lucas-Penrose argument from Gödel's Incompleteness Theorem. This argument is highly debatable, but one simple response in the present context is to

note that it might well be irrelevant: Gödel's Incompleteness Theorem does not say that there are true but unprovable statements of arithmetic (where truth is a matter of semantics and provability is a matter of syntax; see note 3 for what is meant by 'arithmetic'). What it says, rather, is that there is an arithmetic statement S such that neither S nor $\neg S$ can be *proved*. Semantics and truth do not have to enter the picture. Of course, because S can be interpreted as *saying* that S is unprovable, and because Gödel's Incompleteness Theorem proves that, in fact, S *is* unprovable, it follows by a trivial conception of "truth" that S is true. Rosen's second reason is that biological systems are "complex" and hence their "behavior...cannot be fully captured by any syntactic means" (Rosen 1987: 14). But this depends on unprovable, metaphorical analogies between syntax and physical reductionism and between causation and implication. Moreover, Rosen (1987: 2) misunderstands Kleene when he disapprovingly quotes Kleene as saying that, in formal, axiomatic systems, meanings are "left out of account." What Kleene means is, not to *eliminate* meaning altogether, but that the goal of formal axiomatics is to *capture* all meaning syntactically, i.e., to remove it from the *symbols* (i.e., to turn the symbols into "marks," as I used the term above) and to move it (by explicitly re-encoding it) to the (syntactic) axioms. Finally, although Cariani (2001, p. 71) says, "One cannot replace semantics with syntactics," in fact one *can*, *if* the "world states" (Cariani, 2001, p. 70) are *also* symbolized; this is my point about internalis. Oddly, Cariani (2001, p. 77n.13) also says, "Contra Fodor and Putnam, meaning can and does lie in the head." I quite agree; that's part of syntacticism! Fodor 1980 also agrees about the location of meaning, so I fail to understand why Cariani places Fodor and Putnam in the same camp. As Incardona (personal communication, 2011) observed, the real issue might be that I need to explain more clearly than I try above: "how syntacticism justifies the relationship between cognitive agents and the world. [But] internalism does not mean that cognitive agents cannot transform the 'outside world.'" Readers who want such an explanation from me should read my papers on this theme.

50   These ideas were first suggested to me by Shoshana Hardt Loeb (personal communication, ca. 1983.)

51   See discussion and citations in Edelman 2011. Edelman says, "The *phenomenal experience* that arises from th[e] dynamics [of the anatomy and physiology of dreaming] is that of the dream self, situated in a dream world" (p. 3, my italics). So Fetzer could be charitably interpreted as meaning, not that dreams (hence minds) are not computable, but that *dream phenomenology* is not computable, i.e., that the "hard problem" of consciousness is, indeed, hard. But this raises a host of other issues that go beyond the scope of this paper. For some relevant remarks, see Rapaport 2005b.

52   But *simulated* animals *simulatedly* die!—WJR

53   But see my comments above about simulated hurricanes!—WJR

54   Cf. IBM's *Jeopardy*-winning Watson.—WJR

55   See my Web page "What Is Hypercomputation?" for a partial bibliography, [http://www.cse.buffalo.edu/~rapaport/584/hypercompn.html]

APPENDIX

What Is an Algorithm?

Before anyone attempted to define 'algorithm,' many algorithms were in use both by mathematicians as well as by ordinary people (e.g., Euclid's procedures for construction of geometric objects by compass and straightedge—see Toussaint, 1993; Euclid's algorithm for computing the GCD of two integers; the algorithms for simple arithmetic with Hindu-Arabic numerals—see Crossley & Henry, 1990, Wu, 2011; cf. Knuth, 1972). When David Hilbert investigated the foundations of mathematics, his followers began to try to make the notion of algorithm precise, beginning with discussions of "effectively calculable," a phrase first used by Jacques Herbrand in 1931 (Gandy, 1988, p. 68) and later taken up by Church (1936) and Stephen Kleene (1952), but left largely undefined, at least in print.

J. Barkley Rosser (1939, p. 55) made an effort to clarify the contribution of the modifier "effective" (italics and enumeration mine):

"Effective method" is here used in the rather special sense of a method each step of which is [1] precisely predetermined and which is [2] certain to produce the answer [3] in a finite number of steps.

But what, exactly, does 'precisely predetermined' mean? And does 'finite number of steps' mean that the written statement of the algorithm has a finite number of instructions, or that, when executing them, only a finite number of tasks must be performed? (What gets counted: written steps or executed instructions? One written step—"for $i:= 1$ to 100 do $x:= x + 1$"—can result in 100 executed instructions.) Much later, *after* Turing's, Church's, Gödel's, and Post's precise formulations and *during* the age of computers and computer programming, A.A. Markov, Kleene, and Donald Knuth also gave slightly less vague, though still informal, characterizations.

According to Markov (1954/1960, p. 1), an algorithm is a "computational process" satisfying three (informal) properties: (1) being "determined" ("carried out according to a precise prescription...leaving no possibility of arbitrary choice, and in the known sense generally understood"), (2) having "applicability" ("The possibility of starting from original given objects which can vary within known limits"), and (3) having "effectiveness" ("The tendency of the algorithm to obtain a certain result, finally obtained for appropriate original given objects"). These are a bit obscure: Being "determined" may be akin to Rosser's "precisely predetermined." But what about being "applicable"? Perhaps this simply means that an algorithm must not be limited to converting one, specific input to an output, but must be more general. And Markov's notion of "effectiveness" seems restricted to only the second part of Rosser's notion, namely, that of "producing the answer." There is no mention of finiteness, unless that is implied by being computational.

In his undergraduate-level, logic textbook, Kleene (1967) elaborated on the notions of "effective' and "algorithm" that he had left unspecified in his earlier, classic, graduate-level treatise on metamathematics. He continues to identify "effective procedure" with "algorithm" (Kleene, 1967, p. 231), but now he offers a characterization of an algorithm as (1) a "procedure" (i.e., a "finite" "set of rules or instructions") that (2) "in a finite number of steps" answers a question, where (3) each instruction can be "followed" "mechanically, like robots; no insight or ingenuity or invention is required," (4) each instruction "tell[s] us what to do next," and (5) the algorithm "enable[s] us to recognize when the steps come to an end"

(Kleene, 1967, p. 223). Knuth (1973, pp. 1-9) goes into considerably more detail, albeit still informally. He says that an algorithm is "a finite set of rules which gives a sequence of operations for solving a specific type of problem," with "five important features" (Knuth, 1973, p. 4):

1. *"Finiteness.* An algorithm must always terminate after a finite number of steps" (Knuth, 1973, p. 4).

    a. Note the double finiteness: A finite number of rules in the text of the algorithm *and* a finite number of steps to be carried out. Moreover, algorithms must halt. (Halting is not guaranteed by finiteness; see point 5, below.) Interestingly, Knuth also says that an algorithm is a finite "computational method," where a "computational method" is a "procedure" that only has the next four features (Knuth, 1973, p. 4).

2. *"Definiteness.* Each step…must be precisely defined; the actions to be carried out must be rigorously and unambiguously specified…" (Knuth, 1973, p. 5).

    a. This seems to be Knuth's analogue of the "precision" that Rosser and Markov mention.

3. *"Input.* An algorithm has zero or more inputs" (Knuth, 1973, p. 5).

    a. Curiously, only Knuth and Markov seem to mention this explicitly, with Markov's "applicability" property suggesting that there must be at least *one* input. Why does Knuth say *zero* or more? Presumably, he wants to allow for the possibility of a program that simply outputs some information. On the other hand, if algorithms are procedures for computing functions, and if functions are "regular" sets of input-output pairs (regular in the sense that the same input is always associated with the same output), then algorithms would always have to have input. Perhaps Knuth has in mind the possibility of the input being internally stored in the computer rather than having to be obtained from the external environment.

4. *"Output.* An algorithm has one or more outputs" (Knuth, 1973, p. 5).

    a. That there must be at least one output echoes Rosser's property (2) ("certain to produce the answer") and Markov's notion (3) of "effectiveness" ("a certain result"). But Knuth characterizes outputs as "quantities which have a specified relation to the inputs" (Knuth, 1973, p. 5); the "relation" would no doubt be the functional relation between inputs and outputs, but if there is no input, what kind of a relation would the output be in? Very curious!

5. *"Effectiveness.* This means that all of the operations to be performed in the algorithm must be sufficiently basic that they can in principle be done exactly and in a finite length of time by a man [*sic*] using pencil and paper" (Knuth, 1973, p. 6).

    a. Note, first, how the term 'effective' has many different meanings among all these characterizations of "algorithm," ranging from it being an unexplained term, through being synonymous with 'algorithm,' to naming very particular—and very different—properties of algorithms. Second, it is not clear how Knuth's notion of effectiveness differs from his notion of definiteness; both seem to have to do with the preciseness of the operations. Third, Knuth brings in another notion of finiteness: finiteness in time. Note that an instruction to carry out an infinite sequence of steps in a finite time could be accomplished in principle by doing each step twice as fast as the previous step; or each step might only take a finite amount of time, but the number of steps required might take longer than the expected life of the universe (as in computing a perfect, non-losing strategy in chess; see §6.1, above). These may have interesting theoretical implications (see the vast literature on hypercomputation),[55] but do not seem very practical.

Knuth (1973, p. 7) observes that "we want *good* algorithms in some loosely-defined aesthetic sense. One criterion of goodness is the length of time taken to perform the algorithm...." Finally, the "gold standard" of "a [hu]man using pencil and paper" seems clearly to be an allusion to Turing's (1936) analysis.

We can summarize these informal observations as follows: An algorithm (for executor E to accomplish goal G) is:

1. A procedure (or "method")—i.e., a finite set (or sequence) of statements (or rules, or instructions)—such that each statement is:
    a. Composed of a finite number of symbols (or marks) from a finite alphabet
    b. And unambiguous for E—i.e.,
    c. E knows how to do it
    d. E can do it
    e. It can be done in a finite amount of time
    f. And, after doing it, E knows what to do next—
2. And the procedure takes a finite amount of time, i.e., halts,
3. And it ends with G accomplished.

But the important thing to note is that the more one tries to make precise these *informal* requirements for something to be an algorithm, the more one recapitulates Turing's motivation for the formulation of a Turing machine! Turing (1936) describes in excruciating detail what the minimal requirements are for a human to compute:

[T]he computation is carried out on one-dimensional paper, i.e. on a tape divided into squares. ...The behaviour of the computer [i.e., the human who computes!—WJR] at any moment is determined by the symbols which he [sic] is observing, and his "state of mind" at that moment. ...[T]here is a bound B to the number of symbols or squares which the computer can observe at one moment. ...[A]lso...the number of states of mind which need be taken into account is finite. ...[T]he operations performed by the computer...[are] split up into "simple operations" which are so elementary that it is not easy to imagine them further divided. Every such operation consists of some change of the physical system consisting of the computer and his tape. We know the state of the system if we know the sequence of symbols on the tape, which of these are observed by the computer..., and the state of mind of the computer. ...[I]n a simple operation not more than one symbol is altered. ...[T]he squares whose symbols are changed are always "observed" squares. ...[T]he simple operations must include changes of distribution of observed squares. ...[E]ach of the new observed squares is within L squares of an immediately previously observed square. ...The most general single operation must therefore be taken to be one of the following: (A) A possible change...of symbol together with a possible change of state of mind. (B) A possible change...of observed squares, together with a possible change of mind. The operation actually performed is determined... by the state of mind of the computer and the observed symbols. In particular, they determine the state of mind of the computer after the operation is carried out. We may now construct a machine to do the work of this computer" (Turing, 1936, pp. 249-251).

The "machine," of course, is what is now known as a "Turing machine"; it "does the work of this" *human* computer. Hence, my claim, above that the Turing machine was the first AI program. (For further discussion of the role of the goal of an algorithm, see Rapaport 2015. For further discussion of what an algorithm is, see Rapaport 2017b.)

Chapter 8

Signs Conveying Information:
On the Range of Peirce's Notion of Propositions – Dicisigns

Frederik Stjernfelt
Aarhus University, Denmark

ABSTRACT

This chapter introduces Peirce's notion of proposition, "Dicisign." It goes through its main characteristics and argues that its strengths have been overlooked. It does not fall prey to some of the problems in the received notion of propositions (their dependence upon language, upon compositionality, upon human intention). This implies that the extension of Peircean Dicisigns is wider in two respects: they comprise 1) propositions not or only partially linguistic, using in addition gesture, picture, diagrams, etc.; 2) non-human propositions in biology studied by biosemiotics.

INTRODUCTION

Peirce's notion of "Dicisigns" has led a strangely silent life in Peirce's reception. It is, of course, Peirce's notion for propositions, and the central place of that notion in the development of 20th century logic and analytical philosophy probably leading many Peirce scholars to presume that Peirce's notion was merely a forerunner to that development, lacking any intrinsic interest. This is not true, and the goal of this paper is to give an overview over Peirce's notion of Dicisign as well as to highlight those aspects of it which differ form the received notion of propositions in logic and philosophy. Peirce's notion of Dicisign includes logical propositions, and is closely related to Peirce's many discoveries in logic—but due to Peirce's semiotic approach to logic, he is not only, like the logical mainstream, interested in the formalization of propositions and their structures, but also takes a crucial interest in which sign types may carry propositions. This is why Peirce's Dicisign transgresses the narrowly logical notion of propositions in at least two respects.

DOI: 10.4018/978-1-5225-5622-0.ch008

One is that a Peircean Dicisign need not be expressed in language, ordinary or formal. A Dicisign may involve gestures, pictures, diagrams only, or it may involve such devices in combination with language. Thus, the notion of Dicisigns covers a much larger range of human semiotic activity than ordinarily conceived of in the notion of propositions (it is true that most often, logic does not consider that range, focusing instead on logical properties of propositions and propositional content, however expressed)—it gives a much broader idea of which human activities implies claiming something to be the case. The other extension in the notion of Dicisigns as compared to standard conceptions of propositions is that Dicisigns, not being confined to language, also cover animal communication and lower biological sign use studied by biosemiotics. This should not come as a great surprise: communication of any sort couldn't possibly reach any high degree of efficiency if it is not able to indicate things to be the case, which is the central property of Dicisigns. Many of the biological cases, however, do not imply that the signs used correspond to conscious, deliberate claims on the part of a communicating or signifying biological agent—which is probably why many scholars immediately refrain from considering the possibility of Dicisigns in other species.

DICISIGNS

Let's begin by taking a look at the central properties of Peirce's Dicisign doctrine.[1] A basic way of describing Dicisigns is that they are signs which may be ascribed a truth value (Syllabus, 1903; partly reprinted in Peirce, 1998). This is because Dicisigns claim something about something—and this claim may be true or false (or meaningless). The reason why Dicisigns are thus able to claim something is that they have a double structure: they 1) point out an object, and they 2) describe that object in some way. They possess, in some way, a Subject-Predicate structure. The formal part of this analysis of a proposition is largely shared by Frege and Peirce. Frege famously analyzed propositions as consisting of two parts, functions and arguments (corresponding to Predicates and Subjects, respectively), functions indicating properties of the variables indicated by the arguments, and often indicated as $F(a)$—meaning a satisfies the function F; a has the property F. Frege's path-breaking insight was that functions need not have one argument only—properties are not only properties of one object, many properties are relationally shared between objects. Peirce's independent analysis of the Dicisign is quite parallel: a proposition consists of 1) a Predicate, in his terms, an iconic rhema (corresponding to Frege's function) and 2) one or more Subjects, in his terms, indices referring to which objects are claimed to possess the property described by the iconic rhema (corresponding to Frege's arguments). Frege and Peirce count as the founders of relational logic; both of them insist that one and the same rhema (function) may take several indices (arguments). Thus, the rhema "loves" may take two indices: "Peter loves Mary," the rhema "gives" may take three, "Mary gives Peter a rejection." Moreover, both Frege and Peirce invent quantification to give different recipes for the selection of the objects referred to by the indices of a Dicisign. One is existential quantification: "Some berries are red," another is universal quantification "All fruits are colored"—and possibly other quantifiers indicating other modes of selecting objects considered in the Dicisign. Frege and Peirce did not know about each others' discoveries, and it is well known how Frege enjoyed priority in these discoveries because he published them some years before Peirce. It is less well known that the current logical formalism, sometimes referred to as the Peano-Russell formalism, stems from Peirce (via the German philosopher Ernst Schröder) rather than from Frege. Peirce invented the basics of that formalism in his "On the Algebra of Logic" (1885), while Frege's earlier diagrammatical formalism (in the *Begriffsschrift* of 1879) was too cumbersome to be put to common usage.[2]

Thus, there is a large overlap between the basic Fregean and Peircean analyses of the structure of propositions. What differs lies rather in the more or less unspoken assumptions about where and how propositions occur. The Fregean tradition in logic and analytical philosophy (not necessarily Frege himself; the *Begriffsschrift* did indeed present a system for non-linguistic graphical logic representation) has, as a tendency, presumed two things: 1) that propositions must be linguistically expressed, in ordinary language or special, formalized languages, 2) that propositions, in order to be expressed, require a speaking subject taking a "propositional attitude" or a "propositional stance." These two assumptions focus, of course, on the ubiquitous example of propositions expressed by linguistically able human beings. The analytical tradition, having focused more upon the truth-preserving and formal properties of propositions, most often kept such assumptions in the background—but still these ideas form part of what Barry Smith (2005) ridicules as "Fantology," the mistaken doctrine resulting from taking the structure of First Order Predicate Logic (FOPL) as the sole guideline for what should be taken to be the basic entities of ontology. Given the success of FOPL, the Kantian step of reading metaphysical structure directly from logical structure may seem tempting—but then it is important to realize what does indeed belong to logical structure itself, and what are merely artifices of formalization. Smith nicknames the fallacy of disregarding this cautiousness "Fantology" after the F(*a*) structure charted by Frege—"F(*a*)ntology"—and he claims that FOPL formalism has led analytical metaphysics to assume that all generality, not only within the formalism, but also in the world, lies in the predicate and never in the subject. Thus, all existing objects are taken to be particulars, referred to by the *a* (or by the variables *x*'s, *y*'s, and *z*'s of more elaborated formulae), and they may, in turn, share different general properties, referred to by the F's. This, according to Smith, gives rise to a minimalist and nominalist ontology, dispensing with all sorts of natural kinds otherwise central in science (electrons, earthquakes, animal species, revolutions, languages, etc.)—expressed, for instance, in Quine's famous claim that to exist is to be a bound variable of a true proposition, the function or predicate part of the proposition hence having nothing at all to do with real existence.

The ideas that propositions are necessarily 1) linguistic, 2) compositionally constructed from independent parts corresponding to function/argument, and 3) a prerequisite for human beings only, now form another piece of Fantology: we are led to assume these ideas from being accustomed to ordinary and formal languages. Here, Peirce's more general notion of Dicisign, even if formally analogous to Frege's, escapes all of these fantological dangers. Peircean Dicisigns are not confined to be expressed linguistically, they do possess the two aspects of Subject-Predicate but are not for that reason rendered compositional, and they do not require human beings nor indeed conscious intention to appear. These non-fantological properties of Peirce's definition of the Dicisign appear in his basic semiotic description of it: "Every assertion is an assertion that two different signs have the same object" (Short Logic, ca. 1893, CP 2.437)[3]. The Dicisign claims that two signs have the same object—and the two signs form parts or aspects of the Dicisign itself. The two signs are the indexical object reference and the iconic object description, respectively. As Hilpinen says, this notion corresponds to William of Ockham's idea that a proposition is true if its subject and predicate "supposit for the same thing" (Hilpinen, 1992, p. 475). One aspect of the Dicisign indicates its object, another aspect of it describes it, and the Dicisign now claims these two constituent signs have the same object. To be more precise, the iconic part of the Dicisign is not directly a description of the object; rather the iconic part is an icon of the very Dicisign itself, Peirce claims: "every proposition contains a *Subject* and a *Predicate*, the former representing (or being) an Index of the Primary Object, or Correlate of the relation represented, the latter representing (or being) an Icon of the Dicisign in some respect" (Syllabus, 1903, EPII, 279). This, at first sight, surprising doctrine is developed at length in the "Syllabus"—the idea is that the predicate aspect of the Dicisign depicts not only the object, but also depicts the descriptive relation of the Dicisign itself to the object—which is an iconic relation. The reason for this doctrine is to avoid the merely compositional idea that any accidental combination of an index and an icon of the same object could constitute a Dicisign. The icon part is taken to represent not only the description, but also the description claim (indicated by the fact that the copula, in Peirce's analysis, is included into the rheme). This is consistent with Peirce's idea that "The most perfectly thorough analyses throws the whole substance of the Dicisign into the Predicate." (Syllabus, 1903, EPII, 281) which, in turn, is motivated by his analysis that all Predicates function as verbs. In the Dicisign "The sky is blue," thus, the Subject is "the sky," and the Predicate is not only the adjective "blue," but the verbal compound "is blue" with an unsaturated slot which may be filled in by a Subject. Peirce vacillates as to whether these two aspects of the Dicisign must necessarily appear as two independent, explicitly distinguishable parts of the Dicisign: "That is to say, in order to understand the Dicisign, it must be regarded as composed of such two parts whether it be in itself so composed or not. It is difficult to see how this can be, unless it really have two such parts; but perhaps this may be possible." (ibid., 276) His analysis of photographical Dicisigns, however, seems to solve this puzzle: the P-role is here played by the shapes on the photographical plate while the S-role is played by the causal connection of these with the object, granted by the physical effects of focused light rays on the photosensitive surface, chemically or electronically—but without S and P here appearing as autonomous, separate parts of the sign. Thus, a single photograph may act as a Dicisign, provided the observer is capable of connecting it to the Subject indicated, and of understanding its shapes as Predicates of that Subject. So, S and P in general could be characterized as *aspects* of the Dicisign rather than as independent parts—even if they may, in many cases, form such parts.

PROPERTIES OF DICISIGNS

On this basis, let us list some examples of Dicisigns—with their Subject and Predicate aspects indicated by S and P, respectively:

- Linguistic utterances in the indicative: "It (S) rains (P)," "The sky (S) is blue (P)" or "He (S1) loves (P) her (S2)" or "Mary (S1) gives (P) Peter (S2) a present (S3)"
- The double gesture of pointing towards a person (S) and turning the finger to indicate he is crazy (P)
- The gesture of pointing towards a person (S) and then pointing towards a caricature intended to portray him or her (P)
- A picture (P) with the object indicated (S)—specific examples include:
- A cartoon of a well-known figure (Donald Duck, S), acting and expressing graphically represented utterings (P)
- Road signs informing of danger of deer on the road (P), the location of the sign indicating the location of the presence of this danger (S)
- A portrait (P) with a title or label indicating the person portrayed (S)

- A photograph depicting its object by means of shapes and colors on a surface (P), indicating the object by the causal connection via focused light rays (S)
- A photograph (P) with the object made explicit by an accompanying linguistic byline (S)
- A diagram (P) used for describing the structure of an object indicated by accompanying indexical symbols (S)—specific examples include:
- A topographical map (P) equipped with names (S) of countries, cities, landscape features, etc.
- A mathematical graph in a Cartesian plane (P) equipped with specifications of the axis units and a legend indicating the physical phenomenon it describes (S).

A very important upshot of such a list of examples is that linguistically complete, well-formed Dicisigns do not in any way exhaust the category. Many mixed forms, involving gestures, pictures, objects, language in different combinations, may constitute Dicisigns. If you are at a friend's door and he is not home, you may, lacking a pencil, indicate you have been there by leaving a small object you know he knows belongs to you—this object then acting as a minimal Dicisign saying "I (S1) was visiting (P) you (S2)." As to the indexical aspect of the Dicisign, we find very different types of signs realizing it. In very simple cases, the very spatio-temporal context of the sign may indicate the (S)-aspect as in "It rains," where the (P) part of raining is predicated onto the here-and-now of utterance. A bit further away, but still close to the object we find gestural pointing or gaze directions (fundamental in human semiotics according to Michael Tomasello, 2008), potentially further away are symbolic indices like proper names or pronouns, potentially used in the absence of the object. In any case, the Dicisign claims to stand in a direct connection with its object (if it is a true proposition, this claim may be true, referring to existing objects).[4] A very important observation by Peirce is, that in order for a Dicisign to function, the receiver should have "collateral" knowledge about the object indicated—the proposition should not be the only indexical source referring to the object. A photograph of unknown persons taken at an unknown time and place is thus no Dicisign in itself—but it might be used as part of a Dicisign by a person able to identify aspects of the objects caught in the photo: "This is uncle Walter" or "This is how things were in Tübingen." Given the existence of such collateral knowledge, however, a photograph may function as a Dicisign, even without any additional text. If you happen to receive a letter containing recognizable photographs of yourself involved in embarrassing sexual activities, they immediately constitute Dicisigns claiming that you have indeed indulged in such activity.[5]

As to the Predicative or descriptive aspect of the Dicisign, it must contain or involve some kind of iconic representation of aspects of the object. This may be achieved by the direct means of pictures, gestures, diagrams describing aspects of the object's properties, relations or behaviors, or it may be achieved by means of symbolical icons, like conventional gestures, linguistic adjectives, common nouns, verbs, etc.

Finally, the relation between the S and P aspects of the Dicisign must be some sort of syntax: "Finally, our conclusions require that the proposition should have an actual *Syntax*, which is represented to be the Index of those elements of the fact represented that corresponds to the Subject and Predicate" (ibid., 282). This convoluted expression that the syntax is *represented to be* an index of the elements must be taken to mean that the syntax *depicts* the fact involving the object and quality, in turn corresponding to the S and P parts of the Dicisign. It seems simpler and more appropriate to say that the S-P syntax of the Dicisign iconically depicts the combination of object and quality into a fact. This is also in accordance with Peirce's claim elsewhere in the "Syllabus" that "Every informational sign thus involves a Fact, which is its Syntax" (ibid., CP 2.321).

In Peirce's system of trichotomies, Dicisigns form the second level of the triad Rhema-Dicisign-Argument (his version of the traditional logical Term-Proposition-Argument distinction, elsewhere named by him Sumisign-Dicisign-Suadisign, or Seme-Pheme-Delome). We already touched upon the Rhema generalization of the Predicate. Rhemes may form relational predicates with any number of subject slots to be saturated (even if Peirce famously claimed that any *n*-valence Rheme may be analyzed as a composition of at most 3-valence Rhemes), and, as mentioned, they may be linguistic, gestural, pictorial, diagrammatical, etc.[6] Arguments, on the other hand, are the rule-bound combination of Dicisigns so that one Dicisign, the conclusion, logically follows from another, the premise (which may be composite and combined from several Dicisigns). Diagrammatical reasoning, according to Peirce, is the general mode of thus inferring one Dicisign from another, again widening logical inference to embrace diagrammatical and pictorial representations alongside linguistic ones. As to the structure of Rheme-Dicisign-Argument as signs, Peirce gives this beautifully simple description (using the Sumi-Dici-Suadi terminology):

The second trichotomy of representamens is [divided] into: first, simple signs, substitutive signs, or Sumisigns; second, double signs, informational signs, quasi-propositions, or Dicisigns; third, triple signs, rationally persuasive signs, arguments, or Suadisigns (ibid., 275).

To the simple description of the Rheme, claiming nothing, the Dicisign fills in (some of) the empty slots of the Rheme by Indices, making the sign's object relation double. The Argument adds the third claim that the object behaves in the lawful way indicated by the inference from one Dicisign to the next, hence the single-double-triple distinction based on the number of sign-object connections. As to which aspects of their object the three of them represent, Peirce gives the following list:

Or we may say that a Rheme is a sign which is understood to represent its Object in its characters merely; that a Dicisign is a sign which is understood to represent its Object in respect to actual existence; and that an Argument is a sign which is understood to represent its Object in its character as sign. (ibid., 292).

Finally, we may add the description in terms of degrees of explicitness of the signs:

A representamen is either a rhema, a proposition, or an argument. An argument is a representamen which separately shows what interpretant it is intended to determine. A proposition is a representamen which is not an argument, but which separately indicates what object it is intended to represent. A rhema is a simple representation without such separate parts (Lectures on Pragmatism, EPII, 204; CP 5.139)

As to the *meaning* of a proposition, Peirce defines it logically in terms of what may be inferred from it: "what we call the *meaning* of a proposition embraces every obvious necessary deduction from it" (Lectures on Pragmatism, EPII, 214). The final interpretant of a proposition, according to Peirce's later, triadic theory of interpretants, will embrace *everything* which may be deduced from it; the meaning here referred to is rather what he would call the immediate interpretant, the "obvious" implications of it, of course dependent upon the clarity of context as well as the intellectual capacities of the interpreter. This conception of the immediate meaning is thus pragmatic in the sense that it relies upon the information needed by the interpreter and easily extracted from the sign—most often for action possibilities. Even if many facts may be implied by the Dicisign, the immediate meaning is that fact which is immediately relevant to the Interpreter. [7]

We already touched upon the relation between the Dicisign's syntax and Peirce's theory of facts. It is most often overlooked that Peirce developed an ontology of facts which is, to some extent, analogous to the contemporaneous doctrines in Austrian philosophy of *Sachverhalt* (Stumpf, Meinong, later on Wittgenstein). Like in them, the notion of fact as a part of reality is closely connected to that part's representability in propositions, in Dicisigns:

A state of things is an abstract constituent part of reality, of such a nature that a proposition is needed to represent it. There is but one individual, or completely determinate, state of things, namely, the all of reality. A fact is so highly a prescissively abstract state of things, that it can be wholly represented in a simple proposition, and the term "simple," here, has no absolute meaning, but is merely a comparative expression (The Basis of Pragmaticism, 1906, EPII 378, CP 5. 549)

Peirce's theory of the Dicisign as representing a fact is thus a picture theory of semiotics, not unlike the early Wittgenstein's Tractarian picture theory of logic and language. A very important caveat, however, makes Peirce distinguish between the object referred to by a Dicisign (given by the indexical aspect of that Dicisign) on the one hand, and the whole of the fact represented by the Dicisign, on the other. Here we see a strength of the analysis "throwing the whole of the Dicisign into the Predicate:" the object referred to is different from the fact claimed by the Dicisign. It might seem tempting simply to make the fact and the object of the Dicisign one and the same thing—but in that case, the existence of false Dicisigns would become a theoretical problem, if they involve a really existing object. False claims, of course, may refer correctly to existing objects—their falsity lies in the alleged facts they embed those objects in: "Barack Obama is a Muslim." This Dicisign successfully refers to a really existing object—Barack Obama—but the fact that it represents Obama as a Muslim is non-existent, which is why the Dicisign is false. The Russellian example of "The present king of France is bald," on the contrary, does not refer to any existing object and hence cannot (as against Russell) be false, but rather without any truth value.[8] It is an important aspect of Peirce's theory of Dicisigns, highlighted in his Existential Graph representation systems for logic, that Dicisigns are relative to a selected Universe of Discourse in relation to which their truth value should immediately be judged, rather than directly to reality as a whole (as presumed by Russell). The Universe of Discourse is the more or less implicit reference field, agreed upon by the interlocutors, of a given Dicisign. This emancipates Peirce's doctrine from assuming the logic-as-a-universal-language idea with all the consequences in the shape of linguistic relativism, as argued by Jaakko Hintikka (1997)—to see logic instead as consisting of different calculi the plurality of which makes possible the notion of an independent reality to which all of them ultimately refer.

There are many important issues related to Peirce's doctrine of Dicisigns which merit a thorough discussion but which we may only mention briefly in the passing.[9] One is the status of fictive propositions which refer to specific invented Universes of Discourse—and may be true within those Universes ("Donald Duck wears a sailor's shirt").[10] Another is Peirce's meticulous distinction, anticipating Speech Act Theory, of the proposition in itself and 1) the possible mental representation of it, 2) the possible assent to it by some agent, and 3) the possible assertive expression of it in a public sign (plus more uses, as in interrogatives and imperatives). Assertion is the public claim that the Dicisign is true, whereby utterers potentially subject themselves to public consequences (criticism, even court cases if the Dicisign contains threats, libels, etc.). You may publicly assert a Dicisign without yourself believing it (and so stating a lie), you may believe it without expressing it, you may think of it without either assenting to it nor asserting it, etc. The assertion "Osama bin Laden is dead," uttered by a serious participant in a public debate, thus consists of the Dicisign expressed + 1) mental representation + 2) assent + 3) assertion. This embryonic Speech Act theory is very important in order to realize Peirce's fundamentally anti-psychologistic notion of the Dicisign, making the notion of Dicisign or proposition much simpler than the notion of judgment, which involves psychological representation as well as assent and maybe even assertion. A corollary is that, to Peirce, one and the same Dicisign may form part of different types of speech acts such as assertions, questions, imperatives, etc.[11]

ROAD DICISIGNS

Before we proceed to discuss Dicisigns in biology, let us pause to consider a couple of examples of the arch-semiotic category of road signs. In them, the S aspect of the Dicisign is generally played by the very localization of the sign in time and space—not unlike the case in many biological Dicisigns. Consider, for instance, the following road sign:

The immediate interpretation, of course, of this Australian warning sign is that "Camels, koalas, and kangaroos (P) may appear on the road before you for the next 96 kilometers (S)."

In the internet, this same photo has been manipulated for satirical purposes, substituting a stylized missile, plane, and armed person, respectively, for the three animals, and "Baghdad 89 km" for "Next 96 km." This ironic Iraqi warning sign correspondingly realizes the Dicisign that "Falling bombs, Fighter planes and Insurgents (P) may appear before you on the road for the next 89 kilometers to Baghdad (S)" Such signs are, at the same time, descriptive and imperative. The speech act made on this propositional S-P basis describes a real possibility on the road ahead and warns the driver to avoid the general categories of objects depicted if they actualize on the road.

Road signs thus have the peculiarity of being Dicisigns where the S-part is given by the colocalization of the sign interpreter with the sign. The S-part of the sign indexically points out the location where the entities described by the P part may appear. The P part of the sign is the local circumstances described in

Figure 1.

general terms. Categorized by yellow color and the diamond shape as a warning sign, the speech act of the road sign addresses the behavior the driving recipient is advised or ordered to follow (this behavior, of course, may include other general subjects such as the categories of dangerous objects indicated by the Australian and Iraqi road signs).12 So while the subject of the propositional kernel of this Dicisign is the stretch of road ahead of the sign, the subject addressed by the imperative speech act made on this basis is the driver recipient. This structure of Dicisigns—that spatio-temporal whereabouts of the sign vehicle may play a central part for the determination of the S-part of the Dicisign—also occurs in many biosemiotic Dicisigns to which we now turn.

BIOSEMIOTIC DICISIGNS

Above, we made the case that human Dicisigns involve a large semiotic field with different combinations of language, gesture, pictures, drawings, diagrams, moving pictures, etc. This forms one important extension of the Peircean Dicisign category. Another concerns biosemiotics—sign-use in biology. Peirce's definition and description of Dicisigns is emphatically not dependent on human psychology (even if the range of Dicisigns which may be processed by the human brain, of course, vastly surpasses that accessible by simpler biological processes and agents). Peirce only rarely touches upon animal semiotics, but when he does, he leaves little doubt that he conceives of at least higher animals as being capable of processing Dicisigns.[13] Thus, a strong argument can be made that not only do animal semiotics involve pre-linguistic concepts and meanings, but also proto-versions of human propositions.[14] Let us review some classic examples from different levels of biological evolution (with rough translations into human-language Dicisigns added in parentheses):

- E. Coli perceiving sugar and acting by swimming upstream in gradient ("There's more sugar in this direction")
- Fireflies signaling to possible mates with a species-specific flash ("Here is a Photinus female")
- Von Fritsch bee signaling ("Nectar can be found in this distance in that direction")
- Vervet monkey signal calls ("Eagle/ Leopard/ Snake is near!")
- Higher animals trained by humans, able to utter acquired propositions. As an example, to the presentation of a hitherto unseen patch and the question "What matter?" Alex the Grey Parrot answers "Wool."

All of these examples of animal semiosis crucially involve Dicisigns—because all of them may be ascribed truth values, the fundamental Dicisign property. In a very basic sense, this ought not to be surprising: Dicisigns are the only signs which convey information—that is, simpler signs than Dicisigns, or sign aspects like pure icons or pure indices do not convey information and cannot, hence, function efficiently in biological cognition and communication. Still, most biosemioticians discussing biological signification using Peircean terminology prefer to use exactly notions like icon and index. But pure icons are nothing but possibilities, they refer to any object which possess the required similarity—and can not, then, be used in isolation to signify any precise, actually existing object; this requires an additional index. Pure indices, on the other hand, are mere effects of or pointings to actual objects—without the descriptive part necessary to convey information about those objects. Of course, you can isolate icons and indices from their Dicisign contexts and investigate their structure, much like you can isolate single words outside of their sentence context. More complicated indices, like the Dicisign subtypes Dicent Sinsigns or Dicent Indexical Legisigns, may indeed convey such information and have truth values, but they do indeed include, as Dicisigns, iconic information.

The E. Coli example (resumed and discussed in Stjernfelt 2007) pertains to the ability of Coli bacteria to perceive the presence of sugar in their surroundings and act accordingly. They have perceptors for certain selected chemicals (carbohydrates, certain toxins) enabling them to swim towards, resp. away from such compounds. Normally, they swim in a Brownian-motion-like, so-called "random walk" trajectory where directed swimming is interrupted by brief "tumbling" phases changing their direction—not bad as a search strategy looking for nutrients. After the perception of carbohydrate, they change into "biased random walk" where the tumbling phases of their trajectory as a tendency orient their swimming along the carbohydrate gradient. The perception of sugar is made possible by a small, so-called "active site" on the perimeter of the macromolecule—which is why E. Coli, just like human beings, may be fooled by artificial sweeteners displaying the same active site on its molecule, despite being chemically quite different, the active site being literally a surface property only. Full-blown semiotics arguably begins with the possibility of "being fooled," depending on the categorization, within one and the same semantic-behavioral category, of similar but different phenomena (carbohydrates and sweeteners, in this case). So the (fallible) perception of carbohydrate displays the two basic aspects of a Dicisign: the "active site" provides the P aspect of "sugarlike," while the direction indicated by the gradient provides the S aspect—in linguistic paraphrase: "There is sugar (P) in this direction (S)." There is no communication at all present in this case, to be sure, the Dicisign has the same character as when human beings interpret the flag on a pole as indicating the wind coming from a certain direction. Moreover, we have no reason to assume there is any kind of conscious awareness present in the bacterium—the Dicisign structure is realized in the behavioral sequence of perception and action of the single-cell organism. But if the Dicisign had not had its two crucial aspects, P ("sugar") and S ("direction"), the sign would not have been able to convey information enabling the organism to act appropriately.

An early and famous example of semiotic study of animal communication is von Fritsch's "bee dance." Having discovered a group of flowers with nectar, a bee, back in the hive, may communicate this knowledge to other bees by performing a specific, linear wagging dance. The duration of the wagging phase of the dance indicates the distance from the hive to the flowers, and the direction of the wagging, relative to the angle of the direction to the sun, indicates the direction to the nectar-bearing flowers. Here, the wagging behavior constitutes a Dicisign: "There is nectar (P) at this distance in that direction (S)." The predicate part is communicated by the arbitrary symbol of wiggling (as opposed to normal, non-wagging walking or flying), while the subject part is communicated by the direction and duration of the wagging. Here, the most elaborate part of the Dicisign is concerned with indicating a location remote from the locus of communication.

A similarly complicated example is provided by firefly communication. Every firefly species has a specific flashing pattern, enabling males to locate females perched in the grass in order to mate. A whole arms race has taken place, enabling predator fireflies to fool other firefly species by imitating their specific flash patterns and thus lure lovesick males to approach, only to be eaten. (El-Hani, Queiroz, & Stjernfelt, 2010). We shall not go into the intricacies of cross-species code imitating in this context—the basic point is that all the complexities of such mimetic warfare is based on the Dicisign structure of flash patterns in the first place. The temporal shape of the specific pattern forms the P part of the

Dicisign while the S part is provided by the localization of the flashes—so the flashing behavior as a whole expresses the Dicisign "Here (S) is a Photinus female (P)," giving rise to the appropriate mating behavior in the Photinus male. As opposed to the bee example the complicated part of the Dicisign is the predicate part, specifying the species, while the subject part is given directly by the location of the flashes. In both the bee and the firefly cases, again, there is no reason to assume individual consciousness or learning to play any role. The character of the signs is the stable result of evolution, and they are automatized and not subject to individual change. The action interpretation of the signs, however, necessitates individual intelligence—the strategy of approaching the mate and the flowers indicated by the signs must be negotiated with individual perception and mapping of the surroundings. But even if the character of the signs are the result of the long duration of evolution, the use of them is played out in brief, individual, ontogenetic lifetime.

A famous example of monkey communication is the alarm calls of vervet monkeys (Struhsaker 1967). The monkey has, at its disposal, three different calls which it may utter at the sight of an eagle, a leopard, and a snake, something like: [G-l-l-oi], [Arhw-Arhwhehehehe], [K-l-l-HRhr-hr], respectively. These calls cause conspecifics to perch to the ground, flee to a tree or to search the ground around them, respectively. Such calls have s symbolic descriptive aspect (the specific, conventional sound pattern) playing the (P) role, while the time and place for the uttering of them plays, of course, the (S) role of the Dicisign, thus saying, in paraphrase "There is an eagle/leopard/snake (P) around (S)." The origin of the sound patterns is disputed: it seems partially innate, and partially acquired so that young monkey alarm calls are corrected by their parents. Another issue of the dispute is the degree of intention in the monkey emitting the alarm call: does he intend to warn the others, or is here merely expressing an automatized sign which has been selected because it helps the group to survive? Reported cases of exploitative abuse of the sign to lure conspecifics away from fruits (without any actual presence of a predator) might be interpreted as indicating the presence of intention in uttering the calls, but this seems to be an open issue yet. The Dicisign structure, however, is independent of whether a conscious intention is present in the individual uttering the sign—the decisive issue is whether the sign does, in fact, function and lead to the appropriate actions in the receiver monkeys. Conscious awareness, if it is present, may facilitate more complicated behaviors, like the alleged fooling activity. Such activity is indeed found in human beings, and whether or not it is found in vervet monkeys, such more complicated behaviors depend on the simple Dicisign which does not need individual communication intention to be realized.

A last example can be taken from the impressive feats performed by human-trained animals. Irene Pepperberg's famous African Grey Parrot, Alex (recently deceased) was able, after 20 years of training, able to pronounce propositions and correctly answer simple questions pertaining to a series of qualities (number, shape, color, material) of hitherto unseen objects—like answering "Wool" to the question of "What matter?," or "Green" to the question of "What color?." Alex had a vocabulary of around 150 words, among them the names of 50 objects, and was able to count up to seven.[15] Thus, Alex was able to express Dicisigns in a way close to humans—characterizing present objects (S) by some of their qualities (P). Alex was rewarded for the answers, so the communicative goal was immediately food rather than any intentional wish to share information with the trainer—but, again, that does not take away the basic Dicisign characteristic of its utterances, the ability actually to convey information about an object.

These biosemiotic example of Dicisigns vary as to cognition or communication, as to the emphasis on the S or P part of the Dicisign, as to the phylogenetic or acquired character of the sign, etc. They all share, however, the basic pragmatic feature that the communication is more or less closely connected to an action series which is either automatized or leads to an immediate reward in terms of survival or nourishment.[16] Thus, the Dicisigns listed here seem to be simpler than the distinction between Indicatives and Imperatives in human languages; in some sense these biological Dicisigns are descriptive and imperative at one and the same time, leading directly to actions whose success, in most cases, serve as the grant that the sign has been appropriately understood.

The appreciation of the widespread existence of Dicisigns in biology makes, I think, it a bit easier to imagine the crucial transition from animal to human semiotics.[17] The "propositional stance" is not a human, psychological privilege—rather, many biological systems adopt such a stance without necessitating conscious awareness or intention. The Peircean proposal for the specificity of human semiotics places the emphasis on the higher degree of self-control of human beings. About such self-control, Peirce writes: "To return to self-control, which I can but slightly sketch, at this time, of course there are inhibitions and coordinations that entirely escape consciousness. There are, in the next place, modes of self-control which seem quite instinctive. Next, there is a kind of self-control which results from training. Next, a man can be his own training-master and thus control his self-control. When this point is reached much or all the training may be conducted in imagination. When a man trains himself, thus controlling control, he must have some moral rule in view, however special and irrational it may be. But next he may undertake to improve this rule; that is, to exercise a control over his control of control. To do this he must have in view something higher than an irrational rule. He must have some sort of moral principle. This, in turn, may be controlled by reference to an esthetic ideal of what is fine. There are certainly more grades than I have enumerated. Perhaps their number is indefinite. The brutes are certainly capable of more than one grade of control; but it seems to me that our superiority to them is more due to our greater number of grades of self-control than it is to our versatility." (CP, 5.533).

Such levels of self-control enables man to make the sign used explicit—which is what makes possible, in turn, to compare Dicisigns with their constituents, to compare several Dicisigns, to vary the inference from one Dicisign to another experimentally to see how far the inference holds, to invent signs about other signs (Peirce's notion of "hypostatic abstraction" for the creation of second-level thought-objects making reflection upon universals possible). It seems to me a good guess that the very development of consciousness in higher animals with central nervous systems serves to facilitate the growing explicit control, combination and nesting of Dicisigns.[18]

CONCLUSION

To conclude, Peirce's theory of propositions, in the guise of "Dicisigns," makes possible a renewed appreciation of the breadth of empirical phenomena displaying Dicisign structure and behavior—from human signs involving language, gesture, pictures, diagrams, etc. and many crossover forms intermixing them, and to very simple biosemiotic cognition and communication cases. It goes without saying that these Dicisigns vary considerably in terms of constituents, complexity, in degree of conscious access, in generality, and much more. But the unifying perspective of Peirce's Dicisign doctrine seems to me to cast more light upon the appearance of semiotics through the process of evolution: natural selection has had to adapt to Dicisigns, rather than the other way around, because its basic S-P structure is necessary for efficiently conveying information. In that sense, it might be less precise to see Dicisigns as the result of evolution. We should rather see evolution as having increasingly adapted to the structures of Dicisigns.

REFERENCES

Cheney, D. L., & Seyfarth, R. M. (1992). *How monkeys see the world*. Chicago: Chicago University Press.

Deacon, T. (1997). *The symbolic species*. New York: W. W. Norton & Co.

Donald, M. (2001). *A mind so rare*. New York: W.W. Norton & Co.

El-Hani, C., Queiroz, J., & Stjernfelt, F. (2010). Firefly femmes fatales: A case study in the semiotics of deception. *Journal of Biosemiotics*, *3*(1), 33–55. doi:10.1007/s12304-009-9048-2

Eriksen, J.-M., & Stjernfelt, F. (2012). *The democratic contradictions of multiculturalism*. New York: Telos Press.

Hilpinen, R. (1992). On Peirce's philosophical Logic: Propositions and their objects. *Transactions of the Charles S. Peirce Society*, *28*(3), 467.

Hintikka, J. (1997). *Lingua universalis vs. calculus ratiocinator: An ultimate presupposition of twentieth-century philosophy*. Dordrecht, The Netherlands: Kluwer. doi:10.1007/978-94-015-8601-6

Hoffmeyer, J., & Stjernfelt, F. (2015). The great chain of semiosis: Investigating the steps in the evolution of biosemiotic competence. *Biosemiotics*, *9*(1), 7–29. doi:10.1007/s12304-015-9247-y

Hurford, J. (2007). *The origin of meaning*. Oxford, UK: Oxford University Press.

Peirce, C. S. (1992). *The essential Peirce, I*. Bloomington, IN: Indiana University Press.

Peirce, C. S. (1998). *The essential Peirce, II*. Bloomington, IN: Indiana University Press.

Peirce, C. S. (1934-58). *Collected papers, I-VIII*. Cambridge, MA: Belknap Press of the Harvard University Press.

Pepperberg, I. (2002). *The Alex studies: Cognitive and communicative abilities of Grey Parrots*. Cambridge, MA: Harvard University Press.

Short, T. (2007). *Peirce's theory of signs*. Cambridge, UK: Cambridge University Press. doi:10.1017/CBO9780511498350

Smith, B. (2005). Against fantology. In M. Reicher & J. Marek (Eds.), *Experience and analysis*. Vienna: ÖBV & HPT.

Stjernfelt, F. (2007). *Diagrammatology. An investigation on the borderlines of phenomenology, ontology, and semiotics*. Dordrecht, The Netherlands: Springer.

Stjernfelt, F. (2012). The evolution of semiotic self-control. In T. Deacon, T. Schilhab, & F. Stjernfelt (Eds.), *The symbolic species evolved* (pp. 39–63). Dordrecht, The Netherlands: Springer. doi:10.1007/978-94-007-2336-8_3

Stjernfelt, F. (2014). *Natural propositions: The actuality of Peirce's doctrine of Dicisigns*. Boston: Docent Press.

Stjernfelt, F. (2014a). Dicisigns and cognition: The logical interpretation of the ventral-dorsal split in animal perception. *Cognitive Semiotics*, 7(1), 61–82.

Stjernfelt, F. (2015). Iconicity of logic and the roots of the "iconicity" concept. In K. Masako, W. J. Hiraga, K. Z. Herlofsky, & A. Kimi (Eds.), *Iconicity: East meets West* (pp. 35–56). Amsterdam: John Benjamins. doi:10.1075/ill.14.02stj

Stjernfelt, F. (2015a). Dicisigns. *Synthese*, 192(4), 1019–1054. doi:10.1007/s11229-014-0406-5

Stjernfelt, F. (2016). Blocking evil infinites: A note on a note on a Peircean strategy. *Sign Systems Studies*, 42(4), 518–522.

Stjernfelt, F. (2016a). Dicisigns and habits: Implicit propositions and habit-taking in Peirce's pragmatism. In D. West & M. Anderson (Eds.), *Consensus on Peirce's concept of Habit: Before and beyond consciousness* (pp. 241–264). New York: Springer. doi:10.1007/978-3-319-45920-2_14

Struhsaker, T. (1967). Auditory communication among vervet monkeys (*Cercopithecus aethiops*). In S. A. Altmann (Ed.), *Social communication among primates*. Chicago: University of Chicago Press.

Tomasello, M. (2008). *Origins of human communication*. Cambridge, MA: MIT Press.

Von Frisch, K. (1993). *The dance language and orientation of bees*. Cambridge, MA: Harvard University Press. doi:10.4159/harvard.9780674418776

ENDNOTES

[1]   Peirce's main development of the Dicisign doctrine falls in his late period from around 1903, e.g. in the Harvard Lectures of Pragmatism (in the CP as well as the EPII), and in the "Syllabus of Certain Topics of Logic" which has never been published in its entirety, but appears in large parts in the CP and the EPII, etc. I pronounce "Dicisign:" "Dee-see-sign," after the standard way of pronouncing Latin "c" as an "s" when it appears before front vowels like "I" and "e."

[2]   For a discussion of the iconicity of different logic representations, see Stjernfelt 2015.

[3]   Sometimes, Peirce use "assertion" for "proposition" or "Dicisign," despite the fact that he conceived of the Dicisign to be independent of both the mental representation of it, of the assent to it by some agent, and of the assertion of it by some agent (cf. below).

[4]   The Peircean analysis takes Dicisigns claiming to refer to actually existing singular objects to be basic. Hence, even if crucial to logic, quantified Dicisigns form derived cases of vague (existential quantification) and general (universal quantification) Dicisigns.

[5]   Of course, such photographs may be forged or Photoshop manipulated, etc.—but larger or lesser possibilities of falsity will affect all types of Dicisigns.

[6]   Rhemes are often thus identified with isolated Predicates outside of propositional contexts. Strictly speaking, as the Rheme-Dicisign-Argument triad is taken to be exhaustive, also isolated Subjects outside of propositional contexts must be classified as Rhemes.

[7]   The second, "dynamic" interpretant will then be the meaning actually realized in any particular use of the Dicisign.

8   Peirce would claim such a Dicisign is meaningless (his example is "Any Phoenix, when rising from its ashes, sings *Yankee Doodle*") and must be classed with true Dicisigns because no possible perception can make it false. We would rather classify it as without any immediate truth value.

9   The present paper formed, in 2010, my first discussion of Peirce's notion of Dicisign. A broader discussion of Dicisigns and their implications can be found in Stjernfelt 2014 and Stjernfelt 2015a.

10   Peirce's idea that Dicisigns refer to a Universe of Discourse rather than directly to reality makes it easy, it seems to me, the connect his Dicisign doctrine with accounts for fictitious propositions like Roman Ingarden's, cf. Stjernfelt 2007 ch. 16.

11   Another issue which we may only mention in the passing, is Peirce's subdivision of Dicisigns which comprise the following types: 1) Propositions proper—Dicent Symbols (characterized by involving a general idea) - 2) Quasi-Propositions - Informational Indices a) Dicent Sinsigns (a Weathercock, some photographs, some paintings ... but presumably also all sort of empirical occurrences offering information about their particular cause or motivation)   b) Dicent Indexical Legisigns (Peirce's example: a street cry identifying the individual shouting, or expression identifying individuals like "This is him," "That is Napoleon")

These three categories appear as number 9, 7, and 4, respectively, of Peirce's famous ten-sign classification based on the combination of the three trichotomies of the "Syllabus" (EPII, 294ff.) It is an open issue whether the biosemiotic examples discussed below should be categorized as proper Dicent Symbols or rather as Dicent Sinsigns. The apparently simpler status of the latter might immediately tempt one to opt for that choice, but the automatized character of most of the biosemiotic examples rather point in the direction of the generality of Dicent Symbols—E. Coli is probably not concerned with the particularity of this and that lump of sugar, but more interested in its general character of nutrient carbohydrate; very simple signs are also very general.

12   This points to an important issue of plasticity in Peirce's Dicisign doctrine: the exact dividing line between the S and P aspects of one and the same Dicisign may differ with interpretation. This does not indicate the dividing line is arbitrary nor that it may vanish completely—but merely that many Dicisigns, especially nonlinguistic or only partially linguistic Dicisigns may be parsed in S and P parts in different ways, depending upon the immediate use of the interpreter, and that no fixed location of the line could be said to exist. Peirce writes: "Take the proposition "Burnt child shuns fire." Its predicate might be regarded as all that is expressed, or as "has either not been burned or shuns fire" or "has not been burned," or "shuns fire" or "shuns" or "is true"; nor is this enumeration exhaustive. But where shall the line be most truly drawn? I reply that the purpose of this sentence being understood to be to communicate information, anything belongs to the interpretant that describes the quality or character of the fact, anything to the object that, without doing that, distinguishes this fact from others like it; while a third part of the proposition, perhaps, must be appropriated to information about the manner in which the assertion is made, what warrant is offered for its truth, etc. But I rather incline to think that all this goes to the subject. On this view, the predicate is, "is either not a child or has not been burned, or has no opportunity of shunning fire or does shun fire"; while the subject is "any individual object the interpreter may select from the universe of ordinary everyday experience" (Peirce CP 5.473). Hilpinen (1992, p. 476) also discusses this quote and rightly remarks that this idea takes Peirce far away from the logical atomism of Russell and Wittgenstein, according to whom there is only one complete analysis of the proposition (Tractatus). As to the relativity of the S-P, distinction, see also Stjernfelt (2016). Again, this plasticity depends on the Universes of Discourse referred to by the utterer and the interpreter. Another interpreter of

the Australian warning sign might include the sender of the sign and read it as follows: "So, this is the sole three animal species appreciated by the Australian traffic authorities."

[13]   Cf. my arguments in Stjernfelt 2007, ch. 11, pertaining to Peirce's discussions of the abilities of dogs and parrots where he seems to ascribe to them abilities to deal with Dicisigns. Peirce claims that "All thinking is by signs, and the brutes use signs. But they perhaps rarely think of them as signs. To do so is manifestly a second step in the use of language. Brutes use language, and seem to exercise some little control over it. But they certainly do not carry this control to anything like the same grade that we do. They do not criticize their thought logically" (Consequences of Critical Common-Sensism, c. 1905, CP 5.534). So what animals lack, according to him, is rather the ability to subject their own Dicisigns and Argument to explicit and critical scrutiny. See also Stjernfelt (2012) and (2014).

[14]   From another perspective, Hurford (2007) makes a similar argument.

[15]   Critics have argued that Alex was only an example of the "Kluge Hans"-effect where the animal guesses the answer after subtle cues from the trainer. Against this, however, counts the fact that Alex was able to answer questions from other people than the trainer when it could not see anyone knowing the answer.

[16]   The relation of Dicisigns to habits and action, see Stjernfelt (2016a)

[17]   An important corollary is that Dicisign structure as such is not a conventional component of human cultures (even if particular use patterns of Dicisigns may follow such conventions); cf the discussion of "culturalism" as the exaggeration of culture as an explanatory category for human behavior (Eriksen & Stjernfelt, 2012). The sophistication of Dicisign use during evolution must be a central issue for evolutionary biosemiotics, cf. Hoffmeyer & Stjernfelt (2015)

[18]   Cf. the roles of conscious binding, short-term working memory, and medium-term memory discussed by Donald (2001).

Chapter 9
Neologisms:
Semiotic Deconstruction of the New Words "Lizardy," "Staycation," and "Wannarexia" as Peircean Indexes of Culture

Joel West
University of Toronto, Canada

ABSTRACT

While people use language to let others know how and what it is that we think, language is also the means by, and also the substrate within which, humans think. This chapter explores the use of language as the basis for cognition, based on both a chosen word's denotative meaning and also its rhetorical (metaphorical) connotative meanings. The artificial dichotomy between language and speech is deconstructed. Peircean semiotics is used to argue that language is indexical in its primary referential functions, including sociolinguistic functions. Three new words, all of which were coined in the twenty-first century, are examined from a sociolinguistic and a semiotic point of view.

INTRODUCTION: DIFFERENCE AND MEANING

'When I use a word,' Humpty Dumpty said, in rather a scornful tone, 'it means just what I choose it to mean — neither more nor less.'

'The question is,' said Alice, 'whether you **can** *make words mean so many different things.' Lewis Carroll, Through the Looking Glass*

The question of what a verbal message means on a connotative level is truly one of the most human preoccupations. How often have we asked friends, or have friends asked us, "I wonder what he or she meant by that?" It is not that we do not understand the denotative meaning of the words, the oral grunts, whistles and hisses or the written glyphs in which the language was encoded. Instead, by asking this question, we are trying to decode the connotative meaning of the other person's statement. We want to

DOI: 10.4018/978-1-5225-5622-0.ch009

know the meaning of the message beyond its simple denotative meaning. Interestingly, other non-human animals do not seem to have this capacity to create connotative meaning. When a beaver slaps its tail against the water or a bird screams, when a dog barks or a cat meows, the other animals do not stop, reflect and ask, "I wonder what the animal meant by that particular signal?" There is just a denotative signal of some sort, such as a beaver tail slap on the water, followed by some sort of action, usually fight or flight. The point is, though, that there is a direct one-to-one correlation between an animal's signal and what that signal actually means. If the animal is confused by a signal, the animal will often freeze and not continue until it believes that it has fully understood the situation. The animal does not think about the matter but just reacts by following its instincts. Even animals that we humans have taught to speak, like Koko the gorilla (Gorilla Foundation, 2017), do not seem to have the facility to reflect and to wonder about the meaning of the meaning of things. While these gorillas have been taught to speak sign language and while these gorillas even have the capacity to invent new words, their inventions refer only to concrete things or to emotions; things which they already know about (Gorilla Foundation, 2017). The point is that the gorillas can learn words and create new sign language words but all these new sign language words only have simple denotative meanings. Even gorillas do not seem to be capable to use words to connote things or to create metaphors for things that only exist in imagination. The gorillas do not, apparently, have the ability to abstract ideas from the world, to dream up new words, nor do they appear to have capacity to reflect on the words which they know. By "reflect," the authors mean the ability to think about a thing in an abstract sense and to create new meanings based on these reflections which human beings have and which other animals apparently do not. Marcel Danesi discusses this capacity in a Semiotic sense (Danesi, 1993) which he links to the capacity to create metaphor and in fact, he links this to Giambattista Vico's idea of *verum factum,* that human language is a creation based in metaphorical imagination and this capability is at the root of the origins of human language (Danesi, 1993). Such a discussion is, of course, beyond the scope of this work, but we are left with the idea that language is forged by the capacity for metaphor and its connectivity among conceptual domains.

This metaphorical use of language is exactly how humans express themselves. While gorillas use words to denote concrete ideas and concrete objects, human beings use words to connote ideas and things that may not exist concretely. Humans' verbal ideas may exist solely in the imagination. Also, the new words we create are meaning rich, Peircean indexes, signs that indicate not only their very most basic denotative meanings but also that these words indicate meanings which, because they transcend simple denotation, also point to social, socioeconomic and other referential phenomena. Human words work beyond themselves to mean more than simple denotative referents. In Saussure's linguistics we would call the word a "signifier" and its meaning the "signified" and this would be the word's denotative meaning (Saussure, 1959). A re-signified sign, or the connotative meaning of a word, is what Roland Barthes calls a "mythology" (Barthes, 1957). While we will not be using Barthes model of meaning, it important to note that he and the other structuralists (those who follow Saussure's dyadic ideas) are extremely useful in the world of criticism and anthropology, because they limit themselves to binary oppositions that are easy to detect and utilize for analytical purposes. These descriptions are useful but are limited to "one and zero," "yes and no," or "man and woman" descriptors instead of the plethora of other semantic nuances that exist. We will see that imagination transcends these binaries so that a strict one-to-one relationship of signifier and signified might be useful to create a simple deconstruction of meaning but that words are not bound by this binary relationship between a sign and its meaning (Danesi & Perron, 1999).

So instead of structuralist binaries, where words evoke meanings through opposition, studies have shown that words and their meanings are intertwined with themselves as if in a semantic network; it is this network that produces connotation. To this matter, George Lakoff and Mark Johnson demonstrated in their classic work on metaphor that the human capacity to use words, that is, the way that humans use words, always transcends any simple denotative (literal) meanings. People live in a discursive world which is guided by what they call conceptual metaphors (Lakoff & Johnson, 1980). For example, under the rubric of the conceptual metaphor "Life is a journey" we derive specific instantiations via linguistic metaphors such as "She stumbled in life," "He took a wrong turn," and "Her scholarly journey was steep." It is not merely that these metaphors are useful to connote shades of meaning but more importantly, and in fact crucially, that these metaphors, all by themselves, shape the way in which we experience the world, namely, as a journey. This whole line of semantic analysis has its basis in the so-called Whorf-Sapir hypothesis (Whorf, 1956) where the language one speaks shapes how one views the world. As Benjamin Whorf puts it:

The categories and types that we isolate from the world of phenomena we do not find there because they stare every observer in the face. On the contrary the world is presented in a kaleidoscopic flux of impressions which have to be organized in our minds. This means, largely, by the linguistic system in our minds (Whorf, 1956).

Whorf is saying is that the categories and types or denotations are not discovered but rather that linguistic structures are created so that we can interconnect them and organize them cognitively. This means that for Whorf the capacity for organization of information from the world is reflective of the language one speaks and that this capacity is both reflective of what is going on in the mind and also responsible for the capability itself. To be clear, Whorf does not differentiate between language and speech, but sees them as interactive components of cognition. So, let us say instead that a "linguistic system" is how we make sense of the world and that it is also involved in the capacity to convey it via speech. Then, we can say that metaphors are artifacts, human creations, both emanating from, and shaping, linguistic cognition. Our conceptual metaphors, which are systematic of discourse, are a product of our linguistic minds. The upshot of this is that our words do not have a direct one-to-one correlation to their referents but instead are indexical of what they mean—that is, they point to the interconnected network of meanings already present in language.

Jacques Lacan (1985) remarks that "Language pre-exists the infantile subject" suggesting that language itself in the sense that Saussure means (as opposed to speech) exists prior to the infant's sense of self, separate from the world. Saussure makes clear that there is a difference between the language capacity and those utterances, oral or written, which are expressed verbally as speech. He thus differentiates between "*langue,*" which means language, and "*parole*" which means speech; a similar dichotomy exists in the Hebrew words transliterated as "*sapha*" (literally lips) which means speech and "*lashon*" (literally tongue) which means language. Saussure called this delineator, this chasm, between language and expressed speech, as "*difference*" (Saussure, 1959) and in this sense, language is something which, on one hand is a precursor to speech, but also which exists outside of speech. The paradox is, of course, that, language is also that which is expressed in speech. So, as Lacan affirmed, the child's ability to express itself in a subjective sense, as a subject unto itself is based on its ability to express language as speech and also, as a subject, is projected directly into language. To be entirely clear, the ability which exists as a precursor to speech and as the capacity to express words as speech is not what we call speech, but language. This is an innate ability and speech is then an expression of, and only points to or is indexical of, language.

While the difference between language and speech is certainly true at a level of knowledge versus usage of that knowledge, it is also true that the two dimensions are interactive. This implies that the specific words we choose during communication both come from, and modify, communication gradually. The new words that we coin reveal a lot about this dynamic interrelationship. These are interpretant signs, to use Peirce's well-known concept; that is, they are signs that not only stand for something else referentially, but they also interpret it and this system of interpretation thus involves indexing existing meanings. They can be "perhaps more properly thought of a translation or development of the original sign (Atkin, 2013)." To reiterate this point, these words are thus indexical in the Peircean sense, because even the simple referent of these words point beyond their simple denotative dimension by interpreting referent psychologically and culturally at the same time. So, words both point to a referent and interpret it at the same time. Our words do at the very least, double duty.

Part of the reason for this indexical capacity is the fact that all verbal communication is open to multiple interpretations at all times; our words and our meanings are polysemous. One example of this slipperiness is the word "mother." The word "mother" has at a rudimentary semantic level a simple denotative meaning which is "a female parent, usually one who gives birth to a child." But, the phrase "Mother Of All Bombs" (taken from the acronym MOAB for "Massive Ordinance Air Blast," Steinbuch, 2017) does not literally mean that that the military device, the MOAB device, actually is the female parent of all other bombs and that it gives birth to other bombs. Instead, the connotative meaning of the phrase is that the bomb is so devastating that any other bomb is somehow lesser and a "smaller child." So, the denotative meaning of "mother" is transformed into a metaphor, which is made possible because of the inbuilt polysemy of semantic structure. It is because of this constant potential for polysemy alternating between connotation and denotation that we are required to constantly decode possible meanings and to reject ones that do not make sense to us, based on the context. We act as if there is a direct one-to-one correlation between a word and its meaning but, as cognitive linguistics has been showing for at least three decades, there rarely is such a correlation; rather the polysemous potential of words is integrated under the form of conceptual metaphors. Cognitive scientists now describe about how certain circuits work in the brain and how metaphor is evolves from a blending of different circuits (see for example Lakoff, 2012). While the metaphor of circuitry is useful to describe what it is that we mean in terms of how some parts of the brain work and also that this metaphor has in fact entered common usage. This very usage proves the point of the present discussion, since the description of brain structure as consisting of circuits has been adopted by AI as a verity. But metaphor is not certainty or fact; it is an inference. As Umberto Eco says, metaphor "circumscribes cultural units in an asymptotic fashion, without ever allowing one to touch it directly, though making it accessible through other units (Eco, 1976)." (An asymptote is a term that derives from calculus and it means a straight line that approaches a curve without ever actually reaching it. What Eco means is that a sign and its meaning are like a curve and its asymptote in that the two of them will converge closer and closer without ever meeting. This implies that while we can infer (or guess) how cognition works, we will never really grasp it. So, we use the metaphor of circuitry, which is concretely understandable in human terms. It is thus an interpretant for illustrating how cognition is like a computer and vice versa. It is a persuasive one, though, as all indexical interpretants are. So, the various meanings of a word can at very best approach their referent without ever completely imparting the full intended meaning of that referent. Like all metaphors, the meaning of the thing itself only points to the thing it means; it is not that thing. The two dimensions are connected via the layer of interpretation, which unites them and makes them cognitively indistinguishable.

How does all of this work in terms of the sociolinguistics of speech (or linguistic anthropology which we will use as exact synonyms for each other)? How do we create words and what do we mean when we create a neologism based on another word? When we create a new word, how is it that the word means more than what it first appears to encode? For the sake of brevity, I will narrow my focus to three neologisms of the early 21st century, discussing how they are indexical at various levels, in terms of the history of words, starting with a case from the Bible, to show how words do more than encode a referent. I will then will go on to show how, in the twenty-first century, the words "lizardy," "staycation" and "wannarexia" are, aside from simple colloquialisms, semantically and cognitively indexical. While these words have an ostensible denotative (referential) meaning that they evoke senses that are both indexical of socio-economic class and social status relative to the speaker and the culture itself. I will be using sociolinguistic and Peircean semiotic tools to deconstruct the words and show how they are indexical at various levels.

BACKGROUND: PEIRCIAN SEMIOTICS AS SOCIOLINGUISTIC TOOL

While, pragmatist philosopher Charles Sanders Peirce and linguist Ferdinand Saussure, were, separately, the co-founders of the field of semiotics, they differed in several basic theoretical ways. Semiotics is the interdisciplinary study of how humans and animals encode meaning into the artifacts that we create and, furthermore, how meaning may be decoded subsequently from these artifacts. Some human artifacts include art, culture itself, sculpture, architecture and dance but this paper will concentrate on linguistic artifacts. Saussure's semiotic model sees signification as an arbitrary dyadic relation between signifier and signified (Saussure, 1959), for example, a *rose* and the *love* it represents. Peirce, unlike Saussure, claimed that a sign involved a triadic relation whereby a sign (representamen) not only encodes a referent (object) for some pertinent reason, but also interprets it culturally (Danesi & Perron, 1999). For Peirce, it is also significant, not just that the sign represents something in itself but also to the person who uses it. So, agency is always a part of interpretation, unlike in Saussure where the relation between user and sign is arbitrary. Peirce's basic triadic taxonomy of signs, indexical, iconic and symbolic, has become central to both semiotics and linguistics (Atkin, 2013).

Peirce referred to icons as "likenesses", because the sing created stands for its referent by resemblance or likeness (Peirce, 1982). Some examples of icons could include photographs and also, in terms of language onomatopoeic words, which represent by sound likeness their referents. An artistic example of onomatopoeia might be the assonance of the howling sound in Allen Ginsberg's poem "Howl:"

I saw the best minds of my generation destroyed by madness, starving hysterical naked,

dragging themselves through the negro streets at dawn looking for an angry fix,

angelheaded hipsters burning for the ancient heavenly connection to the starry dynamo in the machinery of night (Ginsberg, 1955)

Note the repetition of the phoneme /aɪ/ which can be interpreted, onomatopoeically, as the eponym of howling, or, in this case, an iconic representation of a "howl." So, an icon represents referents by resemblance.

An indexical sign represents by ontology in a relational sense. It points to something and asserts existence, vis-à-vis something (or someone) else: for example a pointing finger, a directional arrow, ownership words such as "My house" and Jenny's dog," locative words such as "the store over there," among others. "All dogs" is also indexical since it points to the set of all of the animals that we call dogs. Even documents, such as biographies, are indexical since they point to the person to whom they refer. Biographies are indexical interpretants of a specific individual (Nye, 1981). *The Autobiography of Benjamin Franklin*, as an example, points to the historical figure Benjamin Franklin and also interprets the man, Benjamin Franklin, from his own point of view. It also indicates a book title, which in turn represents Franklin the man. So, the actual referent of the index is dependent on the context and indeed the referent of the index itself may be equally vague. The index declares the existence of its referent but this referent is also, as has been noted above, an interpretation of the referential object. Even more, because there is an imperfect correlation between the index and its referent we can never be certain if what we see is the actual intended object or if it is just an interpretant. Thus, the movie *Schindler's List* (1993) is an index and to and an interpretation of the book *Schindler's List* (Keneally, 1983) and the man Oskar Schindler. The book is, in turn, a further index and representation of some of the actions of a real human being, Oskar Schindler, 1908 – 1974, (United States Holocaust Museum, 2017) and since the movie was produced, also the movie. The book is indexical of the man and interprets him in the same way that the movie interprets the book which in turn interprets the man. In fact, all three, the book, the movie and the man, all point to and interpret each other. So as we see, even a single index can become, as Anne Urbancic says, "A Lacanian chain of meanings (Urbancic, 1994)," an object which points to an object which in turn points onward to another object *ad infinitum*. At any point in this cascade of meanings, experience and context, guide the interpretive process. Unconventional decoded meanings of words and sentences and the inability to encode verbal meanings in conventional and decodable syntax is regarded, in our culture at least, as symptomatic of pathology and mental illness and is metaphorically called a "word salad."

To continue with Peirce's taxonomy, symbols represent by convention. The onomatopoeic words used to designate a rooster call in English and in French are onomatopoeic—respectively *cock a doodle doo* and *coco ri coco*. But in order to know that they refer to the same referent, we must know the languages involves. This entails symbolic signification—meaning by convention. Many connotations are also based on symbolicity. The words for "red," for example, have culturally based meanings which require a social context for semantic decipherment, based in a cultural milieu, so as to be able to understand the proper meanings which the speaker intends. Even further, the color can have a private meaning for the speaker, which may, or may not, be intended to be understood by others. Obviously then, there is no direct one-to-one correlation between the color sign and its meaning, or the rooster sound words, which should be identical despite the languages, as Saussure claimed. There is obviously some sort of disconnection between the sign and the interpretation which, in this case, is filled in by cultural context.

Signs are rarely meaningful categorically, that is as a pure symbol, index or icon, but often as blends. For example, a souvenir of the Eiffel tower is an icon, index and symbol. The souvenir resembles the tower, it points to the "real" tower in Paris and it is also a keychain, so it a conventional souvenir. Signs are not uni-dimensional, but rather multi-dimensional involving iconicity, indexicality, and symbolicity at once. Words are, for the most part, indexical signs, but this does not preclude their iconic and symbolic functions.

Even how we pronounce words is always indexical of something. One of the first recorded case studies of a culturally established variant of a phoneme being used as a Peircean index is in the Bible. The Israelites had to find a way to differentiate themselves from the surrounding Ephraimites and came up with the following strategy. "They said, 'All right, say 'Shibboleth.' If he said, 'Sibboleth,' because he could not pronounce the word correctly, they seized him and killed him at the fords of the Jordan. Forty-two thousand Ephraimites were killed at that time" (Judges 12:6)."According to the Collins dictionary, *Shibboleth* means an ear of grain; but that literal meaning is moot in this case. While we can presume from the context that the words *Shibboleth* and *Sibboleth* were orthographic variants of same referent, and that the meanings are congruent, the pronunciation of the /ʃ/ phoneme (the "sh" sound) was a cue to a differentiation between two groups of people. This was, in effect, how they differentiated between friend and enemy. Thus the two pronunciations of the same word form a minimal pair with different initial phonemes which are indexical of "Israel" and "Ephraim." The word's pronunciation, not its meaning, was the indexical trigger. More to the point of the present purposes, the word *shibboleth* entered the English lexicon in the 17th Century with the current meaning appearing in 1638 (Merriam-Webster, 2017). Shibboleth in current usage has come to mean either "a word or saying used by adherents of a party, sect, or belief and usually regarded by others as empty of real meaning" or "a use of language regarded as distinctive of a particular group (Merriam-Webster, 2017)." The point here is that an ancient word in a language that was not at all in local use came to mean something else to a group of people because the word was indexical of group appurtenance. The writer took the word *Shibboleth* and presumed that it would have meaning in its indexical sense, pointing to a specific group of people. While at one time in history a variant pronunciation of the word was indicative of an Ephraimite, the word came to mean something completely different.

A contemporary example of pronunciation as indexical of group appurtenance is William Labov's work on the use of the final /r/ phoneme amongst employees in New York (in Danesi, 2016). Labov established that people belonging to a higher socio-economic class, and people who aspired to be a part of that class, pronounced the final /r/ phoneme in words differently than those who were not in that class. Therefore, we can surmise that pronunciation of the /r/ phoneme is indexical in the Peircean sense, indicating a socio-economic class to which one belongs or to which one might desire to belong. This example shows that just by asking "How are you?", for example, we can detect a desire for upward mobility based merely on how this expression is pronounced in terms of the final /r/. Pronunciation is thus not devoid of indexicality; rather it is part of an indexical code that blends in with other codes (verbal and nonverbal) that reveal how people see themselves in relation to others and how they relate to socio-economic status, ethnicity, and the like. All of this information is completely aside from the semantic content of the words themselves. While the mass media may have leveled off pronunciation differences in the mainstream population, these nevertheless persist because of their indexical nature. For example, younger people in North America and even adults who wish to appear younger, tend not to use a so-called "Mid Atlantic" accent which was once indicative of proper English speech (Fallows, 2015), but instead speak using "vocal fry," that is, by using a "pulse register" or "glottalisation of speech" and is found, mostly among young people. This manner of speaking was first popularized by such celebrities as Kim Kardashian and by actresses such as Zoey Deschanel. Someone speaking with "vocal fry" as part of their pronunciation habits reveals a desire to be "in style" with media culture.

The same cultural side-story can be told about any colloquialism or linguistic mannerism that enters speech at any given time. This is the case of the word *cool*, which became indexical of youth culture and the "rebellion of youth" (Danesi, 1994), decades ago. The point here is not that young people rebelled against the prevailing culture but that this rebellion was indexed linguistically. When it became culturally desirable to be a member of that youth culture, only then did *cool*, as its secondary meaning, enter the common lexicon. Words such as *groovy* and *square*, which seem quaint today, entered the lexicon in the same way.

"LIZARDY", "STAYCATION" AND "WANNAREXIA" AS INDEXICAL SIGNS

There is no linguistic authority to govern English language usage, except common usage itself. We are free to say any things in any way we want, presumably, as long as we are understood. Even the Oxford English Dictionary claims that:

The aim for the OED is to record the language as it is, as accurately as possible. It tries to be descriptive rather than prescriptive. The reason for including quotations is quite simply to illustrate how a word is used, because that is what its meaning depends on (Oxford English Dictionary, 2017).

The demise of linguistic authority can even be seen in the fact that the quoted passage ends with the preposition *on*. At one time, ending a sentence with a preposition was considered to be a ungrammatical and "uneducated" English usage. In a previous era, one might have looked to an authoritative source, like, *A dictionary of modern English usage*, by Henry Watson Fowler, for guidance in grammar and vocabulary usage. But today it is unlikely that anyone even knows of its existence. Instead, as John McWhorter notes in the title of his book, we are *Doing our own thing* (McWhorter, 2003). The English lexicon is flowing and ebbing, changing constantly and, instead of traditionally authoritative sources, we are left with Wikis, collaborative websites where users are free to add, remove and revise content at will. Wikipedia (http://www.wikipedia.org) touts itself as an "Online Encyclopedia" (Wikipedia, 2017); but the meaning of encyclopedia has changed drastically. Such are actually good starting off venues to look for information, but they do not hold the reassurance of traditional encyclopedias that the information is genuine or authentic. To be fair, the articles on the Wikipedia site itself often carry notes such as "needs additional citations" and "the neutrality of this article is disputed." That said, Wikis are products of a marketplace form of knowledge. So, while Wikis are useful, there is no assurance of correctness.

The marketplace also is the source of word creation. A 2017 advertisement on the Toronto subway for the Ripley's Museum in Niagara Falls stated that an exhibit of the "Lizard Man is More Lizardy than Before" (Ripley's Believe it or Not Museum, 2017)." The meaning of "lizardy" can be connected to the synesthetic meaning of the exhibit itself via an emphasis of the qualities we would associate with a lizard. The meaning of the advertisement, then, is that the Ripley's exhibit of a "Lizard man" has somehow transcended itself in the measure of whatever it is that we call "lizard." We understand that this measure of "lizardy-ness" is absurd and that the attempt of this advertisement is a humorous attempt to bring people, who may have seen the exhibit years ago, back to see it again by instilling in them a childlike sense of wonder through the infantile nature of the word coinage. Indeed, the word *lizardy* was used in a 2005 children's book called *One Final Firecracker* (Wikitionary, 2017). Given the origin of the word in childhood narrative, the museum evokes a nostalgic sense at the same time that it conveys the sensation of lizardy-ness. A child does not know how to make English words and might create the word "lizardy." An adult would instead use the adjectival phrase "is more like a lizard." Neologisms in advertising are clearly indexical—in this case of a previous childhood state of mind

The fact that we can understand the connotations built into this coinage, implies that we are programmed for connotative (rhetorical) meaning at any time. The fact that a childish neologism was chosen instead of something more formal conveys a sense of childlike informality. More than just informality, the choice of a childish neologism also represents a specific kind of word that would evoke childish wonder, an emotion which might speak to the more jaded urban subway dweller and thus we should not be surprised to find this advertisement in an urban setting, on a subway platform. The word *lizardy* captures our mood and our mode of understanding in an urban setting. When we use words, such as *lizardy*, we are using a childish word, albeit in a mock humorous way, to connote a childish demeanor and attitude. Actually, this can be called "adulating" of word meanings.

Blending of words is a common mode of neologism-construction. For example, *vacation* and *stay* can be amalgamated into *staycation*, a portmanteau that is essentially a pun. The blend comes from the phonemic syllables /vay/ and /stay/. Interestingly, this new construction entered the lexicon via the Canadian TV show *Corner Gas* from the episode "Mail Fraud" (Butt & White, 2005), and is now listed in various (Merriam Webster, 2017, Oxford English Dictionary, 2017). To take a staycation means to go on vacation but instead of travelling to a destination (which one presumes is the point of vacations), one stays at home. The word *vacation* has also blended with *play* to produce *playcation*, which implies a vacation where one plays. The word first appeared on an advertisement for the Ontario Lottery and Gaming Corporation, which proclaimed that "staying home is boring but going to the Casino is "play." The implication is that going to a gambling casino is a vacation of sorts involving play.

Corner Gas takes place in the small fictional community of Dog River in the middle of rural Saskatchewan, Canada. The denizens of the town are average, rural, working class Canadians. They do not aspire to anything much beyond what they already have. Brent, the lead male character, supposedly goes on wonderful vacations, but Lacy, the lead female character and Brent's scripted romantic partner, discovers that Brent's vacations are imaginary. Brent sits and imagines his so-called great vacations. The word *staycation*, from the show's vantage, points to a character's lack of wealth, in that he cannot afford a "real" vacation, so he replaces it with his fertile imagination, as the coinage suggests. Oddly enough, the word *staycation* has spread throughout cyberspace, becoming indicative of a worldwide economic condition. As several websites, including Investopedia note, because of rising costs such as fuel and accommodations, staycations are alternatives to the higher cost of real vacations (Investopedia, 2017). Ostensibly, because fewer people can afford vacations the popularity of staycations rose and so the word itself became a new kind of index, in the Peircean sense, extending Brent's socioeconomic status more broadly in terms of its referential domain.

Finally, the new blend *wannarexia* of *wannabe* and *anorexia* (http://www.urbandictionary.com) indicates that culture does indeed have a leash on words and their referents. As Deborah Cohen notes, in an article in the *British Medical Journal*:

I first heard about "wannarexia" while sitting on the bus. A group of teenage girls on their way into central London to go shopping were dissecting each of their classmates. One girl in particular received marked censure. "She's such a wannarexic," said one. The others all agreed (Cohen, 2007).

Wannarexia, it turns out, is a pejorative term which pokes fun at young females who aspire to developing an eating disorder intentionally, but in the view of their peers have not yet succumbed to the actual condition. A Google search for the word *wannarexia* now shows that it has entered the clinical lexicon: "Wannarexia is a term coined to describe teens that want to develop an eating disorder" (Teen Eating Disorders, 2017). The term is a sardonic one, implying purported body image benefits from becoming anorexic. It is truly an indictment of the times. Incidentally, *wannabe* itself is a contraction of "Want to be" and which was coined, most probably, in the 1980s punk rock scene (Rosenblat, 2000). The word *wannabe* was itself invented as a pejorative epithet and is indexical to those who "want to be" like the real thing. This pejorative meaning is carried into the meaning of the word "wannarexia." The Urban Dictionary appropriately defines *wannarexia* as follows: "In fact, they do not have an eating disorder at all. Most wannarexic people feel that anorexia is a "quick fix" to lose weight and that it is glamorous (Urban Dictionary, 2019)." Deborah Cohen goes on, in her article, to state that there are websites dedicated to wannarexia and how to "achieve" anorexia or at least simulate it (Cohen, 2007). She also explains how "wannabe" eating disorders and other "mock" psychological illnesses are a growing concern in medical practice (Cohen, 2007). One could say that *wannarexia* is indexical not only of a problem of personal body image, but also of a cultural malaise that extols thinness as a virtue or even an icon to be emulated by young females, no matter what the physical and psychological costs.

CONCLUSION

The foregoing discussion was intended to argue by illustration that word creation is based on indexicality—a need to interpret the world ontologically and existentially. Words are more than their denotative referents. They always embody theoretical, connotative, metaphorical, and indexical structure in tandem. They can be viewed as indexical data which point to ourselves as individual people, and to the groups in which we live. The words we use, how we use them, the things we say and the words that we use to say these things are indexical at every level of reference.

While it is important to study the differences in meaning of words in the abstract, as Saussure suggested, it is at least as important to understand what these differences signify indexically, as Peirce certainly knew. This is not to say that the oppositional differences between uses of the words are not significant, but the fact that these differences exist are not just linguistically important, these difference are also semiotically significant. It is worthwhile to remember that while words do have denotative meaning, words are social indices, that point at social, cultural and even individual temperaments and viewpoints. Words are a lodestone which points to not only who we are, but who we think we are and how we see ourselves. Unlike Humpty Dumpty, words always mean more than we intend them to mean. The question for us is to elaborate on and to discover why this is so in human communication.

REFERENCES

Atkin, A. (2013). Peirce's theory of signs. *The Stanford Encyclopedia of Philosophy*. Retrieved July 15, 2017 at 11:01 AM EDT from https://plato.stanford.edu/cgi-bin/encyclopedia/archinfo.cgi?entry=peirce-semiotics

Barthes, R. (1957). *Mythologies. Translated, Lavers, A*. London: Paladin Press.

Blonsky, M. (Ed.). (1985). On signs. Baltimore, MD: John Hopkins University Press.

Butt, B., & White, K. (2005). Mail fraud. *Corner Gas*. First broadcast in Canada, October 24, 2005.

Carroll, L. (2015). *Through the looking glass*. Retrieved July 12, 2017 at 12:23 PM EDT from http://pdfreebooks.org/0-carroll.htm

Cohen, D. (2007). The worrying world of eating disorder wannabes. *The British Medical Journal*. doi: https://doi-org.myaccess.library.utoronto.ca/10.1136/bmj.39328.510880.59

Danesi, M. (1993). *Vico, metaphor and the origin of language*. Bloomington, IN: Indiana University Press.

Danesi, M. (1994). *Cool: The signs and meanings of Adolescence*. Toronto: University of Toronto Press. doi:10.3138/9781442673472

Danesi, M. (2016). *Language, society and new media: Sociolinguistics today*. New York: Routledge.

Danesi, M., & Perron, P. (1999). *Analyzing cultures: An introduction and handbook*. Bloomington, IN: Indiana University Press.

de Saussure, F. (1959). *Course in general linguistics* (W. Baskin, Trans.). Philosophical Library. Retrieved on July 23, 2017 at 12:59 PM EDT from https://archive.org/stream/courseingeneral00saus/courseingeneral00saus_djvu.txt

Eco, U. (1976). *A theory of semiotics*. Bloomington, IN: Indiana University Press. doi:10.1007/978-1-349-15849-2

Eco, U. (1979). *The role of the reader: Explorations in the semiotics of texts*. Bloomington, IN: Indiana University Press.

Fallows, J. (2015). That weirdo announcer-voice accent: Where it came from and why It went away. *The Atlantic*. Retrieved July 23, 2017 at 2:58 EDT from https://www.theatlantic.com/national/archive/2015/06/that-weirdo-announcer-voice-accent-where-it-came-from-and-why-it-went-away/395141/

Ginsberg, A. (1956) *Howl*. Retrieved July 24, 2017 at 10:27 EDT from https://www.poetryfoundation.org/poems/49303/howl

Gorilla Foundation. (2017). Retrieved July 12, 2017 at 12:33 PM EDT from http://www.koko.org/sign-language

Investopedia. (2017). *Staycation*. Retrieved on July 24, 2017 at 4:52 PM from http://www.investopedia.com/terms/s/staycation.asp

Kay, P., & Kempton, W. (1984). *What is the Sapir Whorf Hypothesis?* American Anthropological Association. Retrieved July 23, 2017 at 12:42 PM from http://www1.icsi.berkeley.edu/~kay/Kay&Kempton.1984.pdf

Keneally, T. (1982). *Schindler's List*. Hodder and Stoughton.

Keneally, T., & Zaillian, S. (1993). *Schindler's List*. Dir. Steven Spielberg. Amblin Entertainment. Universal Pictures.

Lakoff, G. (2012). Explaining embodied cognition results. *Topics in Cognitive Science*, *4*(4), 773–785. doi:10.1111/j.1756-8765.2012.01222.x PMID:22961950

Lakoff, G., & Johnson, M. (1980). *Metaphors we live by*. Chicago: University of Chicago Press.

McWhorter, J. (2003). *Doing our own thing: The degradation of language and music and why we should like care*. New York: Gotham Books.

Merriam Webster Dictionary. (2017). Retrieved various dates and times from http://www.merriam-webster.com/dictionary/

Nye, D. (1981). *The Semiotics of Biography: The Lives and Deeds of Thomas Edison*. A Monograph of The Toronto Semiotics Circle.

Oxford English Dictionary. (2017). Retrieved on July 18, 2017 at 11:51 AM EDT from https://en.oxforddictionaries.com/definition/staycation

Oxford English Dictionary. (2017). *A-Level: Introduction to OED online*. Retrieved August 1, 2017 at 1:21 PM EDT from http://public.oed.com/resources/for-students-and-teachers/a-level/

Ripley's Believe it or Not Museum [Billboard]. (2017). Toronto Transit Commission Spadina Station Line 1 Subway, South Bound Platform on July 27 2017 at 9:38 AM EDT.

Rosenblat, D. (2000). *Re: Etymology of wannabe*. Retrieved July 19, 2017 at 4:29 EDT from http://listserv.linguistlist.org/pipermail/linganth/2000-February/000147.html

Steinbuch, Y. (2017). *The New York Times*. Retrieved on July 24, 2017 at 10:12 EDT from http://nypost.com/2017/04/24/this-is-whats-left-after-the-mother-of-all-bombs-hit-afghanistan/

Teen Eating Disorders. (2017). *What is wannarexia*. Retrieved on August 8, 2017 at 11:59 AM EDT from http://www.teeneatingdisorders.us/content/what-is-wannarexia.html

Teflpedia. (2017). Retrieved on July 23, 2017 at 1:36 PM from http://teflpedia.com/IPA_phoneme_/%CA%83/

United States Holocaust Museum. (2017). Retrieved August 1, 2017 at 12:58 PM EDT from https://www.ushmm.org/wlc/en/article.php?ModuleId=10007962

Urbancic, A. (1994). Urban semiotics: Where is downtown? *Signifying Behaviour*, *1*(1), 23–34.

Wikipedia. (2017). Retrieved on August 1, 2017 at 1:50 PM EDT from https://en.wikipedia.org/wiki/Online_encyclopedia

Wikitionary. (2017). Retrieved on August 1, 2017 at 3:17 PM EDT from https://en.wiktionary.org/wiki/lizardy

Chapter 10
Math Talk as Discourse Strategy

Stacy Costa
Ontario Institute for Studies in Education, Canada

ABSTRACT

Mathematical understanding goes beyond grasping numerical values and problem solving. By incorporating visual representation, students can be able to grasp how math can be understood in terms of geometry, which is essentially a visual device. It is important that students be able to incorporate visual representations alongside numerical values to gain meaning from their own knowledge. However, it is also vital that students understand mathematical terminology, via a dialogical-rhetorical pedagogy that now comes under the rubric of "Math Talk," which in turn is part of a system of teaching known as knowledge building, both of which aim to recapture, in a new way, the Socratic method of dialogical interaction. This chapter explores how knowledge building, as a methodology, can assist in furthering student understanding and how math talk leads to a deeper understanding of mathematical principles.

INTRODUCTION

Knowledge Building researchers have identified pedagogical methods that foster idea improvement among learners (Zhang et al., 2009, Scardamalia et al., 2012) via "collective intelligence" development (Broadbent & Gallotti, 2015), which can be described simply as group-based learning (Knowledge Building Gallery, 2016). Studies have examined the validity of Knowledge Building in the area of mathematics (Moss & Beatty, 2006, Moss & Beatty, 2010, Nason, Brett, & Woodruff, 1996, Nason, Woodruff, & Lesh, 2002, Nason & Woodruff, 2004, Knowledge Building Gallery, 2016) finding a common pattern—namely an increase in learning when a derived technique, known as "Math Talk," is utilized in group-based learning contexts. This paper aims to contribute to the growing body of research and theory on Knowledge Building and Math Talk, focusing on elementary school learning.

Mathematics is critical for 21st-century skills. As stated in the Ontario Ministry of Education's document *Growing Success* (2009), "Education directly influences students' life chances and life outcomes" (p. 6). Today's global, knowledge-based economy requires a citizenry with deep understanding of disciplinary knowledge as well as a broad range of what are popularly known as "21st century skills." In the early grades it is a strong predictor of later academic achievement (Duncan et al., 2007). However,

DOI: 10.4018/978-1-5225-5622-0.ch010

with cases of mathematical anxiety and avoidance of math within elementary learning environments, it is hard to engage anxious students in particular in math learning, and allow them to retain information and skills for long-term use and application. Such students do not show an ability to transfer mathematical skills; rather, they rely on mechanical formulas, mathematical "tricks" they might have learned, all of which do not contribute to math competence and fluency. Current evidence of this inability can be gleaned from the Province of Ontario's 2016 EQAO scores, which show only 63% of students meeting the provincial grade three mathematics standards. Bredekamp (2004) argues that educators should examine research and practices on how children actually learn mathematics and develop relevant methods for classroom usage. Knowledge Building pedagogy is grounded in students' natural curiosity and helps expand innate conceptualization skills through a model of learning that encourages the "community dynamics" of the sort found in knowledge-creating organizations. This form of teaching has been shown to boost achievement in literacy, mathematics, engineering, science, across the curriculum (Chen and Hong, 2016). Knowledge Building is built on interactivity models of psychology, encouraging students to design, revise, discuss and apply ideas developed in community (group-based) contexts and spaces so that they can learn in terms of a collective responsibility for knowledge.

The need for mathematics as a tool for creative work in a technology-rich knowledge society is widely recognized (Wagner & Dintersmith, 2015, Ritchhart, 2015). Mathematics is required for problem solving, reasoning, questioning, computational activities and their creative application, all of which reach beyond the typical well-defined classroom problems characterized by a process of "instructor input→student problem-solving→verification." There is growing agreement that this instructional approach is ineffectual (Towers, 1999, Schunk, 2012, Baroody, 2000), although there is no consensus about developing effective alternatives. However, the claim made here is that Knowledge Building is one such alternative (Scardamalia & Bereiter, 2003) and that Math Talk—its methodological derivative—does indeed allow students to learn mathematics effectively through dialogue, discussion, and other interactive strategies. In effect, dialogue, and its rhetorical structure (analogies, allusions, comparisons, and so on) seems to be the best way to learn—a philosophy of education that goes right back to ancient times.

Fostering mathematical thinking is always the goal of curriculum guidelines and professional development workshops for teachers. The *Ontario Mathematics Guidelines*, for example, proclaim that learning involves the ability to control the "relevant facts, skills, and procedures" and develop "the ability to apply the processes of mathematics, and acquire a positive attitude towards mathematics" (Ontario Ministry of Education, 2005, p. 3). This document goes on to state that the main strands that constitute math knowledge are as follows: number sense and numeration, measurement, patterning and algebra, data management and probability, and geometry and spatial sense (Ontario Ministry of Education, 2005). This subdivision applies to math education in all grades and provides a framework for the subject matter to be covered during a school year. Leher & Chazan (1998) and Whiteley (1999) decry, however, the fact indicate that geometry is often neglected or considered of lesser value than numeracy and algebra. So, they suggest adding it to the strands so that math fluency can be developed as an integrated system of numeracy and geometry.

Osta (2014) defines a curriculum as "the pedagogical framework or philosophy underlying the teaching practices and materials, training programs for supporting teachers, and guidelines for assessing students' learning" (p. 417). According to Putnam (2004, p. 33) the educational curriculum is a prescriptive document which dictates an "ordered set of goals in which students are expected to learn, and recommended strategies for teaching material." Among the ever-emerging collaborative-based pedagogies designed to cater to these objectives those that focus on deep learning, inquiry-based learning, project-based learning, and so forth, seem to stand out as efficacious. More generally, the literature on collaborative learning has documented the advantages that can be reaped when students work together. As is well known, Vygotsky (1978) suggested that more challenged learners can be assisted by working with others who are more knowledgeable. The more knowledgeable person need not be the teacher. Every student can contribute to the learning process by injecting his or her talents into the dialogical process.

Knowledge that may be out of reach for someone learning alone may be accessible if the learner has the support of peers or more knowledgeable persons (Van de Walle et al., 2014, p. 6). In Ontario math classrooms, and in many similar educational settings, students come from different cultural backgrounds. This allows them to bring distinctive ideas to the classroom and possibly fill gaps in knowledge by bringing different cultural resources and perspectives to bear on interpreting math ideas or problems. Diversity of this kind enhances learning. As Boaler (2002) argues the view that there is a common learning denominator in all students is erroneous and counterproductive since it leads to mathematics being taught as if students have the same ability, preferred learning style and pace of working. Boaler (1998) found that math students in an open "project-based" environment developed a stronger conceptual understanding of math both within and outside the classroom. The main finding was that collaboration attenuated risks associated with isolating and competitive tasks. Many confuse collaboration and working together with cheating or "passing on the buck" to some in the group. However, this conflation is a misleading view, since it has been found over and over that collaborative learning, which is based on pooling diverse interpretations of the same problem, can allow for a better understanding of complex ideas. This is often called scaffolding, which is a factor in bridging and expanding a student's capability by linking prior knowledge to new knowledge (Wood et al., 1976, Wood and Wood, 1996). Duncan et al. (2007) found that early number knowledge is a key predictor of later academic success and that by allowing number concepts to emerge in a collaborative manner, at an earlier age, students are better able to extend their knowledge to new topics. Students become more comfortable within mathematics and can interpret mathematics learning from a more engaging perspective.

Stigler and Hiebert (1999) reported results showing that students who worked independently risk having insufficient opportunities to translate their work into words, thus hindering their understanding. Students seem to learn best through exploring and emulating how mathematicians tackle unsolved problems. This also seems to be true of mathematics as a profession. The Clay Mathematics Institute is dedicated to encouraging the use of historical artifacts and of getting mathematicians to engage in collaborative tasks. Many of the prizes handed out by the institute are to groups of mathematicians who have worked collaboratively on some project. The underlying idea is that scholars and experts work best and more effectively through collaboration in order to advance mathematical knowledge. In mathematics, it would seem, collaboration wins out over individualized learning. Stahl (2005) argues that revising the mathematics curriculum, orienting it towards a collaborative modality, would allow for a deeper mathematics understanding which comes about when groups tackle problems together and learn from each other in the process. Adapting the mathematics curriculum to the Clay Institute and Knowledge Building model has been found in fact, to enhance classroom learning outcomes (Stahl, 2005).

Pimm (1987) found that when students attempted to express their thoughts out loud in words, other students were helped in the classroom, allowing them to better self-organize their thoughts (p. 23). By having students "turn up the volume on their own self-talk" (Pimm, 1987, p. 25), learning outcomes tend to be more positive and effective. Encouraging student self-talk allows for open-ended multi-step scaffolding derived from collaborative problem solving in classrooms. Through reflective self-talk and engagement with others, students increase their levels of numeracy and procedural knowledge.

Scardamalia and Bereiter (1983) note that "observation, practice and rule learning" (p.63) are the core aspects of acquiring procedural knowledge and thus should not be discarded. But dialogical learning is the system that complements and even directs the flow of rule learning (Sinclair, 2005), since it allows students to share their relative strengths and address their weaknesses. In effect, collaborative learning favors "enculturating" processes in a situation (Towers, 1999), that is, the use of diverse terminologies based on different backgrounds that shed light on some structural problem. Finlayson (2014) found that mathematics structured to produce right or wrong answers from the students encourages formulaic thinking and inhibits the development of higher order learning. Students need to connect previous and new knowledge; simply seeing new forms without understanding does not provide the cognitive foundation for understanding.

"Communication is an essential part of mathematics and mathematics education" (NCTM 2000, p. 60). By becoming familiar with technical mathematical language for expressing new concepts and ideas students can broaden and deepen their overall math competence (Baroody, 2000, Ginsburg et al., 1999, NCTM, 2000). Mathematical communication involves "adaptive reasoning" (Kilpatrick et al., 2001, p. 170) and argumentation (Andriessen, 2006), which give students opportunities to think mathematically as part of a dialogical and social process. Mathematical communication offers students opportunities and encouragement with important consequences, as will be elaborated below.

MATH TALK

The discipline of mathematics consists of symbols, equations, and numbers, a type of language required to interpret and learn mathematical concepts. Mathematics depends fundamentally on language. And this means that the rhetorical strategies inherent in metaphorical comparisons and analogies that characterize common discourse may be at the core of learning. Math Talk is based on this basic premise; it has been examined by the math education literature for over 25 years (NCTM, 2000). Math Talk aims to tap into student curiosity, encouraging them to talk about math in the classroom so that they can relate what they learn to what they know in a discourse-based manner. Teaching and learning mathematics at the elementary level always involves some form of discourse, both between students and between students and the teacher. The advantage of a Math Talk pedagogy lies in its consistent and systematic use of dialogue, rather than a random and sporadic use. As Mountz (2011) characterizes it, "Math Talk is a style of discussion in the classroom that allows students to learn through discourse" (p. 3). Students are encouraged to formulate their views concerning what math ideas are about and how to grasp them, providing valuable input to other students and to teachers as well. While allowing students to figure out solutions after some basic instructional period of time, students need to be able to formulate high-level ideas to work out problem-solving through the ingenious use of discourse to explain them and to relate them to others. Students will this be better able to connect ideas to discourse modalities, thus extending the learning process considerably.

Morgan (2014) notes that some elements of vocabulary is uniquely mathematical, such as, for example, the term "parallelogram;" however mathematics also utilizes everyday language with subtly different meanings as can be seen in terms such as *prime*, *multiple* and *differentiate*. This crisscrossing of semantic domains is very productive for Math Talk pedagogy. Students can thus differentiate between meanings and to understand and apply knowledge via Math Talk, since both semantic systems converge within it. Students will this not feel alienated nor intimidated by math but instead eager to participate in its knowledge system because of its contiguity with language. However, within geometry, not everything can be verbalized so easily; here, figures and geometric diagrams play an essential role. Some mathematical explanations are thus better handled through visualization, rather than verbalization. However, in discussing geometrical notions there is an indirect form of discourse that ties them down semantically as well. Martinovic (2004) refers to this type of learning as "vicarious learning," which refers to learning by overhearing an educational conversation (p. 29). While students may not always be directly involved in Math Talk in such situations, they are nevertheless dialoguing in a "vicarious" way. "Vicarious learning" provides students with a basis of reference when clarifying their understanding or when sharing their answers in the classroom.

Judith Falle (2004) defines the overall goal of Math Talk pedagogy having "students attempt to explore, investigate and solve problems together, resulting in more success their mathematics education" (p. 19). Math Talk is "instructional conversation which references the crucial aspects of: questioning, explanation of math thinking, a source of math ideas and the responsibility of mathematical learning" (NCTM, 2000). Cooke and Adam (1998) emphasize that Math Talk is effective in allowing students to analyze, justify, defend their views, and reason, which are the first steps in developing a stronger mathematical fluency and vocabulary. Falle (2004) examined how Math Talk constitutes a "small-scale strategy of learning that immerses students into speaking patterns which connect to larger-scale forms of learning unconsciously" (p. 19). With students struggling not only to articulate their thoughts, they can also be confused by math symbols. The importance of Math Talk is to impart a "mathematical register" for connecting ideas, symbols, and discourse (Pimm, 1987). "The amount of 'Math Talk' produced by a child's input from the classroom is related to the growth of a child's acquisition of mathematically relevant language" (Boonen et al., 2011, p. 283). This observation implies that conversation should not be constrained to a child-teacher system, but rather to a peer-to-peer one as well. The argument for a systematic type of classroom dialogue that includes everyone is that all, not just those inclined to talk, should be engaged. In a Math Talk model the interconnection of language and mathematics to describe shapes, charts, graphs and to describe how formulas and expressions are constructed produces a supportive context for understanding. In sum, Math Talk classrooms are based on the use of language pedagogically, so as to conduct lessons as clearly as possible. As Resnick (1988) and Martinovic (2004) argue, mathematics is typically taught as a discipline that is distant from language and its rhetorical structure, thus removing learner voices from learning; Math Talk, in the other hand includes debate, argumentation, and gives space to multiple interpretations, allowing students to discuss math ideas fluently and fluidly. Falle (2004) found that "when teachers do listen in mathematics they do so to correct students rather than to reveal possible student cognitive development" (p. 26). Math Talk is also diagnostic; it reveals what students may or may not have grasped, how they value certain concepts on their own terms and thus suggest how they can modify their own responses more advantageously.

Mercer & Sams (2006) emphasize that talk-based group activities develop the individual's mathematical reasoning skills, which serve to enhance the student's problem-solving skills. Hufferd-Ackles et al. (2004) define a Math Talk community aptly as "students who take responsibility for their own learning and the learning of others, which leads to a demonstration of their understanding of problems and of volunteering to assist struggling students" (p. 106). Cooke & Adams (1998) discovered that Math Talk might, however, be difficult to integrate into the classroom because most curricula envision the teacher as being at the center of the learning process. With teachers assuming a direct, authoritative role, there is a tendency for classroom learning to become a rigid competitive process that is hardly conducive to positive learning outcomes. In this situation, constraints are placed upon students by textbook-based learning with no room for spontaneous questions or additional interpretation (Finlayson, 2014), especially in the area o specialized terminology. As Pimm (1988) notes, "a distinctive feature of discourse about mathematics is the widespread use of technical vocabulary" (p. 59). Math Talk provides a space in a classroom to all participants, giving them a voice in the process of grasping the specialized vocabulary through interaction. This situation is defined as encouraging "rich discourse," which Imm & Stylianou (2012) characterize as a learning situation which values the student's ideas as "rich, inclusive, and purposeful" (p. 131). Typically, in the classroom students are not always encouraged to utilize pragmatic conversations about math ideas because the objective in this type of situation is for students to reach an answer to a problem they have been presented individualistically. The reason students struggle is that math conversations for elementary students are rarely casual. Instead, they originate from within the classroom space and are guided by the teacher. Nathan & Knuth (2003) suggest that "productive discourse" can only be engendered in situations that allow teacher-student interactions to build upon ideas through branching out to include other class members.

In Math Talk agreeing and thinking are encouraged to allow students to discover mathematical concepts and to work out their inherent misconceptions (Bruce 2007). Hutton et al. (2013) note that each student must contribute unabashedly to the unfolding discourse, because as Pimm (1987) asserts, mathematical language and everyday language converge semantically and thus every student can feel a part of the conversation without distancing themselves from the mathematic involved. Pimm explores not only the importance of using clear, consistent terminology, or knowledge of the discourse itself, but also understanding the context of words and the context of how to correctly use words mathematically. Martinovic (2004) argues that within the domain of mathematics an important activity for learning is self-explanation, which is different from explanation to others; it is more focused than speaking aloud which may not be motivated by an effort to understand. Such learning strategies can be related to reflection and elaboration. Students need to express their ideas so that they can develop their understanding of mathematical concepts via ordinary language. Klibanoff et al. (2006) found that the more Math Talk is used the greater the acquisition of mathematically relevant language.

KNOWLWDGE BUILDING

Schunk (2012) defines learning theories as attitudes, skills, strategies underlying the process of acquiring and modifying knowledge. Social constructivism (Vygotsky 1978) is an intellectual by-product of this general philosophy; and it is the basis of Knowledge Building pedagogy. Vygotsky emphasized the need for social interaction as a critical vehicle for understanding. In this he included debating, dialoguing, and conversing, all of which enhance understanding because they involve a negotiation of meaning in which the student provides and receives, as in a bartering situation, relevant insights. Students negotiate meaning and refine one another's knowledge as core developmental tools to further learning. As Vygotsky stated, "Every function in the child's cultural development appears twice: first, on the social level, and later, at the individual level; first between people, then inside the child" (Vygotsky 1978, p. 57).

Social constructivism became an area of interest in mathematics education in the 1980s and continues to evolve within it (Artzt and Newman, 1997, Davidson, 1990, Earnest, 1991, 1994, 1998). As elaborated above, collaborative learning in the field of mathematics is grounded in a social constructivist model of learning (Yackel et al. 2011). This envisions students as working together to solve a problem while the teacher stands by as a facilitator. Success depends on contributions from all students. Student accountability forces the community to trust and depend on each other while being respectful and contributing to advance collective and individual understanding in tandem. Overall, this educational model is a form of student-directed learning that engages students in interpretation, synthesis, and investigation through collaboration.

Knowledge Building is a principle-based social constructivist approach in which students reframe and advance ideas in a manner that is conducive to idea improvement. The term "Knowledge Building" was introduced by Bereiter and Scardamalia in the mid 1980s (Scardamalia, Bereiter, Mclean, Swallow, & Woodruff, 1989, Scardamalia & Bereiter, 1991, 1992). Scardamalia (2002) articulated twelve Knowledge Building principles to incorporate in practice. These can be paraphrased as follows:

1. **Real Ideas, Authentic Problems:** Knowledge problems arise from efforts to understand the world. These are problems that learners care about. Thus, a knowledge forum creates a culture for creative work with ideas. Notes, views, and "Rise Aboves" serve as direct reflections of the core work of the organization, and the ideas of its creators.
2. **Improvable Ideas:** All ideas are treated as improvable. Participants work continuously to improve the quality, coherence, and utility of ideas. For such work to prosper, the classroom culture must be one of psychological safety, so that students feel safe in taking risks. Knowledge Building supports all aspects of learning design, but is always prepared to adapt and to revise itself. Its background operations are designed to reflect an ability to change: continual improvement, revision, theory refinement.
3. **Idea Diversity:** Idea diversity is essential to the development of knowledge advancement. To understand an idea is to understand the ideas that surround it, including those that stand in contrast to it. Idea diversity creates a rich environment for ideas to evolve. The technique called Knowledge Forum facilitates students' ability to link ideas, and to bring together different combinations of ideas in different ways, promoting a productive use of diversity.
4. **Epistemic Agency:** Participants set forth their ideas and negotiate a fit between personal ideas and the ideas of others, using contrasts to spark and sustain knowledge advancement rather than depending on others to chart the course. They deal with a problem of goals, motivations, evaluations and long-range planning that are normally left to teachers. Knowledge Forum provides support for theory construction and refinement. Scaffolds for high level knowledge processes are reflected in the use and variety of epistemological terms.
5. **Community Knowledge, Collective Responsibility:** Contributions to shared, top-level goals are prized and rewarded as much as individual achievements. Team members produce ideas of value to others and share responsibility for the overall advancement of knowledge in the community. Knowledge Forum's open, collaborative workspace holds conceptual artefacts, which are contributed by community members. Community membership is defined in terms of reading and building on the notes of others, ensuring views are informative and helpful for the entire community. Participants share responsibility for the highest levels of the knowledge attained.
6. **Democratizing Knowledge:** All participants are legitimate contributors to the shared goals of the community; all take pride in knowledge advances achieved by the group. The diversity and divisional differences represented in any group do not lead to separations along have/have-not or innovator/non-innovator lines. All are empowered to engage in knowledge innovation. Analytic tools allow participants to assess evenness of contributions and other indicators of the extent to which all members do their part in a join enterprise.

7. **Symmetric Knowledge Advancement:** Expertise is distributed within and between communities. Symmetry in knowledge advancement results from knowledge exchange and from the fact that to give knowledge is to get knowledge.

8. **Pervasive Knowledge Building:** Knowledge Building is not confined to particular occasions or subjects but pervades mental life, in and out of school.

9. **Constructive Uses of Authoritative Sources:** To know a discipline is to be in touch with the present state and growing edge of knowledge in the field. This requires respect and understanding of authoritative sources, combined with a critical stance toward them. Knowledge Building encourages participants to use authoritative sources, along with other information sources as the basis for their own Knowledge Building and idea-improving processes.

10. **Knowledge Building Discourse:** The discourse of Knowledge Building communities results in more than the sharing of knowledge; the knowledge itself is refined and transformed through the discursive practices of the community—practices that have the advancement of knowledge as their explicit goal. Knowledge Forum supports rich intertextual and inter-team notes and views and emergent rather than predetermined goals and workspaces. Revision, reference, and annotation further encourage participants to identify shared problems and gaps in understanding and to advance understanding beyond the level of the most knowledgeable individual.

11. **Embedded, Concurrent and Transformative Assessment:** Assessment is part of the effort to advance knowledge—it is used to identify problems as the work proceeds and is embedded in the day-to-day workings of the group. The community engages in its own internal assessment, which is both more fine-tuned and rigorous than external assessment, and serves to ensure that the community's work will exceed the expectations of external assessors. Standards and benchmarks are objects of discourse in Knowledge Forum, to be annotated, built on, and risen above. Increases in literacy, twenty-first-century skills, and productivity are by-products of mainline knowledge work, and advance in parallel.

12. **Rise Above:** Creative Knowledge Building entails working toward more inclusive principles and higher level formulations of problems. It means learning to work with diversity, complexity and messiness, to achieve new syntheses. Knowledge Builders transcend trivialities and oversimplifications and move beyond current best practices. Conditions to which people adapt are a result of the success of other people in the environment. Notes and views support unlimited embedding of ideas and support emergent rather than fixed goals.

In essence, Knowledge Building pedagogy supports students in posing questions, defining goals, acquiring and building knowledge and assuming collective responsibility for knowledge advancement. Collective responsibility is more complex than collaboration, requiring individual contributions to the community and individual reflective thought to enable individuals to achieve something greater than they could by working exclusively on their own, but never suppressing independent action. Knowledge Building should achieve a productive blend of individual and teamwork encouraging students to examine their own ideas while advancing the community's knowledge. Knowledge Building discourse is supported through "epistemological markers," which support a "process towards revelation and special problems" that can be easily identified and solved pedagogically (Guba & Lincoln, 2005, p.196). Co-creation assists students by allowing them to explore solutions in a space suited to their needs. In Knowledge Building pedagogy, the classroom shifts focus from providing pre-validated answers, and instead focuses on thinking as a consequence of group interaction, facilitated through blended face-to-face and online discourse.

Mathematics education must resonate with a new generation of students who experience the need for mathematics through many technological facets of their lives. The classroom should be able to incorporate technology as a complementary form of pedagogy. One area of direct relevance is in computer-assisted collaborative pedagogy. While learning is acquired ultimately individualistically, learning is enhanced through collaboration (Stahl 2005). Confrey and Kazak (2006) and Steffe and Kieren (1994) argue that a curriculum focussed on only word problems is a complicated learning method for students, which may be riddled with misconceptions. Mathematical knowledge becomes subjective knowledge through construction in a collaborative framework; individual student learning is understood as undergoing reformation and reconstruction by peer interactions. When combined with the computer-assisted collaborative model, a process called "Distributed Cognition" emerges whereby knowledge can be spread across a group of people through the digital tools that they use to solve problems (Norman, 1993, Hutchins 1996). This technological model thus not only supports student collaboration but also promotes it, encouraging a process of mutual understanding, respect, and openness to the ideas shared and explored.

Before the current technological possibilities, Brown (1982) referred to the creation of an "immediate environment" needed to incorporate written symbols, diagrams on paper and utilizing written accounts. In all these techniques, there is awareness that face-to-face communication is rendered more effective through written and technological support. Mathematics classrooms should consist of blended face-to-face communications with technological environments to enhance and complement one another. Students should not have the sole goal to regurgitate content but to grow their ideas. Teachers should ensure a balance between the use of paper-and-pencil learning with the digital tools. In this way, the student can come to see the abstract principles involved, independently of the medium used. Knowledge Building pedagogy and technology has been used to support mathematics education in this blended fashion (Beatty & Moss, 2006a, 2006b, 2009, Hurme & Jarvela, 2005, Nason & Woodruff, 2002, Moss & Beatty, 2010, Moss, Beatty, McNab, & Eisenband, 2006, Hutton, et al., 2013). Nason & Woodruff (2002) noted the need for Knowledge Forum to provide symbolic math representations to enhance the discourse space. The most recent version of Knowledge Forum (KF6) allows, in response, a use of symbols and graphical representations of math ideas; students can now hand-write their equations using a blended method of typed and handwritten text, utilize voice-over command software to transform their spoken ideas into typed text, including equations, import pictures, and work more productively with other software devices. By uploading an image from a smartphone or tablet, students can now share graphical representations of their ideas across platforms and across writing methods. In addition to the capacity to add pictures, students can build representations of mathematical structures and designs and other multimedia components can be attached and uploaded onto the group view, allowing students to build on ideas in various formats. The capacity to manipulate and present data in different forms supports students to constructing their own representations, comparing them, and extrapolating principles of mathematics. Some students benefit from showcasing their knowledge in a visual manner, others in a text-based manner; the technology is flexible as to cognitive learning style. Moreover, all uploads and representations can be co-authored, allowing for collaborative production with potential for stronger motivation and engagement. Boaler (2002) has put forth the common critique that "math is often taught as if students have the same ability, and pace of working" (p.101). Knowledge Building pedagogy, on the contrary, is designed to accommodate diverse needs. Lazarus et al. (2013) argued that the Ontario mathematics curriculum (Ontario Ministry of Education, 2005) does not expand on what communication looks like in online environments, but identifies what it means to communicate within mathematics (p. 34). Sinclair (2005) found that students need additional sessions at which they can discuss their mathematical ideas, explore other mathematical definitions, and find further information to support the use of mathematical terminology. Students need to organize and consolidate their thinking in which Math Talk does not only occur within the math classroom but throughout all classrooms. Moss & Beatty (2006) found that students utilizing Knowledge Forum "interwove words, numbers, and formal symbols in their interactions with one another" (p. 460) but that more research was needed to analyze the significance of such discourse.

Jacobsen, Lock & Friesen (2013) argue, overall, that Knowledge Building environments promote intellectual engagement, which provides ongoing learning opportunities to ensure a richer learning experience for students. This was evident within a study conducted by the author where Math Talk provided an opportunity for grade 2 students to engage in collaborative mathematical inquiry within a Knowledge Building framework. Students were asked to work together in generating questions, ideas, and designing methods to foster their learning. By exploring geometrical topics, such as *the nature of shapes* students were incorporating and advancing their ideas, creating community knowledge by curating mathematical cognitive artefacts.

The Ontario Mathematics Curriculum defines the Big Ideas for geometry as "intuition about the characteristics of two and three-dimensional shapes and the effects and changes of shapes in relation to spatial sense. To recognize basic shapes and figures, to distinguish between the attributes of an object, while understanding and appreciating the geometric aspects of our world" (Ontario Ministry of Education, 2005, p. 9). The purpose of these Big Ideas is to guide the formulation of essential questions to help educators explore topics more in depth with their students. Within the grade 2 math curriculum two Key Ideas are relevant in the teaching and learning of geometry:

Key Idea 1: Geometric objects have properties that allow them to be classified and described in a variety of different ways;

Key Ideas 2: Understanding relationships between geometric objects allows us to create any geometric object by composing and decomposing other geometric objects.

These are certainly encompassed by the model of Knowledge Building, and especially the Knowledge Forum, which is an entry point for ideas to become refined by students. Students make contributions in which they consider their bodies as geometric objects, explore shape properties and dimensions, and generate and explore both conceptual and plastic two-dimensional and three-dimensional shapes. By engaging their own bodies in the learning process, students can explore geometric ideas and thus extend their mathematical reasoning. Students also challenge the ideas of the community, their own ideas, and those of the experts they reference. Students understand the importance of an expert's work, by providing their own justifications.

Face-to-face and online discourse interactions demonstrate what Knowledge Builders call "Collaborative Justification." This theoretical phrase was coined by Kopp and Mandl (2011), who use it to refer to a learner's justification for arguments during a collaborative task. Their work only explores undergraduate students, and no further studies have considered its validity at the elementary level. Collaborative justification utilizes communities' contributions to allow for students to incorporate similar or opposing ideas to justify their understanding and definition of a term. Sinclair and Bruce (2015) argued that the applications of this approach to geometric learning in classrooms requires broader applications of learning through computer based tools and models, especially visual models. The use of Knowledge Forum is not meant to replace the creation and exchange of ideas, but instead assist the process. As Alavi et al. (2002) have argued, the amount of effort that students invest in learning may decrease as technology or a task becomes more complex, hence simpler systems may outperform complex systems in certain contexts.

CONCLUSION

In sum, Knowledge Building and Math Talk are effective in getting students to build on their own ideas math students and thus to progress effectively within the expected time frames of pre-established curricula, via a powerful dialogic form of inquiry, in line with Mikhail Bakhtin's theory of knowledge (Zack et Graves, 2001), which suggests that one learns through others, not by oneself. Students' collective thoughts and ideas can be easily integrated individualistically to the advantage of the student. Through the construction and exchange of collective thoughts each single learner has the opportunity to develop ideas in effective ways.

Knowledge Building allows students to enhance the way they talk about, and thus learn, mathematics—hence the validity of Math Talk as a rhetorical-discursive pedagogical strategy. Students can develop mathematical ways to reason and become reflective in regards to anything that could be a numerical of geometric concept. This leads to further student mathematical discourse for initiating new ideas and work together to understand them. Knowledge Building and Math Talk are clearly part of the "collective intelligence" movement within math education (Broadbent & Gallotti, 2015).

REFERENCES

Alavi, M., Marakas, G. M., & Yoo, Y. (2002). A comparative study of distributed learning environments on learning outcomes. *Information Systems Research*, *13*(4), 404–415. doi:10.1287/isre.13.4.404.72

Andriessen, J. (2006). Arguing to learn. In R. K. Sawyer (Ed.), *The Cambridge Handbook of the Learning Sciences*. New York: Cambridge University Press.

Artzt, A., & Newman, C. M. (1997). *How to use cooperative learning in the mathematics class* (2nd ed.). Reston: National Council of Teachers of Mathematics.

Baroody, A. J. (2000). Does mathematics instruction for three-to-five-year-olds really make sense? *Young Children*, *55*(4), 61–67.

Beatty, R., & Moss, J. (2006a). Connecting grade 4 students from diverse urban classrooms: Virtual collaboration to solve generalizing problems. *Proceedings of the Seventeenth ICMI Study: Technology Revisited.*

Beatty, R., & Moss, J. (2006b). Multiple vs. numeric approaches to developing functional understanding through patterns: Affordances and limitations for grade 4 students. In S. Alatorre, J. L. Cortina, M. Sáiz, & A. Méndes (Eds.), *Proceedings of the 28th Annual Meeting of the North American Chapter of the International Group for the Psychology of Mathematics Education*. Mérida, México: Universidad Pedagógica Nacional.

Beatty, R., & Moss, J. (2009). *I think for every problem you can solve it…for every pattern there is a rule: Tracking young students' developing abilities to generalize in the context of patterning problems.* Paper presented at the Annual Meeting of the American Educational Research Association, San Diego, CA.

Boaler, J. (1998). Open and closed mathematics: Student experiences and understandings. *Journal for Research in Mathematics Education, 29*(1), 41–62. doi:10.2307/749717

Boaler, J. (2002). Paying the price for "sugar and spice:" Shifting the analytical lens in equity research. *Mathematical Thinking and Learning, 4*(2/3), 127–144. doi:10.1207/S15327833MTL04023_3

Boonen, A., Kolkman, M., & Kroesbergen, E. (2011). The relation between teachers' math talk and the acquisition of number sense within kindergarten classrooms. *Journal of School Psychology, 49*(1), 281–299. doi:10.1016/j.jsp.2011.03.002 PMID:21640245

Bredekamp, S. (2004). Standards for preschool and kindergarten mathematics education. In D. H. Clements & J. Sarama (Eds.), *Engaging young children in mathematics: Standards for early childhood mathematics education* (pp. 77–87). Mahwah, NJ: Lawrence Erlbaum Associates.

Broadbent, S., & Gallotti, M. (2015). *Collective intelligence: How does it emerge?* Nesta School of Advanced Study, University of London.

Brown, A. L. (1992). Design experiments: Theoretical and methodological challenges in creating complex interventions in classroom settings. *Journal of the Learning Sciences, 2*(2), 141–178. doi:10.1207/s15327809jls0202_2

Bruce, C.D (2007). *Student interaction in the math class: Stealing ideas or building understanding.* OISE Presentation.

Chen, B., & Hong, H. Y. (2016). Schools as knowledge-building organizations: Thirty years of design research. *Educational Psychologist, 51*(2), 266–288. doi:10.1080/00461520.2016.1175306

Confrey, J., & Kazak, S. (2006). A thirty-year reflection on constructivism in mathematics education in PME. In A. Gutiérrez & P. Boero (Eds.), *Handbook of research on the psychology of mathematics education: past, present and future* (pp. 305–345). Rotterdam: Sense Publications.

Cooke, L., & Adams, V. (1998). "Math talk" in the classroom. *Middle School Journal, 29*(5), 35–40. doi:10.1080/00940771.1998.11495918

Davidson, N. (1990). *Cooperative learning in mathematics: A handbook for teachers.* Menlo Park, CA: Addison-Wesley.

Duncan, G., Dowsett, C., Claessens, A., Magnuson, K., Huston, A., Klebanov, P., & Japel, C. et al. (2007). School readiness and later achievement. *Developmental Psychology, 43*(6), 428–1446. doi:10.1037/0012-1649.43.6.1428 PMID:18020822

Ernest, P. (1991). *The philosophy of mathematics education.* London: Falmer Press.

Ernest, P. (1994). Social constructivism and the psychology of mathematics education. In P. Ernest (Ed.), *Constructing mathematical knowledge: Epistemology and mathematics education* (pp. 68–77). London: Falmer Press.

Falle, J. (2004). Let's Talk Maths: A model for teaching to reveal student understandings. *Australian Senior Mathematics Journal*, *18*(2), 17–27.

Finlayson, M. (2014). Addressing math anxiety in the classroom. *Improving Schools*, *17*(1), 99–115. doi:10.1177/1365480214521457

Ginsburg, H. P., Inoue, N., & Seo, K. H. (1999). Young children doing mathematics: Observations of everyday activities. In J. V. Copley (Ed.), *Mathematics in the early years* (pp. 88–99). Reston, VA: NCTM.

Guba, E. G., & Lincoln, Y. S. (2005). Paradigmatic controversies, contradictions and emerging confluences. In N. K. Denzin & Y. S. Lincoln (Eds.), *The SAGE handbook of qualitative research* (2nd ed.; pp. 191–215). Thousand Oaks, CA: Sage.

Hufferd-Ackles, K., Fuson, K. C., & Sherin, M. G. (2004). Describing levels and components of a math-talk learning community. *Journal for Research in Mathematics Education*, *35*(2), 81–116. doi:10.2307/30034933

Hurme, T., & Jarvela, S. (2005). Students' activity in computer-supported collaborative problem solving in mathematics. *International Journal of Computers for Mathematical Learning*, *10*(1), 49–73. doi:10.1007/s10758-005-4579-3

Hutchins, E. (1996). *Cognition in the wild*. Cambridge, MA: MIT Press.

Hutton, K., Chen, B., & Moss, J. (2013). *A knowledge building discourse analysis of proportional reasoning in grade 1*. Paper presented at the 2013 Knowledge Building Summer Institute, Puebla, Mexico.

Imm, K., & Stylianou, A. (2012). Talking mathematically: An analysis of discourse. *The Journal of Mathematical Behavior*, *31*(1), 130–148. doi:10.1016/j.jmathb.2011.10.001

Jacobsen, M., Lock, J., & Friesen, S. (2013). *Strategies for engagement: Knowledge building and intellectual engagement in participatory learning environments*. Toronto: Education Canada.

Kilpatrick, J., Swafford, J., & Findell, B. (2001). *Adding it up: Helping children learn mathematics*. Washington, DC: National Academies Press.

Klibanoff, R. S., Levine, S., Huttenlocher, J., Hedges, L. V., & Vasilyeva, M. (2006). Preschool children's mathematical knowledge: The effect of teacher "Math Talk." *Developmental Psychology*, *42*(1), 59–69. doi:10.1037/0012-1649.42.1.59 PMID:16420118

Knowledge Building Gallery. (in press). Knowledge building in math: How do we help students to think like mathematicians? In *Leading Student Achievement: Networks for Learning* (pp. 51-79). Academic Press.

Kopp, B., & Mandl, H. (2011). Fostering argument justification using collaboration scripts and content schemes. *Learning and Instruction*, *21*(5), 636–649. doi:10.1016/j.learninstruc.2011.02.001

Lazarus, J., & Roulet, G. (2013). *Blended math-talk community: Extending the boundaries of classroom collaboration*. OAME/AOEM Gazette.

Lehrer, R., & Chazan, D. (1998). *Designing learning environments for developing understanding of geometry and space*. Mahwah, NJ: Lawrence Erlbaum Associates.

Martinovic, D. (2004). *Communicating mathematics online: The case of online help* (PhD thesis). University of Toronto.

Mercer, N., & Sams, C. (2006). Teaching children how to use language to solve maths problems. *Language and Education, 20*(6), 507–528. doi:10.2167/le678.0

Morgan, C. (2014). Mathematical language. In S. Lerman (Ed.), *Encyclopedia of Mathematics Education* (pp. 388–391). New York: Springer.

Moss, J., & Beatty, R. (2006). Knowledge building in mathematics: Supporting collaborative learning in pattern problems. *International Journal of Computer-Supported Collaborative Learning, 1*(4), 441–465. doi:10.1007/s11412-006-9003-z

Moss, J., & Beatty, R. (2010). Knowledge building and mathematics: Shifting the responsibility for knowledge advancement and engagement. *Canadian Journal of Learning and Technology, 36*(1). doi:10.21432/T24G6B

Moss, J., Beatty, R., McNab, S. L., & Eisenband, J. (2006). The potential of geometric sequences to foster young students ability to generalize in mathematics. Paper Presented at the *Annual Meeting of the American Educational Research Association*, San Francisco, CA.

Mountz, R. (2011) *Let's talk math: Investigating the use of mathematical discourse, or "math talk," in a lower-achieving elementary class* (Dissertation). Penn State University.

Nason, R., & Woodruff, E. (2002). New ways of learning mathematics: Are we ready for it? *Proceedings of the International Conference on Computers in Education*. doi:10.1109/CIE.2002.1186334

Nason, R. A., Brett, C., & Woodruff, E. (1996). Creating and maintaining knowledge-building communities of practice during mathematical investigations. In P. Clarkson (Ed.), *Technology in mathematics education* (pp. 20–29). Melbourne: Mathematics Education Research Group of Australia.

Nason, R. A., & Woodruff, E. (2004). Online collaborative learning in mathematics: Some necessary innovations. In T. Roberts (Ed.), *Online learning: Practical and theoretical considerations* (pp. 103–131). Hershey, PA: Idea Group Inc. doi:10.4018/978-1-59140-174-2.ch005

Nason, R. A., Woodruff, E., & Lesh, R. (2002). Fostering authentic, sustained and progressive mathematical knowledge-building activity in CSCL communities. In B. Barton, C. Irwin, M. Pfannkuch, & M. O. J. Thomas (Eds.), *Mathematics education in the South Pacific: Proceedings of the Annual Conference of the Mathematics Education Research Group of Australasia* (pp. 504-511). Sydney, Australia: MERGA.

Nathan, M. J., & Knuth, E. J. (2003). A study of whole classroom mathematical discourse and teacher change. *Cognition and Instruction, 21*(2), 175–207. doi:10.1207/S1532690XCI2102_03

National Council of Teachers of Mathematics. (2000). *Principles and standards for school mathematics*. Reston, VA: National Council of Teachers of Mathematics.

Norman, D. A. (1993). *Things that make us smart*. Reading: Addison-Wesley Publishing.

Nuns, C. A. A., Nunes, M. M. R., & Davis, C. (2003). Assessing the inaccessible: Metacognition and attitudes. *Assessment in Education, 10*(3), 375-388.

Ontario Education Quality and Accountability Office. (2016). Toronto: Ontario Government.

Ontario Ministry of Education and Training. (2005). *The Ontario curriculum, grades 1-8: Mathematics, revised.* Retrieved from http://www.edu.gov.on.ca/eng/curriculum/elementary/math18curr.pdf

Ontario Ministry of Education and Training. (2009). *Growing success document.* Retrieved from www.edu.gov.on.ca/eng/policyfunding/growsuccess.pdf

Osta, I. (2014). Mathematics curriculum evaluation. In S. Lerman (Ed.), *Encyclopedia of Mathematics Education* (pp. 417–423). New York: Springer.

Pimm, D. (1987). *Speaking mathematically: Communications in mathematics classrooms.* London: Routledge & Kegan Paul.

Pimm, D. (1988). Mathematical metaphor. *For the Learning of Mathematics, 8*(1), 30–34.

Putnam, R. (1987). Structuring and adjusting content for students: A study of live and simulated tutoring of addition. *American Educational Research Journal, 24*(1), 13–28. doi:10.3102/00028312024001013

Resnick, L. B. (1988). Treating mathematics as an ill-structured discipline. In R. I. Charles & E. A. Silver (Eds.), *The teaching and assessing of mathematical problem solving* (pp. 32–60). Hillsdale, NJ: Lawrence Erlbaum Associates, Inc.

Ritchhart, R. (2015). *Creating Cultures of Thinking: The 8 forces We Must Master to Truly Transform our Schools.* Hoboken, NJ: Wiley.

Scardamalia, M., & Bereiter, C. (1983). Child as coinvestigator: Helping children gain insight into their own mental processes. In S. Paris, G. Olson, G., & H. Stevenson (Eds.), Learning and motivation in the classroom (pp. 61-82). Hillsdale, NJ: Lawrence Erlbaum Associates.

Scardamalia, M., & Bereiter, C. (1991). Higher levels of agency for children in knowledge-building: A challenge for the design of new knowledge media. *Journal of the Learning Sciences, 1*(1), 37–68. doi:10.1207/s15327809jls0101_3

Scardamalia, M., & Bereiter, C. (2003). Knowledge building. Encyclopedia of Education, 2, 1370-1373.

Scardamalia, M., Bereiter, C., McLean, R. S., Swallow, J., & Woodruff, E. (1989). Computer supported intentional learning environments. *Journal of Educational Computing Research, 5*(1), 51–68. doi:10.2190/CYXD-6XG4-UFN5-YFB0

Scardamalia, M., Bransford, J., Kozma, B., & Quellmalz, E. (2012). New Assessments and Environments for Knowledge Building. In P. Griffin, P., B. McGaw, & E. Care (Eds.), Assessment and teaching of 21st century skills (pp. 231-300). New York: Springer. doi:10.1007/978-94-007-2324-5_5

Schunk, D. H. (2012). *Learning theories: An educational perspective.* New York, NY: Macmillan Publishing Co, Inc.

Sinclair, M. P. (2005). Peer interactions in a computer lab: Reflections on results of a case study involving web-based dynamic geometry sketches. *The Journal of Mathematical Behavior, 24*(1), 89–107. doi:10.1016/j.jmathb.2004.12.003

Sinclair, N., & Bruce, C. (2015). New opportunities in geometry education at the primary school. *ZDM Mathematics Education, 47*(3), 319–329. doi:10.1007/s11858-015-0693-4

Stahl, G. (2005). Group cognition in computer-assisted collaborative learning. *Journal of Computer Assisted Learning, 21*(2), 79–90. doi:10.1111/j.1365-2729.2005.00115.x

Steffe, L. P., & Kieren, T. (1994). Radical constructivism and mathematics education. *Journal for Research in Mathematics Education, 25*(6), 711–733. doi:10.2307/749582

Stigler, J., & Hiebert, J. (1999). *The teaching gap: Best ideas from the world's teachers for improving education in the classroom*. New York: The Free Press.

Towers, J. (1999). *In what ways do teachers' interventions interact with and occasion the growth of students' mathematical understanding* (Unpublished doctoral dissertation). University of British Columbia.

Van den Akker, J., Graveemeijer, K., McKenney, S., & Nieveen, N. (2006). *Educational design research*. London: Routledge.

Vygotsky, L. S. (1978). *Mind in society: The development of higher psychological processes*. Cambridge, MA: Harvard University Press.

Wagner, T., & Dintersmith, T. (2015). *Most likely to succeed: Preparing our kids for the innovation era*. New York: Scribner's.

Whiteley, W. (1999). The decline and rise of Geometry in 20th century North America. In J. Grant (Ed.), *Proceedings of CMESG 1999 Annual Meeting* (pp. 7–30). Academic Press.

Wood, D., Bruner, J. S., & Ross, G. (1976). The role of tutoring in problem solving. *Journal of Child Psychology and Psychiatry, and Allied Disciplines, 17*(2), 89–100. doi:10.1111/j.1469-7610.1976.tb00381.x PMID:932126

Wood, D., & Wood, H. (1996). Vygotsky, tutoring and learning. *Oxford Review of Education, 22*(1), 5–16. doi:10.1080/0305498960220101

Yackel, E., Gravemeijer, K., & Sfard, A. (2011). *A journey in mathematics education research: Insights from the work of Paul Cobb*. Dordrecht: Springer.

Zack, V., & Graves, B. (2001). Making mathematical meaning through dialogue: Once you think of it, the z minus three seems pretty weird. *Educational Studies in Mathematics, 46*(1/3), 229–271. doi:10.1023/A:1014045408753

Zhang, J., Scardamalia, M., Reeve, R., & Messina, R. (2009). Designs for collective cognitive responsibility in knowledge building communities. *Journal of the Learning Sciences, 18*(1), 7–44. doi:10.1080/10508400802581676

Chapter 11
The Reincarnation of the Aura:
Challenging Originality With Authenticity in Plaster Casts of Lost Sculptures

Victoria Bigliardi
University of Toronto, Canada

ABSTRACT

In 1935, Walter Benjamin introduced the aura as the abstract conceptualization of uniqueness, authenticity, and singularity that encompasses an original art object. With the advent of technological reproducibility, Benjamin posits that the aura of an object deteriorates when the original is reproduced through the manufacture of copies. Employing this concept of the aura, the author outlines the proliferation of plaster casts of sculptures in eighteenth- and nineteenth-century Europe, placing contextual emphasis on the cultural and prestige value of originals and copies. Theories of authenticity in both art history and material culture are used to examine the nature of the aura and to consider how the aura transforms when an original object is lost from the material record. Through an object biography of a fifteenth-century sculpture by Francesco Laurana, the author proposes that the aura does not disappear upon the loss of the original, but is reincarnated in the authentic reproduction.

INTRODUCTION

In his seminal essay on "The Work of Art in the Age of its Technological Reproducibility" (1935), Walter Benjamin argues that the reproduction of an art object leads to the "decay of the aura" of the original work (p. 15). The aura, as Benjamin defines it, is "the here and now" of the object; it is the quality of a unique existence, "a strange tissue of space and time" that gives the appearance of being slightly distant or removed, "however near it may be" (1935, p. 15). It is the *here-and-nowness* of the original that "constitutes the concept of its authenticity" (1935, p. 13). The way we seek to perceive history in the present day means that we desire to become closer to objects while "overcoming the uniqueness of every reality through its reproducibility" (1935, p. 16). This concept of the aura that is put forth by Benjamin is seen as an inherent, magical property that is present in every original art object. I endeavour here to apply Benjamin's conceptualization of the aura to a specific context that persists throughout our material

DOI: 10.4018/978-1-5225-5622-0.ch011

culture histories in a very pertinent way. Can an object have authenticity if it does not have originality? Do copies have a life beyond their prescribed role as mere representation or replication? Can the concept of the aura be maintained in a world of copies and replicas? These key questions guide my analysis of Benjamin's theoretical approach to reproducibility in the technological age.

With the foundations of my research built upon "The Work of Art in the Age of its Technological Reproducibility" (1935), I explore the implications of Benjamin's work through a case study of a particular example of artistic reproduction. Looking specifically at the casting and replication of sculpture in 18th- and 19th-century Europe, I outline the social, historical, and cultural forces that led to the proliferation of casts and copies in the Western world. Situating the originals in relation to their copies, I consider differences in function, value, meaning, and prestige. To illustrate the prevalence and significance of casting practices in Europe during this era, a historical overview of the Cast Courts at the Victoria & Albert Museum in London contextualizes the focal point of this paper: a plaster cast of a bust by Renaissance sculptor Francesco Laurana, entitled *Portrait of a Woman*. It is the investigation of this piece in particular that highlights my exploration of the tension between originality and authenticity. Considering Benjamin's concept of the aura, along with further critical commentaries on authenticity in material culture and on reproducibility, I use Laurana's *Portrait of a Woman* as a prime example of *the vanishing original*. In examining this work, I suggest that the aura may not always decay as Benjamin says. Using object biography to consider the historical relevance and iconic meaning of a reproduction, I posit that the value and prestige of the original are not lost through the process of casting. Instead, Benjamin's aura may be temporarily concealed or displaced. I argue that in instances of an absent original, the aura is reincarnated within the authentic reproduction. Put simply, the iconic sign becomes the very object to which it refers.

AUTHENTICITY IN MATERIAL CULTURE

I begin by asking a simple question: what makes an object authentic? In light of Benjamin's contentious essay, a number of critical commentaries and responses have emerged, revealing the complicated resonances that are tied up in the meaning of authenticity. Jaworski (2013) prefaces his argument that skilful reproductions of art objects are "just as good" (p. 392) as the original objects by clarifying that he does not believe they are of the same value. His description of historical originality as a value of the authentic object appears closely aligned with Benjamin's conceptualization of the aura. "We seem to care about 'survival value'," which is "the value objects have in virtue of having survived throughout the years" (Jaworski, 2013, p. 393). In this sense, the way that an object is situated in time gives a good indication of its authenticity. This temporal quality is connected to causal history, which Goodman explains as follows: knowing the history of production of an art object provides a link back "to the hand of the artist," and because of this history, "the original […] will be authentic in virtue of its causal history" (as cited in Jaworski, 2013, p. 397). A reproduction or a forgery, which is bereft of a causal history as intricate as the original, "lacks authenticity" (p. 397). A rich causal history implies that an object also has a strong survival value, and it is the uniqueness of this narrative and its position in time that bring it closer to originality and authenticity. This temporal individuality, with a complex object biography, points to the aura that is imbued in the original object via its unique existence, as Benjamin says.

Others suggest that authenticity is not so complicated, especially when understood directly through Benjamin's lens. Zeller (2012) posits that the aura, which is "a notion of aesthetic experience," is "the utopian concept of authenticity" (p. 75). In this sense, the aura and authenticity are inextricably linked, and both are opposed to the inherently dystopian advent of technological advancement.

George Kubler (2008) is less direct in his assessment of the authentic object, largely symptomatic of his structural approach to the relationships between originals and copies. In his mind, there cannot be one without the other. Originals are referred to as prime objects, and reproductions in the face of technological development become part of the replica-mass (Kubler, 2008, p. 35). He emphasizes that "the number of prime objects is distressingly small" (p. 38), and that copies are vital by way of their "adhesive properties," as they "[hold] together the present and the past" (p. 66). There is a sequential relationship between the authentic original and the authentic copy in Kubler's world. It is the creation of copies in this sequence that leads to the generation of symbolic associations, because symbols emerge from repetitions (p. 67). "Its identity among its users depends upon their shared ability to attach the same meaning to a given form" (p. 67). While most evidence points to the notion that prestige and value reside in the authentic original, and that the reproduction of these originals reduces their prestige, Kubler's concept offers the reverse. If reproductions actually serve to create symbolic meaning for an art object among its collective viewers, then the copies themselves seem to be promoting the original rather than detracting from its unique aura. This reliance on copies to mediate between time periods and to develop symbolic value for the form of the art object reveals the sense of hierarchical dependency that Kubler says exists between prime objects and their replicas. Philip Fisher (1991) notes that Kubler's work tries to balance "the authority of invention, originality, and breakthrough with a portrait of the social process of repetition or duplication" (p. 99). This element of the social, in many ways, is left out of Benjamin's conceptualization of the aura and authenticity. Fisher also makes a connection to temporality by saying that "copying aims at providing a successor in time" (1991, p. 100). Similar to Jaworski's discussion of survival value, it seems straightforward to assume that copies themselves can assume a place in time that leads to the development of their own historical originality. In the causal history of the original object, the copy is situated as a successor with the sole purpose of continuing with a pre-existing material culture legacy. This familial undertone seems to indicate that the authentic nature of the original (perhaps its aura) can be passed down to succeeding objects. While this contradicts Benjamin's notion of the withering aura, it adds an interesting layer to my later discussion of the aura prevailing beyond the original.

In the realm of art history, which is particularly relevant to this paper, authenticity is not without controversy. "In most discourses of Art History, the art object is present only through representations" (Dam Christensen, 2010, p. 200). In many cases, scholars and art historians may study art objects whose original forms they have never actually encountered. The collective knowledge of art historians is largely comprised of responses to authentic representations rather than originals (p. 213). Dam Christensen (2010) argues that there is a repressive logic to it. Somewhat discounting the weight of Benjamin's thesis in the process, he suggests that when we are privy to the aura of the object and "the immediacy of its location in time and space" without considering "the multiplied discursive representations of the object," we are dismissing the significance and intellectual resonances of the reproductions as they exist in the collective knowledge of art history (p. 213). This instance is a strong example of how social values alter the way that authenticity and originality would be encountered objectively; the educational and accessibility agendas of copying art objects carry an important cultural connection that cannot be ignored, particularly since the study of these representations has been vital to the development of art history as a discipline. In this way, authenticity and auras can be pushed aside to sometimes reveal lasting social implications of reproducibility that are not easily erased.

Some commentaries on authenticity in material culture are openly critical of Benjamin's theories. Ian Knizek, when revisiting "The Work of Art in the Age of its Technological Reproducibility" (Benjamin, 1935) argues that the aura is "extra-aesthetic" and that it detracts from the formal and compositional properties of an art object (Knizek, 1993, p. 358). He is most critical of Benjamin for not persuasively clarifying *why* and *how* the aura disintegrates (p. 360). "There is then no very good reason why even reproductions cannot appropriate for themselves the features composing the world's 'authenticity'. In other words, these features can be transmitted or transferred to reproductions by imagination" (p. 361). It is this criticism from Knizek (1993) that bears the most resemblance to my argument here in this paper. Adopting a similar perspective in proposing that the aura does not truly disappear, and can in some cases be transferred from original to copy, I move forward with a critical eye toward exploring the roles that authenticity and originality have played in the practice of casting sculptures in 18th- and 19th-century Europe.

TASTE AND THE ANTIQUE

Francis Haskell and Nicholas Penny in their book *Taste and the Antique: The Lure of Classical Sculpture 1500-1900* (1981) discuss the myriad social and historical contexts that were entwined with the prestige associated with the ownership of Classical sculpture in the Renaissance and the centuries that followed. Kubler's (2008) comment on the distressingly small number of prime objects in the world is made manifest through this discussion, particularly in the retelling of Bernini being unimpressed "by the number of original antique sculptures to be found in France" (Haskell & Penny, 1981, p. 37). This shortage of originals from antiquity led to the proliferation of casts and copies, which dominates much of the history of sculpture in Europe. Even Charles I of England "knew that the best statues could only be obtained in casts and copies" (p. 31).

The desire for reproductions of original Classical sculpture was rooted in a number of causes. There was the educational purpose, where casts of pieces from antiquity were viewed as essential models for drawing in European art academies. Bernini emphasized the necessity for "young art students to copy from casts taken from 'all the most beautiful statues, bas-reliefs, and busts of antiquity' before learning to draw from nature" (Haskell & Penny, 1981, p. 37). There was also the decorative purpose: royalty commissioned complete sets of plaster and bronze casts to decorate their palaces and gardens. These casts and copies were "made for kings" (p. 35); imitation was not enough, as "the work produced by students at the [academies] must [...] be 'more perfect than the antique'" (p. 39). The prestige and beauty of the casts often meant that royalty ended up holding on to collections that were meant for the academies, losing them within the "cult of grandeur" (p. 42). By the mid-1700s, "copies and casts of ancient statues had become generally associated with the refinement of good taste and good breeding" (McNutt, 1990, p. 159). A reputable and authentic sculpture cast required a skilful craftsman who was also often an artist, especially since students at European art academies worked so closely with casts from the start.

In a way, the unbelievable quantity of these beautiful casts and copies in Europe during this time meant that, as Fisher said, the reproductions couldn't help but become successors for the originals. As time passes, material culture becomes more vulnerable. Unintentionally, some of the casts of Classical sculpture produced during this time became the only remnant of the original art object. With time and political turmoil, casts were often all that remained of vital European history, at least in its complete form. This significant quality of casting during this era has had a lasting impact not only on art history, but also on the way reproductions are viewed in the present day. To contextualize this notion and explore the prevalence of casting culture, I will here provide a historical overview of the Cast Courts at the Victoria and Albert Museum in London.

THE CAST COURTS

In 1873, the Cast Courts were opened at what was then known as the South Kensington Museum in London (Trench, 2010, p. 13). Originally called the Architectural Courts (Trusted, 2007, p. 167), the purpose of the gallery was "to make available a selection of the finest sculpture and architectural fragments from around Europe" (Trench, 2010, p. 13). This notion of accessibility is a big part of why casts were made in the first place; the reproducibility of original sculpture meant that casts could be made and shared widely for educational and decorative purposes. "Casts were absolutely reliable replicas" that could be "easily taken to all countries" because of their comparable lightness next to original marbles and bronzes (Haskell & Penny, 1981, p. 16). The Cast Courts at the Victoria and Albert Museum (formerly the South Kensington) served this purpose of making Classical art accessible to the public; "with access to [casts], there was no excuse for anyone anywhere failing to follow 'the good and ancient path'" (Haskell & Penny, 1981, p. 16). The intellectual prestige associated with this "good and ancient path" points to the profound sociocultural underpinnings of casting as a practice. Undoubtedly, the appreciation and ownership of casts was an important social practice for anyone of great learning or power. Casts were tools of self-improvement and self-fashioning, and in this way they bore a cultural function similar to that of their original counterparts, particularly from the Renaissance.

The collection in the Cast Courts at the Victoria and Albert Museum is "the largest single group of plasters" in world, acquired and displayed since the 19th-century (Trusted, 2007, p. 153). These reproductions were initially added to the museum "to fill gaps in the collections of so-called 'original' works of art," but in many cases they were collected as "important examples in their own right" (p. 160). Students increasingly sought access to these sculptural and architectural masterworks to further their studies, which, along with the aesthetic desires of royalty to have casts as ornaments, led to a rapid increase in plaster reproductions in the 19th-century (p. 161). Casting at this time was considered to be a very lucrative practice, with a number of artisanal manufacturers existing throughout Europe (p. 166). The famous workshop of Domenico Brucciani, a prolific *formatore* (cast manufacturer), was originally located in Covent Garden in the mid-19th century. Following Brucciani's death, his workshop was taken over by the Board of Education and brought to the Victoria and Albert Museum, which at the time already housed a number of Brucciani's masterpieces (p. 168). This led to the establishment of a casting workshop within the museum itself, which represents the true pervasiveness of casting as a practice with mass-appeal.

The Cast Courts have not been without controversy. Leading into the 20th-century, opinions surrounding the value of casts began to change; "casts were considered by many as inferior copies that had no place in a museum with original works of art" (Trusted, 2007, p. 170). In some cases, even a damaged fragment of an original was highly valued over an exact replication of a monument in plaster (p. 170). The Courts are now more of a historical record than they are a growing collection. The museum still emphasizes a stronger desire for original art objects, which reinforces Benjamin's work on the allure of the aura and the power of the unique existence of the original. However, the prevalence of casting in 18th- and 19th-century Europe cannot be disputed, and the proliferation of casts and copies in the art world forms a body of work that is now situated at a distinct moment in the past. Providing "a unique record of what is intrinsically a Victorian collection," casts serve to preserve "the appearance of an object or monument as it was" a couple of centuries prior, while also continuing to make art objects of antiquity available to students and artists (Trusted, 2007, p. 170-1). Importantly, as I will discuss further, some casts "record objects that no longer exist" (p. 170).

While Benjamin's aura of unique existence may not be present in plaster reproductions, it can be said, however, that the reproductions themselves are unique in a different sense. They provide a window to the past that can provide us with valuable information—information that may otherwise be lost to the vulnerabilities of antique materialities, including the deterioration, modification, and destruction of original art objects through changing environmental, social, and political forces. In the case study that follows, I explore the implications of the loss of original art objects, and work to connect this absence with the theoretical frameworks I outlined earlier in this paper.

FRANCESCO LAURANA'S *PORTRAIT OF A WOMAN*

Francesco Laurana, a Croatian-Italian sculptor from the Renaissance, produced work for foreign royal courts in the mid-to-late 15th century (Royal Academy of Arts, "After Francesco Laurana," 2015, para. 3). He is best known "for his series of portrait busts of the royal ladies of the Aragonese court at Naples, which he executed […] in the 1470s" (para. 3). His work is noted for being sensitive, depicting "smooth features and calm expressions" in all of his portrait busts (para. 5). One such example of these marble busts is his *Portrait of a Woman* (sometimes referred to as *Bust of a Woman;* see Figure 1). Sculpted in Naples in 1472, the sitter is suspected to be Ippolita Maria Sforza, the wife of King Alfonso II of Naples (V&A Collections, *Bust of a Woman*, "Physical Description" section). The original sculpture bore rich gilded ornamentation on the dress, and showed "the remains of extensive pigmentation" (*Bust of a Woman*, "Historical context note" section). A hole at the breast indicates that there may have once been a metal brooch attached to the bust. *Portrait of a Woman* once resided in the collection of the Berlin Staatliche Museum (Royal Academy of Arts, "After Francesco Laurana," 2015, para. 6). Two closely related busts exist to this day in their original form in the Frick Collection in New York and the National Gallery of Art in Washington; the similarly decorated bases and facial resemblance suggest that all three depict the same sitter (V&A Collections, *Bust of a Woman*, "Historical context note" section).

Formatore Domenico Brucciani, mentioned earlier in this paper, produced a plaster cast of this bust in 1889 (Royal Academy of Arts, "After Francesco Laurana," para. 2; see Figure 2). It is this cast that was acquired by the Victoria and Albert Museum in 1889 from the Berlin Museum for 15 marks (V&A Collections, *Bust of a Woman*, "Object history note" section). The bust, 51cm in height and 46cm in width, has resided in the Cast Courts collection ever since. Originally praised for its beauty and included in the collection as an example of exceptional sculptural talent on the part of Laurana, the cast was included first as an educational tool. However, as discussed here, the cast of *Portrait of a Woman* has, over time, taken on new meaning.

The original *Portrait of a Woman* bust has a complicated and violent history. During World War Two, the marble bust, along with hundreds of other works of art, was removed from the collection of the Berlin Staatliche Museum and housed in the Freidrichshain flak tower for protection (Royal Academy of Arts, "After Francesco Laurana," para. 6). "In the closing days of the war, two fires occurred at the tower, causing devastating damage to the artworks" (para. 6). The original Laurana bust was left completely destroyed as a result of these fires in 1945. The authentic original no longer exists, and the plaster cast

Figure 1. Original marble bust, destroyed in Berlin during World War Two. Portrait of a Woman, 1472, by Francesco Laurana ([Untitled photograph], n.d.).

Figure 2. Plaster cast of Francesco Laurana's Portrait of a Woman, now residing in the Cast Courts at the Victoria and Albert Museum in London. Cast in 1889 by Domenico Brucciani (widdowquinn, 2010).

in the Cast Courts collection becomes the only complete and authentic replica of the marble. At face value, the cast is now historically important as a representation of an art object that is now lost. From the reproduction, one can still gather information about the time period within which the original was produced. For example, the physical representation of the sitter can reveal information about Italian ideals of womanhood in the 15ᵗʰ century and standards of art in royal courts, showing "appropriate female comportment and decorum" (para. 5). The hair of the sitter as depicted in the bust shows a style that was "reserved only for the elite social classes" and that communicates aristocratic standing and status (para. 5). The cast here acts as a complete record of some of these historical resonances in Renaissance art works.

On another level, the cast can be explored through the framework of Benjamin's theories, particularly in the case of the vanishing original. Fisher's (1991) notion that copies become the successors of originals seems particularly apt in this situation; the copy of the Laurana bust acts almost as a descendant of the authentic marble, carrying its material culture legacy forward in time. Kubler's (2008) conceptualization of the systematic connections between prime objects and replicas is deeply tied up in Fisher's proposal of succession.

But, as mentioned before, what happens to the aura? Benjamin (1935) suggests that the aura of the original disintegrates as copies are produced. This aura, the unique existence of the original, is what gives it its authenticity. The survival value and historical originality that Jaworski (2013) addresses are not present in the reproduction; the temporal biography of the authentic original concludes when the life of the original ends. This begs other questions: Does the life of the original object end if it has been reproduced? Do reproductions not offer some extension of life to the originals? Again, Fisher's notion of succession comes into play here with resonances of lineage and genealogy. The parent prime object may die, but it may persist in some form through its familial ties to reproductions as successors. It is this connection that I have used to frame my argument in this paper.

CONCLUSION

I agree with Benjamin when he posits that the value and prestige of the original decrease when the object loses its uniqueness. This loss of uniqueness is, in essence, the withering of the aura. The ethereal distance we keep between ourselves and the unattainable original is what creates that aura, and when art objects are made more accessible through reproductions, the magic of that separation is diminished. I suggest, however, that the loss of this distance, the loss of the original object, changes the status of the aura once again. While the uniqueness of the original may be eclipsed by the creation of copies, I argue that it does not completely disappear—it is temporarily concealed. In the event of the loss of the original, as is the case with Laurana's *Portrait of a Woman*, this hidden aura may re-emerge in the presence of the authentic reproduction. I suggest here that the authentic quality of the surviving plaster cast allows for the reincarnation of the original's aura in the body of the reproduction. The absence of the original means then that since 1945, the cast of the bust in the Victoria and Albert Museum has come to embody the authenticity and cultural meaning that was once found within the original bust. Like a family heirloom, the unique quality of the original has the potential to be passed down over time, ensuring the continued existence of the art object.

In essence, I suggest that authenticity can survive without originality, and that it is the value of the authentic more than the value of the original that allows for the transference of the aura. This alternative reading of authenticity brings an additional, critical perspective to the existing analyses of Benjamin's "The Work of Art in the Age of its Technological Reproducibility" (1935). More broadly, it seeks to address the shifts in values and cultural meanings that occur within the reproduced art object upon the loss of its original. With the above case study, I have endeavoured to show that the authenticity of an object is not reduced when it is copied. Walter Benjamin's conceptualization of the aura is not always destroyed by the reproducible, but can instead function as a mediating force that connects past masterpieces with present representations in the historical record.

REFERENCES

Benjamin, W. (2010). The work of art in the age of its technological reproducibility (M. W. Jennings, Trans.). *Grey Room, 39*, 11-37. (Original work published 1935)

Dam Christensen, H. (2010). The repressive logic of a profession: On the use of reproductions in art history. *Konsthistorik tidskrift/Journal of Art History, 79*(4), 200-215.

Fisher, P. (1991). *Making and effacing art: Modern American art in a culture of museums*. Cambridge, MA: Harvard University Press.

Haskell, F., & Penny, N. (1981). *Taste and the antique: The lure of classical sculpture 1500-1900*. London, UK: Yale University Press.

Jaworski, P. (2013). In defense of fakes and artistic treason: Why visually-indistinguishable duplicates of paintings are just as good as the originals. *The Journal of Value Inquiry, 47*(4), 391–405. doi:10.1007/s10790-013-9383-z

Knizek, I. (1993). Walter Benjamin and the mechanical reproducibility of art works revisited. *British Journal of Aesthetics, 33*(4), 357–366. doi:10.1093/bjaesthetics/33.4.357

Kubler, G. (2008). *The shape of time: Remarks on the history of things*. New Haven, CT: Yale University Press. (Original work published 1962)

McNutt, J. K. (1990). Plaster casts after antique sculpture: Their role in the elevation of public taste and in American art instruction. *Studies in Art Education, 31*(3), 158–167. doi:10.2307/1320763

Untitled photograph of the original bust of *Portrait of a Woman*. (n.d.). Retrieved from http://www.croatia.org/crown/articles/10219/

Royal Academy of Arts Collections. (2015). *After Francesco Laurana: Bust of a woman, possibly Ippolita Maria Sforza, c. 1473, 19th-century plaster cast*. Retrieved from http://www.racollection.org.uk/ixbin/indexplus?record=ART13883

Trench, L. (2010). *The Victoria and Albert Museum*. London, UK: V&A Publishing.

Trusted, M. (Ed.). (2007). *The making of sculpture: The materials and techniques of European sculpture.* London, UK: V&A Publishing.

V&A Collections. (n.d.). *Bust of a woman* (artifact record). Retrieved from http://collections.vam.ac.uk/item/O39862/

widdowquinn (Photographer). (2010). *Portrait of a woman* [Photograph]. Retrieved from https://flic.kr/p/8kfZP6

Zeller, C. (2012). Language of immediacy: Authenticity as a premise in Benjamin's *The Work of Art in the Age of Its Technological Reproducibility. Monatshefte, 104*(1), 70–85. doi:10.1353/mon.2012.0015

Chapter 12
Orders of Experience:
The Evolution of the Landscape Art-Object

Aaron Rambhajan
University of Toronto, Canada

ABSTRACT

Does art tend towards immersion? Positing James Turrell's Roden Crater (2015) as the modern epitome of the landscape art-object, the evolution of the medium is traced through prominent examples its transformations: Titian's Venus and the Organist with Dog (1550), De Loutherbourg's Eidophusikon (1781), and Barker's Panorama (1792). Discussion regarding Roden Crater's predecessors serve to illustrate distinct innovations that greatly influenced its construction of sensory experience, spanning the use of dialogue to the integration of physicality. This chronology is used to demonstrate an overarching tendency of media towards immersion, and to reflect how the development of contemporary culture evolves towards progressively psychological experiences.

INTRODUCTION

When a medium can evolve and adapt to capture more attention, it will. This is an inclination pervasive in all realms of media—from smartphones to augmented reality, when we consider how the mediating layers between people and technology have diminished over time, this notion becomes obvious. Throughout the course of history, we have continually bridged the gap between the physical and technological, as we now cross the final layers between the asceticism of pre-Industrial society and the hyper-connectivity of the future. In this paper, I argue this phenomenon is symptomatic of an overarching tendency of media towards immersion. I seek to understand how and why the notion of 'experiential' has become so salient in modernity, and will discuss these phenomena through the evolution of landscape art-objects. I have chosen to examine the landscape genre because, unlike any other, it embodies a visual rhetoric that necessitates the physical, where meaning is *created,* not observed—unlike the didactic practices of traditional art that solely encompass the act of *seeing* (Jelić, 2015). I will posit James Turrell as the culmination of this tendency, because of his synthesis of the *viewing* and *sensing* spaces: *viewing* space, wherein one merely 'sees' something (as with most exhibited art), and *sensing* space, wherein one 'feels'

DOI: 10.4018/978-1-5225-5622-0.ch012

something and engages it on a sensory level ("The Wolfsburg Project," 2009). The bridging of these spaces is necessary to sensory experience, and is the boundary wherein art is either observed, or is felt. To understand Turrell's work, I will reconcile *Roden Crater* (2015) with what I propose are three historical antecedents: the dialogue of Titian's *Venus and The Organist with Dog* (1550), the phenomenology of Philip James De Loutherbourg's *Eidophusikon* (1781), and the physicality of Robert Barker's 19th century *Panorama* (1792). I will examine these innovations in the landscape art-object to demonstrate how the amalgamation of the *viewing* and *sensing* space have come to define landscape art as the visual rhetoric of 'experience'—demonstrating art's overarching tendency towards immersion.

TURRELL'S RODEN CRATER (2015)

James Turrell is one of few to utilize the low-level coding processes behind the brain's interpretation of the world in order to manipulate experience. Where the art historical canon is devoted to the *viewing* of art, Turrell focuses on the *experience*, the *perception* of the art. This begets the question, what *is Roden Crater?* This piece, which I purport to be the modern paragon of immersion in art, is James Turrell's magnum opus. Having created an expansive body of installation work centered around perceptual experimentation, Turrell's influence in art has been far-reaching. His oeuvre is distinctly different from most artists—he identifies his material as light and his medium as perception (Hylton, 2013). Located in the Painted Desert, Northern Arizona, *Roden Crater* is a volcanic cinder cone he purchased four decades ago with a Guggenheim grant. The epitome of the landscape art-object, he has meticulously tailored it into a naked-eye observatory with a total of 21 viewing spaces and 6 tunnels in order to create an experience of light and shadow in geological and celestial time (Matts & Tynan, 2012). It is an art that elicits surrender—he first isolates his audience geographically through its remote location, and then isolates with the distinctly unitary space. Borrowing theory from Kengo Kuma, a renown Japanese architect, it is an architecture of erasure, one that through naturalistic, psychological vernacular, achieves unification with its environment (Bognar et al., 2009). With *Roden Crater,* Turrell sought to humanize astronomical phenomena, but contrary to what its description might suggest, it is not merely a curation of views, nor a framing device for these objects ("About Roden Crater," n.d). With profound understanding of human perception, Turrell creates a space that one must *discover*. This requisite interaction is unprecedentedly dialogical and is highly distinct from the act of viewing because there is no hiding nor revealing—it is *created* through discovery. This happens because Turrell's design is a carefully constructed framework, revealing the unobstructed totality and wholeness of the sky and its celestial bodies. It is overwhelmingly physical, meant to be touched and experienced. It eludes the photograph, there is no element of presentation—*Roden Crater* demonstrates knowing "perceptual and bodily responses precede conscious awareness…", noting that "pre-reflective judgment of architectural space is delivered by perceptual experience…" (Jelić, 2015). This speaks to Turrell's ability to engage perception—by allowing participants the agency for exploration in a designed environment, they are actually able to *create* the reality themselves—akin to the natural affect discovering the sublime ("James Turrell's Roden Crater," 2013).

These phenomenological elements are what Turrell uses to bridge the *viewing* and the *sensing* spaces; it is made with specific attention to perception so there is never just *one* sensory experience occurring at any time—it is always a multitude. *Roden Crater* borrows a number of key elements which amalgamate to represent the tendency of art towards immersion—as a statement for what it takes to affect people.

The sublimity of *Roden Crater* expresses a physical understanding of the emotional, not offering viewers an experience of language, but "an experience of the *absence* of language." (Lunberry, 2009).

TITIAN'S VENUS AND THE ORGANIST WITH DOG (1550)

The face of the 16th century Venetian school, Titian is a relevant case for this narrative as a revered innovator in early landscape painting. His expression of dimension, color, and brushwork influenced Western art for centuries to come—and his ability to integrate the viewer into painting to create experience set him apart. Early in the chronology of the landscape art-object, Titian's work exhibited an innate understanding of landscape's sensorial capacities. At that time, landscape works were few and far between, largely depicted idealistically. While he was no exception to this, Titian never failed to engender tension between the *viewing* and the *sensing* space, as demonstrated in his 1550 work, *Venus and The Organist with Dog*. Unlike most work of this era, Titian is unprecedentedly dialogical. Seen equally in his other nudes, *Venus of Urbino* (1510) and *Venus and the Lute Player* (1570), he manages to capture the notion of aural experience in the form of visual echo. The gestures his characters make are atypically forward, even *inviting*. They acknowledge the viewer's presence, reverberating an echo that elucidates the painting's circulation (Eberhart, 2012). This is evidenced by the observable disconnect in gestural communication between the Organist and Venus. There is reason behind Titian's decision to create this tension—what might otherwise be a simple interaction, instead permits the audience to participate in this vulnerability. By allowing such vulnerability in circulation, Titian effectively reflects the echo towards the audience, connecting them to the work, and becoming a part of it. This gesture begets interaction in place of worship; *sensing* instead of *viewing,* and juxtaposes the prescribed relationship audience was to have with painting. At the time, the stature of the visual arts meant all other senses were too ephemeral to ever be divine—the work of paint-on-canvas were thought to birth works of immortality (Eberhart, 2012). Titian, whilst engaging these ideals, delicately subverted them— whether Titian's intent was to challenge the pedestalled visual art through the visual echo of unreciprocated gesture is subject for debate, but the innuendos are difficult to contest. Titian's work is more filmic than it is a still image, carrying with it phenomenological connotations whose resonances are heard as far as Turrell's work. He ingeniously brings this dialogue to life, embellished by the openness of Venus' expression.

The use of visual echo and circulation is called upon in *Roden Crater,* as an index to Titian's interest in the fullest experience possible, previously unbeknownst to canvas (Eberhart, 2012). *Venus and The Organist with Dog* was seminal in the development of dialogue in the landscape art-object, as one of the first steps down from the pedestal that was the artistic canon. It is the same of how Turrell manipulates of light, space, and material to connect his work to people's perception. He aims to engage through experience, to create frameworks for perception instead of didactic artwork which demand the viewer to *look* in a certain way. It is highly subjective in that *Roden Crater* does not ask its participant to explicitly understand—it asks for engagement, for interaction. To experience its sublimity requires no particular knowledge, just as Titian's work denotes a change in the relationship between viewing and sensing spaces. Through the invitation of dialogue, visual echo, and its circulation, Titian suggests the work is to be *felt* rather than *seen,* bridging the viewing and the sensing space to immerse his audience just as *Roden Crater* has done.

DE LOUTHERBOURG'S EIDOPHUSIKON (1781)

In the chronology of the art-object, the *Eidophusikon* is where the transition to the art-object prolifer-ated—and where it adopted phenomenological properties. Nearly two centuries after the work of Titian, Philip James de Loutherbourg invented the *Eidophusikon*, or the 'image of nature'. A painter by profes-sion, he was recruited by David Garrick, then-manager of Drury Lane Theatre, to occupy a brand new role as set designer (which had once been the responsibility of the actors) (Kornhaber, 2009). There were a number of significant changes that led to the *Eidophusikon,* the first of which was De Loutherbourg's creation of the backdrop. This was pivotal because prior to this, visual art and theatre were foreign. By intertwining the realms, Garrick then supplanted what was then homogeneous theatre lighting and instead used lighting to distinguish audience and stage—allowing the audience to focus their attention on the backdrop. The mobility of Garrick's lighting would allow De Loutherbourg to forge the third *dimension* of theatre, pulling from the sublime, vast landscapes that his paintings were famous for and separating theatre from the static morbidity of the still image (Kornhaber, 2009). After working for some time to make mechanized tools for animating scenery, colors, and movement to create a dynamic stage, De Loutherbourg made his foray into miniatures. It was at this point that he built the *Eidophusikon*, a miniature theatre that could be entirely operated by one individual. It was entirely mechanized and did not require actors, allowing him to singlehandedly *become* a theatre—De Loutherbourg held shows for smaller crowds, and created a much more intimate theatre experience (Kornhaber, 2009).

De Loutherbourg's quotations of traditional landscape art defined the next evolution of the land-scape art-object. In retrospect, the addition of a third dimension to the stage sounds painfully obvious, but this only speaks to the gravity of the innovation. The ability to exacerbate the drama of theatre is a notion that has never left the medium, and few innovations have been as significant. Akin to Turrell, the *Eidophusikon* is a form of spectacle—as a miniature theatre, it is a space to be engaged in, one that privileges the experience of few. It did something theatre had not done before: directly integrated the experiential to create a new *dimension* of interaction. As a result of his background as a painter, De Loutherbourg was able to create a way to bridge *viewing* and *sensing* spaces by literally engaging dialogue with his audience, by creating an environment where the phenomenology of temporality and intimacy were prime. *Roden Crater* engages temporality and intimacy, too—it is intended for very few people—because the artwork directly interacts with people and requires their involvement to achieve af-fect. Unlike the *Eidophusikon,* however, *Roden Crater* is *created* by the viewer, where the *Eidophusikon* merely requires the interaction of the viewer, though it would not be a 'show' per se, without the viewer. As a landscape artist, De Loutherbourg was known for his use of the Burkian sublime—the evocation of universal terror through scenes near some grandiose landscape facing death. This is central to the *Eidophusikon*—by bringing the 'image of nature' *to* people, he could affect them by virtue of the medium itself—resulting in an art much more immersive than it is isolating, subverting the static theatre. The *viewing* and *sensing* spaces here begin to amalgamate—it provoked sensations beyond visual experience, merging sight with the experience of space and situating it as the first phenomenological development of the landscape art-object. The *Eidophusikon* still requires viewing, but sensing is created where the interaction between its operator and the audience begins. *Roden Crater* similarly expresses an impetus towards sensation in the landscape art-object, the difference with being that Turrell's work goes *beyond* physicality, into perception. *Roden Crater's viewing* and *sensing* spaces are amalgamated, indistinguishable and unified—but De Loutherbourg's willingness to dialogue with his audience and experience *with* them created a new dimension that reflects a necessary step towards this unity. Where Titian's use of the visual echo and circulation insinuated a relationship with the audience, De Loutherbourg's introduction of a new *dimension* facilitated the introduction of phenomenology to the landscape art-object, taking its place in history as the experience of modern theatre, and in the bridging of *viewing* and *sensing* spaces that led to *Roden Crater.*

BARKER'S PANORAMA (1792)

Not long after the *Eidophusikon* came the invention of the *Panorama,* inspired by De Loutherbourg. Created by Robert Barker in 1792, the *Panorama* was a large-scale painting that surrounded the viewer 360°. To do this, Barker placed the works *exclusively* in circular buildings set to a very specific kind of lighting, then using immersive paint techniques to hide the borders of the painting (Ellis, 2008). It was a meticulously curated sensory experience, very similar to the process of selection and depiction inherent in creating a landscape, and is undoubtedly the most intentionally dialogical of the preceding works.

Barker, discussing this new medium, lamented, "[a] mere description is inadequate to impress a just idea of the performance, which, from the entire novelty of the thought, is not perfectly understood until seen" (Ellis, 2008). The premise that one must see it, must *be* there to understand it, describes precisely what makes the *Panorama* so irrevocably phenomenological—it does not merely *express* physicality so much as it *is* physical. It had to be experienced to be understood, and this put the *Panorama* in juxtaposition to the proliferation of print media of the time. *Roden Crater* is derivative of this thinking—its priority is affect, to arouse *feeling* and not just *seeing,* and it achieves this by engaging as many sensations as is physically possible. The *Panorama* manipulated perception through physicality, and in doing so truly bridged the *viewing* and *sensing* spaces—Barker took a medium designed specifically for viewing, and altered its environment to make the environment a part of the art. *Roden Crater* achieves this not through its physicality, but through its allusion to the physicality of space. Through highly perceptual manipulations of light, space, and material, Turrell ubiquitously contorts sensation—one cannot *view* a reproduction of *Roden Crater,* it *requires* you to experience it. Keeping in mind the significance of their respective environments, both the *Panorama* and *Roden Crater* explore notions of controlling their locale: *Roden Crater* exists in a singular geography and the *Panorama* was housed exclusively in rotundas. As Peter Zumthor describes, both are demonstrably "…closed architectural bod[ies] that isolate space within [themselves]…", opening to an intimate, physical dialogue with its audience (Zumthor, 2006). It was at the center of a perceptual shift from "viewing something that one knows is an illusion, yet feels like reality" (Byerly, 2007). The *Panorama* was a transitory point for how time and space were understood in art, demarcating a shift wherein interaction is required from the viewer. Much like in *Roden Crater,* the medium actually *guides* experience, prompting the viewer to discover and create the space they occupy. In the essay, *Spectacle, Landscape, and the Visual Demands of Panorama Painting,* Ethan Robey describes the *Panorama* "[a]s a mode of representation that had to be assembled by the viewer's own experience…[had] become emblematic of a type of visual experience new to the nineteenth century: the subjective experience of perception was not necessarily a means of comprehending an external reality" (Robey, 2014). This could not more succinctly describe Turrell's voice in *Roden Crater*—he never *displays* anything for his viewer, but allows them to discover it themselves. Where the *Panorama* creates a physical fiction to express a bodily phenomenology, Turrell's work appropriates this framework to reimagine the experience of celestial realities, a realm light years away from our own. This is the brilliance of the *Panorama*—its function as a 360-degree work is profound as a form of perceptual contrivance, ultimately affecting viewers with its spectacle and amalgamating of the lines between the *viewing* and *sensing* spaces through its sheer physicality.

CONCLUSION

Roden Crater defines what landscape has become in modernity: totality. Encompassing the principles of its antecedents to create a profoundly psychological experience, it utilizes phenomenological thinking to *exceed* experience—for the viewer to discover to *create*. From Titian's work it gains its dialogical facets, from the *Eidophusikon* it derives sensation, and from the *Panorama* it manifests physicality, and Turrell uses these frameworks to close the gap between the *viewing* and *sensing* space. He elicits participation and dialogue in his audience such that the act observation is not possible to begin with—*Roden Crater* can only be discovered and created. Viewers become participants, the art wields the ability to *affect,* and the distance between the medium and the person diminishes even further. I have demonstrated how different dimensions in the chronology of the landscape art-object have culminated in James Turrell's seminal work, *Roden Crater*, capitalizing on innovations in the landscape art-object medium to immerse his audience. Having bridged the *viewing* and *sensing* spaces to reach sensory perception, *Roden Crater* expresses no dialogue, no experience nor viewing—in Turrell's hands, it all becomes *one*. Turrell serves as a metaphorical parallel to the evolution of contemporary culture, serving as an exemplar for the tendency of art, and media as a whole, towards immersion.

REFERENCES

About Roden Crater. (n.d.). Retrieved from http://rodencrater.com/about/

Bognar, B., Bognár, B., & Kuma, K. (2009). *Material immaterial: The new work of Kengo Kuma*. New York: Princeton Architectural Press.

Byerly, A. (2007). A prodigious map beneath his feet: Virtual travel and the panoramic perspective. *Nineteenth-Century Contexts, 29*(2-3), 151–168. doi:10.1080/08905490701584643

Eberhart, M. (2012). Sensing, time, and the aural imagination in Titian's 'Venus with Organist and Dog.'. *Artibus et Historiae, 33*(65), 79–95. Retrieved from http://www.jstor.org/stable/23509712

Ellis, M. (2008). Spectacles within doors: Panoramas of London in the 1790s. *Romanticism, 14*(2), 133–148. doi:10.3366/E1354991X0800024X

Hylton, W. S. (2013, June 13). *How James Turrell knocked the art world off its feet*. Retrieved from http://www.nytimes.com/2013/06/16/magazine/how-james-turrell knocked-the-art-world-off-its-feet.html

James Turrell's Roden Crater. (2013). Los Angeles County Museum of Art. Retrieved from https://vimeo.com/67926427

Jelić, A. (2015). Designing "pre-reflective" architecture. *Ambiances, Experimental Simulation*. Retrieved from http://ambiances.revues.org/628

Kornhaber, D. (2009). Regarding the Eidophusikon: Spectacle, scenography, and culture in eighteenth century England. *TAJ - Theatre Arts Journal: Studies in Scenography and Performance, 1*(1). Retrieved from http://myaccess.library.utoronto.ca/login?url=http://search.proquest.com/docview/1086486005?accountid=14771

Kunstmuseum Wolfsburg. (2009, November 25). *James Turrell: The Wolfsburg Project*. Retrieved from https://www.youtube.com/watch?v=QWekIcZaKns

Lunberry, C. (2009). Soliloquies of silence: James Turrell's theatre of installation. *Mosaic, 42*(1), 33–50. Retrieved from http://myaccess.library.utoronto.ca/login?url=http://search.proquest.com/docview/205371831?accountid=14771

Matts, T., & Tynan, A. (2012). Earthwork and eco-clinic: Notes on James Turrell's 'Roden Crater' Project. *Design Ecologies, 2*(2), 188–211. doi:10.1386/des.2.2.188_1

Robey, E. (2014). John Vanderlyn's view of Versailles: Spectacle, landscape, and the visual demands of panorama painting. *Early Popular Visual Culture, 12*(1), 1–21. doi:10.1080/17460654.2013.876922

Zumthor, P. (2006). *Thinking architecture*. Basel: Birkhäuser.

Related References

To continue our tradition of advancing media and communications research, we have compiled a list of recommended IGI Global readings. These references will provide additional information and guidance to further enrich your knowledge and assist you with your own research and future publications.

Aas, B. G. (2012). What's real? Presence, personality and identity in the real and online virtual world. In N. Zagalo, L. Morgado, & A. Boa-Ventura (Eds.), *Virtual worlds and metaverse platforms: New communication and identity paradigms* (pp. 88–99). Hershey, PA: IGI Global. doi:10.4018/978-1-60960-854-5.ch006

Aceti, V., & Luppicini, R. (2013). Exploring the effect of mhealth technologies on communication and information sharing in a pediatric critical care unit: A case study. In J. Tan (Ed.), *Healthcare information technology innovation and sustainability: Frontiers and adoption* (pp. 88–108). Hershey, PA: IGI Global. doi:10.4018/978-1-4666-2797-0.ch006

Acilar, A. (2013). Factors affecting mobile phone use among undergraduate students in Turkey: An exploratory analysis. In I. Lee (Ed.), *Strategy, adoption, and competitive advantage of mobile services in the global economy* (pp. 234–246). Hershey, PA: IGI Global. doi:10.4018/978-1-4666-1939-5.ch013

Adams, A. (2013). Situated e-learning: Empowerment and barriers to identity changes. In S. Warburton & S. Hatzipanagos (Eds.), *Digital identity and social media* (pp. 159–175). Hershey, PA: IGI Global. doi:10.4018/978-1-4666-1915-9.ch012

Adeoye, B. F. (2013). Culturally different learning styles in online learning environments: A case of Nigerian university students. In L. Tomei (Ed.), *Learning tools and teaching approaches through ICT advancements* (pp. 228–240). Hershey, PA: IGI Global. doi:10.4018/978-1-4666-2017-9.ch020

Agarwal, N., & Mahata, D. (2013). Grouping the similar among the disconnected bloggers. In G. Xu & L. Li (Eds.), *Social media mining and social network analysis: Emerging research* (pp. 54–71). Hershey, PA: IGI Global. doi:10.4018/978-1-4666-2806-9.ch004

Aiken, M., Wang, J., Gu, L., & Paolillo, J. (2013). An exploratory study of how technology supports communication in multilingual groups. In N. Kock (Ed.), *Interdisciplinary applications of electronic collaboration approaches and technologies* (pp. 17–29). Hershey, PA: IGI Global. doi:10.4018/978-1-4666-2020-9.ch002

Aikins, S. K., & Chary, M. (2013). Online participation and digital divide: An empirical evaluation of U.S. midwestern municipalities. In *Digital literacy: Concepts, methodologies, tools, and applications* (pp. 63–85). Hershey, PA: IGI Global. doi:10.4018/978-1-4666-1852-7.ch004

Al Disi, Z. A., & Albadri, F. (2013). Arab youth and the internet: Educational perspective. In F. Albadri (Ed.), *Information systems applications in the Arab education sector* (pp. 163–178). Hershey, PA: IGI Global. doi:10.4018/978-1-4666-1984-5.ch012

Al-Dossary, S., Al-Dulaijan, N., Al-Mansour, S., Al-Zahrani, S., Al-Fridan, M., & Househ, M. (2013). Organ donation and transplantation: Processes, registries, consent, and restrictions in Saudi Arabia. In M. Cruz-Cunha, I. Miranda, & P. Gonçalves (Eds.), *Handbook of research on ICTs for human-centered healthcare and social care services* (pp. 511–528). Hershey, PA: IGI Global. doi:10.4018/978-1-4666-3986-7.ch027

Al-Khaffaf, M. M., & Abdellatif, H. J. (2013). The effect of information and communication technology on customer relationship management: Jordan public shareholding companies. In R. Eid (Ed.), *Managing customer trust, satisfaction, and loyalty through information communication technologies* (pp. 342–350). Hershey, PA: IGI Global. doi:10.4018/978-1-4666-3631-6.ch020

Al-Nuaim, H. A. (2012). Evaluation of Arab municipal websites. In *Wireless technologies: Concepts, methodologies, tools and applications* (pp. 1170–1185). Hershey, PA: IGI Global. doi:10.4018/978-1-61350-101-6.ch505

Al-Nuaim, H. A. (2013). Developing user profiles for interactive online products in practice. In M. Garcia-Ruiz (Ed.), *Cases on usability engineering: Design and development of digital products* (pp. 57–79). Hershey, PA: IGI Global. doi:10.4018/978-1-4666-4046-7.ch003

Al Omoush, K. S., Alqirem, R. M., & Shaqrah, A. A. (2013). The driving internal beliefs of household internet adoption among Jordanians and the role of cultural values. In A. Zolait (Ed.), *Technology diffusion and adoption: global complexity, global innovation* (pp. 130–151). Hershey, PA: IGI Global. doi:10.4018/978-1-4666-2791-8.ch009

Al-Shqairat, Z. I., & Altarawneh, I. I. (2013). The role of partnership in e-government readiness: The knowledge stations (KSs) initiative in Jordan. In A. Mesquita (Ed.), *User perception and influencing factors of technology in everyday life* (pp. 192–210). Hershey, PA: IGI Global. doi:10.4018/978-1-4666-1954-8.ch014

AlBalawi, M. S. (2013). Web-based instructions: An assessment of preparedness of conventional universities in Saudi Arabia. In M. Khosrow-Pour (Ed.), *Cases on assessment and evaluation in education* (pp. 417–451). Hershey, PA: IGI Global. doi:10.4018/978-1-4666-2621-8.ch018

Alejos, A. V., Cuiñas, I., Expósito, I., & Sánchez, M. G. (2013). From the farm to fork: Information security accomplishment in a RFID based tracking chain for food sector. In P. Lopez, J. Hernandez-Castro, & T. Li (Eds.), *Security and trends in wireless identification and sensing platform tags: Advancements in RFID* (pp. 237–270). Hershey, PA: IGI Global. doi:10.4018/978-1-4666-1990-6.ch010

Alkazemi, M. F., Bowe, B. J., & Blom, R. (2013). Facilitating the Egyptian uprising: A case study of Facebook and Egypt's April 6th youth movement. In N. Azab (Ed.), *Cases on web 2.0 in developing countries: Studies on implementation, application, and use* (pp. 256–282). Hershey, PA: IGI Global. doi:10.4018/978-1-4666-2515-0.ch010

Almutairi, M. S. (2012). M-government: Challenges and key success factors – Saudi Arabia case study. In Wireless technologies: Concepts, methodologies, tools and applications (pp. 1698-1717). Hershey, PA: IGI Global. doi:10.4018/978-1-61350-101-6.ch611

Alyagout, F., & Siti-Nabiha, A. K. (2013). Public sector transformation: Privatization in Saudi Arabia. In N. Pomazalová (Ed.), *Public sector transformation processes and internet public procurement: Decision support systems* (pp. 17–31). Hershey, PA: IGI Global. doi:10.4018/978-1-4666-2665-2.ch002

Amirante, A., Castaldi, T., Miniero, L., & Romano, S. P. (2013). Protocol interactions among user agents, application servers, and media servers: Standardization efforts and open issues. In D. Kanellopoulos (Ed.), *Intelligent multimedia technologies for networking applications: Techniques and tools* (pp. 48–63). Hershey, PA: IGI Global. doi:10.4018/978-1-4666-2833-5.ch003

Andres, H. P. (2013). Shared mental model development during technology-mediated collaboration. In N. Kock (Ed.), *Interdisciplinary applications of electronic collaboration approaches and technologies* (pp. 125–142). Hershey, PA: IGI Global. doi:10.4018/978-1-4666-2020-9.ch009

Andrus, C. H., & Gaynor, M. (2013). Good IT requires good communication. In S. Sarnikar, D. Bennett, & M. Gaynor (Eds.), *Cases on healthcare information technology for patient care management* (pp. 122–125). Hershey, PA: IGI Global. doi:10.4018/978-1-4666-2671-3.ch007

Annafari, M. T., & Bohlin, E. (2013). Why is the diffusion of mobile service not an evolutionary process? In I. Lee (Ed.), *Mobile services industries, technologies, and applications in the global economy* (pp. 25–38). Hershey, PA: IGI Global. doi:10.4018/978-1-4666-1981-4.ch002

Anupama, S. (2013). Gender evaluation of rural e-governance in India: A case study of E-Gram Suraj (e-rural good governance) scheme1. In *Digital literacy: Concepts, methodologies, tools, and applications* (pp. 1059–1074). Hershey, PA: IGI Global. doi:10.4018/978-1-4666-1852-7.ch055

Ariely, G. (2013). Boundaries of socio-technical systems and IT for knowledge development in military environments. In J. Abdelnour-Nocera (Ed.), *Knowledge and technological development effects on organizational and social structures* (pp. 224–238). Hershey, PA: IGI Global. doi:10.4018/978-1-4666-2151-0.ch014

Arsenio, A. M. (2013). Intelligent approaches for adaptation and distribution of personalized multimedia content. In D. Kanellopoulos (Ed.), *Intelligent multimedia technologies for networking applications: Techniques and tools* (pp. 197–224). Hershey, PA: IGI Global. doi:10.4018/978-1-4666-2833-5.ch008

Artail, H., & Tarhini, T. (2013). Runtime discovery and access of web services in mobile environments. In I. Lee (Ed.), *Mobile services industries, technologies, and applications in the global economy* (pp. 193–213). Hershey, PA: IGI Global. doi:10.4018/978-1-4666-1981-4.ch012

Asino, T. I., Wilder, H., & Ferris, S. P. (2013). Innovative use of ICT in Namibia for nationhood: Special emphasis on the Namibian newspaper. In H. Rahman (Ed.), *Cases on progressions and challenges in ICT utilization for citizen-centric governance* (pp. 205–216). Hershey, PA: IGI Global. doi:10.4018/978-1-4666-2071-1.ch009

Atici, B., & Bati, U. (2013). Identity of virtual supporters: Constructing identity of Turkish football fans on digital media. In S. Warburton & S. Hatzipanagos (Eds.), *Digital identity and social media* (pp. 256–274). Hershey, PA: IGI Global. doi:10.4018/978-1-4666-1915-9.ch018

Azab, N., & Khalifa, N. (2013). Web 2.0 and opportunities for entrepreneurs: How Egyptian entrepreneurs perceive and exploit web 2.0 technologies. In N. Azab (Ed.), *Cases on web 2.0 in developing countries: Studies on implementation, application, and use* (pp. 1–32). Hershey, PA: IGI Global. doi:10.4018/978-1-4666-2515-0.ch001

Bainbridge, W. S. (2013). Ancestor veneration avatars. In R. Luppicini (Ed.), *Handbook of research on technoself: Identity in a technological society* (pp. 308–321). Hershey, PA: IGI Global. doi:10.4018/978-1-4666-2211-1.ch017

Baporikar, N. (2013). Critical review of academic entrepreneurship in India. In A. Szopa, W. Karwowski, & P. Ordóñez de Pablos (Eds.), *Academic entrepreneurship and technological innovation: A business management perspective* (pp. 29–52). Hershey, PA: IGI Global. doi:10.4018/978-1-4666-2116-9.ch002

Barroca, L., & Gimenes, I. M. (2013). Computing postgraduate programmes in the UK and Brazil: Learning from experience in distance education with web 2.0 support. In N. Azab (Ed.), *Cases on web 2.0 in developing countries: Studies on implementation, application, and use* (pp. 147–171). Hershey, PA: IGI Global. doi:10.4018/978-1-4666-2515-0.ch006

Barton, S. M. (2013). Facilitating learning by going online: Modernising Islamic teaching and learning in Indonesia. In E. McKay (Ed.), *ePedagogy in online learning: New developments in web mediated human computer interaction* (pp. 74–92). Hershey, PA: IGI Global. doi:10.4018/978-1-4666-3649-1.ch005

Barzilai-Nahon, K., Gomez, R., & Ambikar, R. (2013). Conceptualizing a contextual measurement for digital divide/s: Using an integrated narrative. In *Digital literacy: Concepts, methodologies, tools, and applications* (pp. 279–293). Hershey, PA: IGI Global. doi:10.4018/978-1-4666-1852-7.ch015

Bénel, A., & Lacour, P. (2012). Towards a participative platform for cultural texts translators. In C. El Morr & P. Maret (Eds.), *Virtual community building and the information society: Current and future directions* (pp. 153–162). Hershey, PA: IGI Global. doi:10.4018/978-1-60960-869-9.ch008

Bentley, C. M. (2013). Designing and implementing online collaboration tools in West Africa. In N. Azab (Ed.), *Cases on web 2.0 in developing countries: Studies on implementation, application, and use* (pp. 33–60). Hershey, PA: IGI Global. doi:10.4018/978-1-4666-2515-0.ch002

Berg, M. (2012). Checking in at the urban playground: Digital geographies and electronic flâneurs. In F. Comunello (Ed.), *Networked sociability and individualism: Technology for personal and professional relationships* (pp. 169–194). Hershey, PA: IGI Global. doi:10.4018/978-1-61350-338-6.ch009

Bers, M. U., & Ettinger, A. B. (2013). Programming robots in kindergarten to express identity: An ethnographic analysis. In *Industrial engineering: Concepts, methodologies, tools, and applications* (pp. 1952–1968). Hershey, PA: IGI Global. doi:10.4018/978-1-4666-1945-6.ch105

Binsaleh, M., & Hassan, S. (2013). Systems development methodology for mobile commerce applications. In I. Khalil & E. Weippl (Eds.), *Contemporary challenges and solutions for mobile and multimedia technologies* (pp. 146–162). Hershey, PA: IGI Global. doi:10.4018/978-1-4666-2163-3.ch009

Bishop, J. (2013). Cooperative e-learning in the multilingual and multicultural school: The role of "classroom 2.0" for increasing participation in education. In P. Pumilia-Gnarini, E. Favaron, E. Pacetti, J. Bishop, & L. Guerra (Eds.), *Handbook of research on didactic strategies and technologies for education: Incorporating advancements* (pp. 137–150). Hershey, PA: IGI Global. doi:10.4018/978-1-4666-2122-0.ch013

Bishop, J. (2013). Increasing capital revenue in social networking communities: Building social and economic relationships through avatars and characters. In J. Bishop (Ed.), *Examining the concepts, issues, and implications of internet trolling* (pp. 44–61). Hershey, PA: IGI Global. doi:10.4018/978-1-4666-2803-8.ch005

Bishop, J. (2013). Lessons from the emotive project for increasing take-up of big society and responsible capitalism initiatives. In P. Pumilia-Gnarini, E. Favaron, E. Pacetti, J. Bishop, & L. Guerra (Eds.), *Handbook of research on didactic strategies and technologies for education: Incorporating advancements* (pp. 208–217). Hershey, PA: IGI Global. doi:10.4018/978-1-4666-2122-0.ch019

Blau, I. (2013). E-collaboration within, between, and without institutions: Towards better functioning of online groups through networks. In N. Kock (Ed.), *Interdisciplinary applications of electronic collaboration approaches and technologies* (pp. 188–203). Hershey, PA: IGI Global. doi:10.4018/978-1-4666-2020-9.ch013

Boskic, N., & Hu, S. (2013). Blended learning: The road to inclusive and global education. In E. Jean Francois (Ed.), *Transcultural blended learning and teaching in postsecondary education* (pp. 283–301). Hershey, PA: IGI Global. doi:10.4018/978-1-4666-2014-8.ch015

Botero, A., Karhu, K., & Vihavainen, S. (2012). Exploring the ecosystems and principles of community innovation. In A. Lugmayr, H. Franssila, P. Näränen, O. Sotamaa, J. Vanhala, & Z. Yu (Eds.), *Media in the ubiquitous era: Ambient, social and gaming media* (pp. 216–234). Hershey, PA: IGI Global. doi:10.4018/978-1-60960-774-6.ch012

Bowe, B. J., Blom, R., & Freedman, E. (2013). Negotiating boundaries between control and dissent: Free speech, business, and repressitarian governments. In J. Lannon & E. Halpin (Eds.), *Human rights and information communication technologies: Trends and consequences of use* (pp. 36–55). Hershey, PA: IGI Global. doi:10.4018/978-1-4666-1918-0.ch003

Brandão, J., Ferreira, T., & Carvalho, V. (2012). An overview on the use of serious games in the military industry and health. In M. Cruz-Cunha (Ed.), *Handbook of research on serious games as educational, business and research tools* (pp. 182–201). Hershey, PA: IGI Global. doi:10.4018/978-1-4666-0149-9.ch009

Brost, L. F., & McGinnis, C. (2012). The status of blogging in the republic of Ireland: A case study. In T. Dumova & R. Fiordo (Eds.), *Blogging in the global society: Cultural, political and geographical aspects* (pp. 128–147). Hershey, PA: IGI Global. doi:10.4018/978-1-60960-744-9.ch008

Burns, J., Blanchard, M., & Metcalf, A. (2013). Bridging the digital divide in Australia: The potential implications for the mental health of young people experiencing marginalisation. In *Digital literacy: Concepts, methodologies, tools, and applications* (pp. 772–793). Hershey, PA: IGI Global. doi:10.4018/978-1-4666-1852-7.ch040

Cagliero, L., & Fiori, A. (2013). News document summarization driven by user-generated content. In G. Xu & L. Li (Eds.), *Social media mining and social network analysis: Emerging research* (pp. 105–126). Hershey, PA: IGI Global. doi:10.4018/978-1-4666-2806-9.ch007

Camillo, A., & Di Pietro, L. (2013). Managerial communication in the global cross-cultural context. In B. Christiansen, E. Turkina, & N. Williams (Eds.), *Cultural and technological influences on global business* (pp. 397–419). Hershey, PA: IGI Global. doi:10.4018/978-1-4666-3966-9.ch021

Canazza, S., De Poli, G., Rodà, A., & Vidolin, A. (2013). Expressiveness in music performance: Analysis, models, mapping, encoding. In J. Steyn (Ed.), *Structuring music through markup language: Designs and architectures* (pp. 156–186). Hershey, PA: IGI Global. doi:10.4018/978-1-4666-2497-9.ch008

Carrasco, J. G., Ovide, E., & Puyal, M. B. (2013). Closing and opening of cultures. In F. García-Peñalvo (Ed.), *Multiculturalism in technology-based education: Case studies on ICT-supported approaches* (pp. 125–142). Hershey, PA: IGI Global. doi:10.4018/978-1-4666-2101-5.ch008

Carreras, I., Zanardi, A., Salvadori, E., & Miorandi, D. (2013). A distributed monitoring framework for opportunistic communication systems: An experimental approach. In V. De Florio (Ed.), *Innovations and approaches for resilient and adaptive systems* (pp. 220–236). Hershey, PA: IGI Global. doi:10.4018/978-1-4666-2056-8.ch013

Carter, M., Grover, V., & Thatcher, J. B. (2013). Mobile devices and the self: developing the concept of mobile phone identity. In I. Lee (Ed.), *Strategy, adoption, and competitive advantage of mobile services in the global economy* (pp. 150–164). Hershey, PA: IGI Global. doi:10.4018/978-1-4666-1939-5.ch008

Caruso, F., Giuffrida, G., Reforgiato, D., & Zarba, C. (2013). Recommendation systems for mobile devices. In I. Lee (Ed.), *Mobile services industries, technologies, and applications in the global economy* (pp. 221–242). Hershey, PA: IGI Global. doi:10.4018/978-1-4666-1981-4.ch014

Casamassima, L. (2013). eTwinning project: A virtual orchestra. In P. Pumilia-Gnarini, E. Favaron, E. Pacetti, J. Bishop, & L. Guerra (Eds.) Handbook of research on didactic strategies and technologies for education: Incorporating advancements (pp. 703-709). Hershey, PA: IGI Global. doi:10.4018/978-1-4666-2122-0.ch061

Caschera, M. C., D'Ulizia, A., Ferri, F., & Grifoni, P. (2012). Multiculturality and multimodal languages. In G. Ghinea, F. Andres, & S. Gulliver (Eds.), *Multiple sensorial media advances and applications: New developments in MulSeMedia* (pp. 99–114). Hershey, PA: IGI Global. doi:10.4018/978-1-60960-821-7.ch005

Catagnus, R. M., & Hantula, D. A. (2013). The virtual individual education plan (IEP) team: Using online collaboration to develop a behavior intervention plan. In N. Kock (Ed.), *Interdisciplinary applications of electronic collaboration approaches and technologies* (pp. 30–45). Hershey, PA: IGI Global. doi:10.4018/978-1-4666-2020-9.ch003

Ch'ng, E. (2013). The mirror between two worlds: 3D surface computing for objects and environments. In D. Harrison (Ed.), *Digital media and technologies for virtual artistic spaces* (pp. 166–185). Hershey, PA: IGI Global. doi:10.4018/978-1-4666-2961-5.ch013

Chand, A. (2013). Reducing digital divide: The case of the 'people first network' (PFNet) in the Solomon Islands. In *Digital literacy: Concepts, methodologies, tools, and applications* (pp. 1571–1605). Hershey, PA: IGI Global. doi:10.4018/978-1-4666-1852-7.ch083

Chatterjee, S. (2013). Ethical behaviour in technology-mediated communication. In J. Bishop (Ed.), *Examining the concepts, issues, and implications of internet trolling* (pp. 1–9). Hershey, PA: IGI Global. doi:10.4018/978-1-4666-2803-8.ch001

Chen, C., Chao, H., Wu, T., Fan, C., Chen, J., Chen, Y., & Hsu, J. (2013). IoT-IMS communication platform for future internet. In V. De Florio (Ed.), *Innovations and approaches for resilient and adaptive systems* (pp. 68–86). Hershey, PA: IGI Global. doi:10.4018/978-1-4666-2056-8.ch004

Chen, J., & Hu, X. (2013). Smartphone market in China: Challenges, opportunities, and promises. In I. Lee (Ed.), *Mobile services industries, technologies, and applications in the global economy* (pp. 120–132). Hershey, PA: IGI Global. doi:10.4018/978-1-4666-1981-4.ch008

Chen, Y., Lee, B., & Kirk, R. M. (2013). Internet use among older adults: Constraints and opportunities. In R. Zheng, R. Hill, & M. Gardner (Eds.), *Engaging older adults with modern technology: Internet use and information access needs* (pp. 124–141). Hershey, PA: IGI Global. doi:10.4018/978-1-4666-1966-1.ch007

Cheong, P. H., & Martin, J. N. (2013). Cultural implications of e-learning access (and divides): Teaching an intercultural communication course online. In A. Edmundson (Ed.), *Cases on cultural implications and considerations in online learning* (pp. 82–100). Hershey, PA: IGI Global. doi:10.4018/978-1-4666-1885-5.ch005

Chhanabhai, P., & Holt, A. (2013). The changing world of ICT and health: Crossing the digital divide. In *Digital literacy: Concepts, methodologies, tools, and applications* (pp. 794–811). Hershey, PA: IGI Global. doi:10.4018/978-1-4666-1852-7.ch041

Chuling, W., Hua, C. M., & Chee, C. J. (2012). Investigating the demise of radio and television broadcasting. In R. Sharma, M. Tan, & F. Pereira (Eds.), *Understanding the interactive digital media marketplace: Frameworks, platforms, communities and issues* (pp. 392–405). Hershey, PA: IGI Global. doi:10.4018/978-1-61350-147-4.ch031

Ciaramitaro, B. L. (2012). Introduction to mobile technologies. In B. Ciaramitaro (Ed.), *Mobile technology consumption: Opportunities and challenges* (pp. 1–15). Hershey, PA: IGI Global. doi:10.4018/978-1-61350-150-4.ch001

Cicconetti, C., Mambrini, R., & Rossi, A. (2013). A survey of wireless backhauling solutions for ITS. In R. Daher & A. Vinel (Eds.), *Roadside networks for vehicular communications: Architectures, applications, and test fields* (pp. 57–70). Hershey, PA: IGI Global. doi:10.4018/978-1-4666-2223-4.ch003

Ciptasari, R. W., & Sakurai, K. (2013). Multimedia copyright protection scheme based on the direct feature-based method. In K. Kondo (Ed.), *Multimedia information hiding technologies and methodologies for controlling data* (pp. 412–439). Hershey, PA: IGI Global. doi:10.4018/978-1-4666-2217-3.ch019

Code, J. (2013). Agency and identity in social media. In S. Warburton & S. Hatzipanagos (Eds.), *Digital identity and social media* (pp. 37–57). Hershey, PA: IGI Global. doi:10.4018/978-1-4666-1915-9.ch004

Comunello, F. (2013). From the digital divide to multiple divides: Technology, society, and new media skills. In *Digital literacy: Concepts, methodologies, tools, and applications* (pp. 1622–1639). Hershey, PA: IGI Global. doi:10.4018/978-1-4666-1852-7.ch085

Consonni, A. (2013). About the use of the DMs in CLIL classes. In F. García-Peñalvo (Ed.), *Multiculturalism in technology-based education: Case studies on ICT-supported approaches* (pp. 9–27). Hershey, PA: IGI Global. doi:10.4018/978-1-4666-2101-5.ch002

Constant, J. (2012). Digital approaches to visualization of geometric problems in wooden sangaku tablets. In A. Ursyn (Ed.), *Biologically-inspired computing for the arts: Scientific data through graphics* (pp. 240–253). Hershey, PA: IGI Global. doi:10.4018/978-1-4666-0942-6.ch013

Cossiavelou, V., Bantimaroudis, P., Kavakli, E., & Illia, L. (2013). The media gatekeeping model updated by R and I in ICTs: The case of wireless communications in media coverage of the olympic games. In M. Bartolacci & S. Powell (Eds.), *Advancements and innovations in wireless communications and network technologies* (pp. 262–288). Hershey, PA: IGI Global. doi:10.4018/978-1-4666-2154-1.ch019

Cropf, R. A., Benmamoun, M., & Kalliny, M. (2013). The role of web 2.0 in the Arab Spring. In N. Azab (Ed.), *Cases on web 2.0 in developing countries: Studies on implementation, application, and use* (pp. 76–108). Hershey, PA: IGI Global. doi:10.4018/978-1-4666-2515-0.ch004

Cucinotta, A., Minnolo, A. L., & Puliafito, A. (2013). Design and implementation of an event-based RFID middleware. In N. Karmakar (Ed.), *Advanced RFID systems, security, and applications* (pp. 110–131). Hershey, PA: IGI Global. doi:10.4018/978-1-4666-2080-3.ch006

D'Andrea, A., Ferri, F., & Grifoni, P. (2013). Assessing e-health in Africa: Web 2.0 applications. In N. Azab (Ed.), *Cases on web 2.0 in developing countries: Studies on implementation, application, and use* (pp. 442–467). Hershey, PA: IGI Global. doi:10.4018/978-1-4666-2515-0.ch016

de Guinea, A. O. (2013). The level paradox of e-collaboration: Dangers and solutions. In N. Kock (Ed.), *Interdisciplinary applications of electronic collaboration approaches and technologies* (pp. 166–187). Hershey, PA: IGI Global. doi:10.4018/978-1-4666-2020-9.ch012

Dhar-Bhattacharjee, S., & Takruri-Rizk, H. (2012). An Indo-British comparison. In C. Romm Livermore (Ed.), *Gender and social computing: Interactions, differences and relationships* (pp. 50–71). Hershey, PA: IGI Global. doi:10.4018/978-1-60960-759-3.ch004

Díaz-Foncea, M., & Marcuello, C. (2013). ANOBIUM, SL: The use of the ICT as niche of employment and as tool for developing the social market. In T. Torres-Coronas & M. Vidal-Blasco (Eds.), *Social e-enterprise: Value creation through ICT* (pp. 221–242). Hershey, PA: IGI Global. doi:10.4018/978-1-4666-2667-6.ch013

Díaz-González, M., Froufe, N. Q., Brena, A. G., & Pumarola, F. (2013). Uses and implementation of social media at university: The case of schools of communication in Spain. In B. Pătruţ, M. Pătruţ, & C. Cmeciu (Eds.), *Social media and the new academic environment: Pedagogical challenges* (pp. 204–222). Hershey, PA: IGI Global. doi:10.4018/978-1-4666-2851-9.ch010

Ditsa, G., Alwahaishi, S., Al-Kobaisi, S., & Snášel, V. (2013). A comparative study of the effects of culture on the deployment of information technology. In A. Zolait (Ed.), *Technology diffusion and adoption: Global complexity, global innovation* (pp. 77–90). Hershey, PA: IGI Global. doi:10.4018/978-1-4666-2791-8.ch006

Donaldson, O., & Duggan, E. W. (2013). Assessing mobile value-added preference structures: The case of a developing country. In I. Lee (Ed.), *Strategy, adoption, and competitive advantage of mobile services in the global economy* (pp. 349–370). Hershey, PA: IGI Global. doi:10.4018/978-1-4666-1939-5.ch019

Douai, A. (2013). "In YouTube we trust": Video exchange and Arab human rights. In J. Lannon & E. Halpin (Eds.), *Human rights and information communication technologies: Trends and consequences of use* (pp. 57–71). Hershey, PA: IGI Global. doi:10.4018/978-1-4666-1918-0.ch004

Dromzée, C., Laborie, S., & Roose, P. (2013). A semantic generic profile for multimedia document adaptation. In D. Kanellopoulos (Ed.), *Intelligent multimedia technologies for networking applications: Techniques and tools* (pp. 225–246). Hershey, PA: IGI Global. doi:10.4018/978-1-4666-2833-5.ch009

Drucker, S., & Gumpert, G. (2012). The urban communication infrastructure: Global connection and local detachment. In *Wireless technologies: Concepts, methodologies, tools and applications* (pp. 1150–1169). Hershey, PA: IGI Global. doi:10.4018/978-1-61350-101-6.ch504

Drula, G. (2013). Media and communication research facing social media. In M. Pătruţ & B. Pătruţ (Eds.), *Social media in higher education: Teaching in web 2.0* (pp. 371–392). Hershey, PA: IGI Global. doi:10.4018/978-1-4666-2970-7.ch019

Dueck, J., & Rempel, M. (2013). Human rights and technology: Lessons from Alice in Wonderland. In J. Lannon & E. Halpin (Eds.), *Human rights and information communication technologies: Trends and consequences of use* (pp. 1–20). Hershey, PA: IGI Global. doi:10.4018/978-1-4666-1918-0.ch001

Dumova, T. (2012). Social interaction technologies and the future of blogging. In T. Dumova & R. Fiordo (Eds.), *Blogging in the global society: Cultural, political and geographical aspects* (pp. 249–274). Hershey, PA: IGI Global. doi:10.4018/978-1-60960-744-9.ch015

Dunn, H. S. (2013). Information literacy and the digital divide: Challenging e-exclusion in the global south. In *Digital literacy: Concepts, methodologies, tools, and applications* (pp. 20–38). Hershey, PA: IGI Global. doi:10.4018/978-1-4666-1852-7.ch002

Elias, N. (2013). Immigrants' internet use and identity from an intergenerational perspective: Immigrant senior citizens and youngsters from the former Soviet Union in Israel. In R. Luppicini (Ed.), *Handbook of research on technoself: Identity in a technological society* (pp. 293–307). Hershey, PA: IGI Global. doi:10.4018/978-1-4666-2211-1.ch016

Elizabeth, L. S., Ismail, N., & Tun, M. S. (2012). The future of the printed book. In R. Sharma, M. Tan, & F. Pereira (Eds.), *Understanding the interactive digital media marketplace: Frameworks, platforms, communities and issues* (pp. 416–429). Hershey, PA: IGI Global. doi:10.4018/978-1-61350-147-4.ch033

Erne, R. (2012). Knowledge worker performance in a cross-industrial perspective. In S. Brüggemann & C. d'Amato (Eds.), *Collaboration and the semantic web: Social networks, knowledge networks, and knowledge resources* (pp. 297–321). Hershey, PA: IGI Global. doi:10.4018/978-1-4666-0894-8.ch015

Ertl, B., Helling, K., & Kikis-Papadakis, K. (2012). The impact of gender in ICT usage, education and career: Comparisons between Greece and Germany. In C. Romm Livermore (Ed.), *Gender and social computing: Interactions, differences and relationships* (pp. 98–119). Hershey, PA: IGI Global. doi:10.4018/978-1-60960-759-3.ch007

Estapé-Dubreuil, G., & Torreguitart-Mirada, C. (2013). ICT adoption in the small and medium-size social enterprises in Spain: Opportunity or priority? In T. Torres-Coronas & M. Vidal-Blasco (Eds.), *Social e-enterprise: Value creation through ICT* (pp. 200–220). Hershey, PA: IGI Global. doi:10.4018/978-1-4666-2667-6.ch012

Eze, U. C., & Poong, Y. S. (2013). Consumers' intention to use mobile commerce and the moderating roles of gender and income. In I. Lee (Ed.), *Strategy, adoption, and competitive advantage of mobile services in the global economy* (pp. 127–148). Hershey, PA: IGI Global. doi:10.4018/978-1-4666-1939-5.ch007

Farrell, R., Danis, C., Erickson, T., Ellis, J., Christensen, J., Bailey, M., & Kellogg, W. A. (2012). A picture and a thousand words: Visual scaffolding for mobile communication in developing regions. In W. Hu (Ed.), *Emergent trends in personal, mobile, and handheld computing technologies* (pp. 341–354). Hershey, PA: IGI Global. doi:10.4018/978-1-4666-0921-1.ch020

Fidler, C. S., Kanaan, R. K., & Rogerson, S. (2013). Barriers to e-government implementation in Jordan: The role of wasta. In A. Mesquita (Ed.), *User perception and influencing factors of technology in everyday life* (pp. 179–191). Hershey, PA: IGI Global. doi:10.4018/978-1-4666-1954-8.ch013

Filho, J. R. (2013). ICT and human rights in Brazil: From military to digital dictatorship. In J. Lannon & E. Halpin (Eds.), *Human rights and information communication technologies: Trends and consequences of use* (pp. 86–99). Hershey, PA: IGI Global. doi:10.4018/978-1-4666-1918-0.ch006

Fiordo, R. (2012). Analyzing blogs: A hermeneutic perspective. In T. Dumova & R. Fiordo (Eds.), *Blogging in the global society: Cultural, political and geographical aspects* (pp. 231–248). Hershey, PA: IGI Global. doi:10.4018/978-1-60960-744-9.ch014

Fischer, G., & Herrmann, T. (2013). Socio-technical systems: A meta-design perspective. In J. Abdelnour-Nocera (Ed.), *Knowledge and technological development effects on organizational and social structures* (pp. 1–36). Hershey, PA: IGI Global. doi:10.4018/978-1-4666-2151-0.ch001

Fleury, M., & Al-Jobouri, L. (2013). Techniques and tools for adaptive video streaming. In D. Kanellopoulos (Ed.), *Intelligent multimedia technologies for networking applications: Techniques and tools* (pp. 65–101). Hershey, PA: IGI Global. doi:10.4018/978-1-4666-2833-5.ch004

Freeman, I., & Freeman, A. (2013). Capacity building for different abilities using ICT. In T. Torres-Coronas & M. Vidal-Blasco (Eds.), *Social e-enterprise: Value creation through ICT* (pp. 67–82). Hershey, PA: IGI Global. doi:10.4018/978-1-4666-2667-6.ch004

Gallon, R. (2013). Communication, culture, and technology: Learning strategies for the unteachable. In R. Lansiquot (Ed.), *Cases on interdisciplinary research trends in science, technology, engineering, and mathematics: Studies on urban classrooms* (pp. 91–106). Hershey, PA: IGI Global. doi:10.4018/978-1-4666-2214-2.ch005

García, M., Lloret, J., Bellver, I., & Tomás, J. (2013). Intelligent IPTV distribution for smart phones. In D. Kanellopoulos (Ed.), *Intelligent multimedia technologies for networking applications: Techniques and tools* (pp. 318–347). Hershey, PA: IGI Global. doi:10.4018/978-1-4666-2833-5.ch013

García-Plaza, A. P., Zubiaga, A., Fresno, V., & Martínez, R. (2013). Tag cloud reorganization: Finding groups of related tags on delicious. In G. Xu & L. Li (Eds.), *Social media mining and social network analysis: Emerging research* (pp. 140–155). Hershey, PA: IGI Global. doi:10.4018/978-1-4666-2806-9.ch009

Gerpott, T. J. (2013). Attribute perceptions as factors explaining mobile internet acceptance of cellular customers in Germany: An empirical study comparing actual and potential adopters with distinct categories of access appliances. In I. Lee (Ed.), *Strategy, adoption, and competitive advantage of mobile services in the global economy* (pp. 19–48). Hershey, PA: IGI Global. doi:10.4018/978-1-4666-1939-5.ch002

Giambona, G. J., & Birchall, D. W. (2012). Collaborative e-learning and ICT tools to develop SME managers: An Italian case. In *Wireless technologies: Concepts, methodologies, tools and applications* (pp. 1606–1617). Hershey, PA: IGI Global. doi:10.4018/978-1-61350-101-6.ch605

Giannakos, M. N., Pateli, A. G., & Pappas, I. O. (2013). Identifying the direct effect of experience and the moderating effect of satisfaction in the Greek online market. In A. Scupola (Ed.), *Mobile opportunities and applications for e-service innovations* (pp. 77–97). Hershey, PA: IGI Global. doi:10.4018/978-1-4666-2654-6.ch005

Giorda, M., & Guerrisi, M. (2013). Educating to democracy and social participation through a "history of religion" course. In P. Pumilia-Gnarini, E. Favaron, E. Pacetti, J. Bishop, & L. Guerra (Eds.), *Handbook of research on didactic strategies and technologies for education: Incorporating advancements* (pp. 152–161). Hershey, PA: IGI Global. doi:10.4018/978-1-4666-2122-0.ch014

Goggins, S., Schmidt, M., Guajardo, J., & Moore, J. L. (2013). 3D virtual worlds: Assessing the experience and informing design. In B. Medlin (Ed.), *Integrations of technology utilization and social dynamics in organizations* (pp. 194–213). Hershey, PA: IGI Global. doi:10.4018/978-1-4666-1948-7.ch012

Gold, N. (2012). Rebels, heretics, and exiles: Blogging among estranged and questioning American Hasidim. In T. Dumova & R. Fiordo (Eds.), *Blogging in the global society: Cultural, political and geographical aspects* (pp. 108–127). Hershey, PA: IGI Global. doi:10.4018/978-1-60960-744-9.ch007

Görgü, L., Wan, J., O'Hare, G. M., & O'Grady, M. J. (2013). Enabling mobile service provision with sensor networks. In I. Lee (Ed.), *Mobile services industries, technologies, and applications in the global economy* (pp. 175–192). Hershey, PA: IGI Global. doi:10.4018/978-1-4666-1981-4.ch011

Gregory, S. J. (2013). Evolution of mobile services: An analysis. In I. Lee (Ed.), *Mobile services industries, technologies, and applications in the global economy* (pp. 104–119). Hershey, PA: IGI Global. doi:10.4018/978-1-4666-1981-4.ch007

Grieve, G. P., & Heston, K. (2012). Finding liquid salvation: Using the Cardean ethnographic method to document second life residents and religious cloud communities. In N. Zagalo, L. Morgado, & A. Boa-Ventura (Eds.), *Virtual worlds and metaverse platforms: New communication and identity paradigms* (pp. 288–305). Hershey, PA: IGI Global. doi:10.4018/978-1-60960-854-5.ch019

Guha, S., Thakur, B., Konar, T. S., & Chakrabarty, S. (2013). Web enabled design collaboration in India. In N. Kock (Ed.), *Interdisciplinary applications of electronic collaboration approaches and technologies* (pp. 96–111). Hershey, PA: IGI Global. doi:10.4018/978-1-4666-2020-9.ch007

Gulati, G. J., Yates, D. J., & Tawileh, A. (2013). Explaining the global digital divide: The impact of public policy initiatives on e-government capacity and reach worldwide. In *Digital literacy: Concepts, methodologies, tools, and applications* (pp. 39–62). Hershey, PA: IGI Global. doi:10.4018/978-1-4666-1852-7.ch003

Gupta, J. (2013). Digital library initiatives in India. In T. Ashraf & P. Gulati (Eds.), *Design, development, and management of resources for digital library services* (pp. 80–93). Hershey, PA: IGI Global. doi:10.4018/978-1-4666-2500-6.ch008

Gururajan, R., Hafeez-Baig, A., Danaher, P. A., & De George-Walker, L. (2012). Student perceptions and uses of wireless handheld devices: Implications for implementing blended and mobile learning in an Australian university. In *Wireless technologies: Concepts, methodologies, tools and applications* (pp. 1323–1338). Hershey, PA: IGI Global. doi:10.4018/978-1-61350-101-6.ch512

Gwilt, I. (2013). Data-objects: Sharing the attributes and properties of digital and material culture to creatively interpret complex information. In D. Harrison (Ed.), *Digital media and technologies for virtual artistic spaces* (pp. 14–26). Hershey, PA: IGI Global. doi:10.4018/978-1-4666-2961-5.ch002

Hackley, D. C., & Leidman, M. B. (2013). Integrating learning management systems in K-12 supplemental religious education. In A. Ritzhaupt & S. Kumar (Eds.), *Cases on educational technology implementation for facilitating learning* (pp. 1–22). Hershey, PA: IGI Global. doi:10.4018/978-1-4666-3676-7.ch001

Hale, J. R., & Fields, D. (2013). A cross-cultural measure of servant leadership behaviors. In M. Bocarnea, R. Reynolds, & J. Baker (Eds.), *Online instruments, data collection, and electronic measurements: Organizational advancements* (pp. 152–163). Hershey, PA: IGI Global. doi:10.4018/978-1-4666-2172-5.ch009

Hanewald, R. (2012). Using mobile technologies as research tools: Pragmatics, possibilities and problems. In *Wireless technologies: Concepts, methodologies, tools and applications* (pp. 130–150). Hershey, PA: IGI Global. doi:10.4018/978-1-61350-101-6.ch108

Hanewald, R. (2013). Professional development with and for emerging technologies: A case study with Asian languages and cultural studies teachers in Australia. In J. Keengwe (Ed.), *Pedagogical applications and social effects of mobile technology integration* (pp. 175–192). Hershey, PA: IGI Global. doi:10.4018/978-1-4666-2985-1.ch010

Hayhoe, S. (2012). Non-visual programming, perceptual culture and mulsemedia: Case studies of five blind computer programmers. In G. Ghinea, F. Andres, & S. Gulliver (Eds.), *Multiple sensorial media advances and applications: New developments in MulSeMedia* (pp. 80–98). Hershey, PA: IGI Global. doi:10.4018/978-1-60960-821-7.ch004

Henschke, J. A. (2013). Nation building through andragogy and lifelong learning: On the cutting edge educationally, economically, and governmentally. In V. Wang (Ed.), *Handbook of research on technologies for improving the 21st century workforce: Tools for lifelong learning* (pp. 480–506). Hershey, PA: IGI Global. doi:10.4018/978-1-4666-2181-7.ch030

Hermida, J. M., Meliá, S., Montoyo, A., & Gómez, J. (2013). Developing rich internet applications as social sites on the semantic web: A model-driven approach. In D. Chiu (Ed.), *Mobile and web innovations in systems and service-oriented engineering* (pp. 134–155). Hershey, PA: IGI Global. doi:10.4018/978-1-4666-2470-2.ch008

Hernández-García, Á., Agudo-Peregrina, Á. F., & Iglesias-Pradas, S. (2013). Adoption of mobile video-call service: An exploratory study. In I. Lee (Ed.), *Strategy, adoption, and competitive advantage of mobile services in the global economy* (pp. 49–72). Hershey, PA: IGI Global. doi:10.4018/978-1-4666-1939-5.ch003

Hesapci-Sanaktekin, O., & Somer, I. (2013). Mobile communication: A study on smart phone and mobile application use. In I. Lee (Ed.), *Strategy, adoption, and competitive advantage of mobile services in the global economy* (pp. 217–233). Hershey, PA: IGI Global. doi:10.4018/978-1-4666-1939-5.ch012

Hill, S. R., Troshani, I., & Freeman, S. (2013). An eclectic perspective on the internationalization of Australian mobile services SMEs. In I. Lee (Ed.), *Mobile services industries, technologies, and applications in the global economy* (pp. 55–73). Hershey, PA: IGI Global. doi:10.4018/978-1-4666-1981-4.ch004

Ho, V. (2013). The need for identity construction in computer-mediated professional communication: A Community of practice perspective. In R. Luppicini (Ed.), *Handbook of research on technoself: Identity in a technological society* (pp. 502–530). Hershey, PA: IGI Global. doi:10.4018/978-1-4666-2211-1.ch027

Hudson, H. E. (2013). Challenges facing municipal wireless: Case studies from San Francisco and Silicon Valley. In A. Abdelaal (Ed.), *Social and economic effects of community wireless networks and infrastructures* (pp. 12–26). Hershey, PA: IGI Global. doi:10.4018/978-1-4666-2997-4.ch002

Humphreys, S. (2012). Unravelling intellectual property in a specialist social networking site. In A. Lugmayr, H. Franssila, P. Näränen, O. Sotamaa, J. Vanhala, & Z. Yu (Eds.), *Media in the ubiquitous era: Ambient, social and gaming media* (pp. 248–266). Hershey, PA: IGI Global. doi:10.4018/978-1-60960-774-6.ch015

Iglesias, A., Ruiz-Mezcua, B., López, J. F., & Figueroa, D. C. (2013). New communication technologies for inclusive education in and outside the classroom. In D. Griol Barres, Z. Callejas Carrión, & R. Delgado (Eds.), *Technologies for inclusive education: Beyond traditional integration approaches* (pp. 271–284). Hershey, PA: IGI Global. doi:10.4018/978-1-4666-2530-3.ch013

Igun, S. E. (2013). Gender and national information and communication technology (ICT) policies in Africa. In B. Maumbe & J. Okello (Eds.), *Technology, sustainability, and rural development in Africa* (pp. 284–297). Hershey, PA: IGI Global. doi:10.4018/978-1-4666-3607-1.ch018

Ikolo, V. E. (2013). Gender digital divide and national ICT policies in Africa. In *Digital literacy: Concepts, methodologies, tools, and applications* (pp. 812–832). Hershey, PA: IGI Global. doi:10.4018/978-1-4666-1852-7.ch042

Imran, A., & Gregor, S. (2013). A process model for successful e-government adoption in the least developed countries: A case of Bangladesh. In *Digital literacy: Concepts, methodologies, tools, and applications* (pp. 213–241). Hershey, PA: IGI Global. doi:10.4018/978-1-4666-1852-7.ch012

Ionescu, A. (2013). ICTs and gender-based rights. In J. Lannon & E. Halpin (Eds.), *Human rights and information communication technologies: Trends and consequences of use* (pp. 214–234). Hershey, PA: IGI Global. doi:10.4018/978-1-4666-1918-0.ch013

Iyamu, T. (2013). The impact of organisational politics on the implementation of IT strategy: South African case in context. In J. Abdelnour-Nocera (Ed.), *Knowledge and technological development effects on organizational and social structures* (pp. 167–193). Hershey, PA: IGI Global. doi:10.4018/978-1-4666-2151-0.ch011

Jadhav, V. G. (2013). Integration of digital reference service for scholarly communication in digital libraries. In T. Ashraf & P. Gulati (Eds.), *Design, development, and management of resources for digital library services* (pp. 13–20). Hershey, PA: IGI Global. doi:10.4018/978-1-4666-2500-6.ch002

Jäkälä, M., & Berki, E. (2013). Communities, communication, and online identities. In S. Warburton & S. Hatzipanagos (Eds.), *Digital identity and social media* (pp. 1–13). Hershey, PA: IGI Global. doi:10.4018/978-1-4666-1915-9.ch001

Janneck, M., & Staar, H. (2013). Playing virtual power games: Micro-political processes in inter-organizational networks. In B. Medlin (Ed.), *Integrations of technology utilization and social dynamics in organizations* (pp. 171–192). Hershey, PA: IGI Global. doi:10.4018/978-1-4666-1948-7.ch011

Januska, I. M. (2013). Communication as a key factor in cooperation success and virtual enterprise paradigm support. In P. Renna (Ed.), *Production and manufacturing system management: Coordination approaches and multi-site planning* (pp. 145–161). Hershey, PA: IGI Global. doi:10.4018/978-1-4666-2098-8.ch008

Jayasingh, S., & Eze, U. C. (2013). Consumers' adoption of mobile coupons in Malaysia. In I. Lee (Ed.), *Strategy, adoption, and competitive advantage of mobile services in the global economy* (pp. 90–111). Hershey, PA: IGI Global. doi:10.4018/978-1-4666-1939-5.ch005

Jean Francois, E. (2013). Transculturality. In E. Jean Francois (Ed.), *Transcultural blended learning and teaching in postsecondary education* (pp. 1–14). Hershey, PA: IGI Global. doi:10.4018/978-1-4666-2014-8.ch001

Jensen, S. S. (2012). User-driven content creation in second life a source of innovation? Three case studies of business and public service. In N. Zagalo, L. Morgado, & A. Boa-Ventura (Eds.), *Virtual worlds and metaverse platforms: New communication and identity paradigms* (pp. 1–15). Hershey, PA: IGI Global. doi:10.4018/978-1-60960-854-5.ch001

Johnston, W. J., Komulainen, H., Ristola, A., & Ulkuniemi, P. (2013). Mobile advertising in small retailer firms: How to make the most of it. In I. Lee (Ed.), *Strategy, adoption, and competitive advantage of mobile services in the global economy* (pp. 283–298). Hershey, PA: IGI Global. doi:10.4018/978-1-4666-1939-5.ch016

Kadas, G., & Chatzimisios, P. (2013). The role of roadside assistance in vehicular communication networks: Security, quality of service, and routing issues. In R. Daher & A. Vinel (Eds.), *Roadside networks for vehicular communications: Architectures, applications, and test fields* (pp. 1–37). Hershey, PA: IGI Global. doi:10.4018/978-1-4666-2223-4.ch001

Kale, S. H., & Spence, M. T. (2012). A trination analysis of social exchange relationships in e-dating. In C. Romm Livermore (Ed.), *Gender and social computing: Interactions, differences and relationships* (pp. 257–271). Hershey, PA: IGI Global. doi:10.4018/978-1-60960-759-3.ch015

Kamoun, F. (2013). Mobile NFC services: Adoption factors and a typology of business models. In I. Lee (Ed.), *Mobile services industries, technologies, and applications in the global economy* (pp. 254–272). Hershey, PA: IGI Global. doi:10.4018/978-1-4666-1981-4.ch016

Kaneda, K., & Iwamura, K. (2013). New proposals for data hiding in paper media. In K. Kondo (Ed.), *Multimedia information hiding technologies and methodologies for controlling data* (pp. 258–285). Hershey, PA: IGI Global. doi:10.4018/978-1-4666-2217-3.ch012

Kastell, K. (2013). Seamless communication to mobile devices in vehicular wireless networks. In O. Strobel (Ed.), *Communication in transportation systems* (pp. 324–342). Hershey, PA: IGI Global. doi:10.4018/978-1-4666-2976-9.ch012

Kaye, B. K., & Johnson, T. J. (2012). Net gain? Selective exposure and selective avoidance of social network sites. In F. Comunello (Ed.), *Networked sociability and individualism: Technology for personal and professional relationships* (pp. 218–237). Hershey, PA: IGI Global. doi:10.4018/978-1-61350-338-6.ch011

Kaye, B. K., Johnson, T. J., & Muhlberger, P. (2012). Blogs as a source of democratic deliberation. In T. Dumova & R. Fiordo (Eds.), *Blogging in the global society: Cultural, political and geographical aspects* (pp. 1–18). Hershey, PA: IGI Global. doi:10.4018/978-1-60960-744-9.ch001

Kefi, H., Mlaiki, A., & Peterson, R. L. (2013). IT offshoring: Trust views from client and vendor perspectives. In J. Wang (Ed.), *Perspectives and techniques for improving information technology project management* (pp. 113–130). Hershey, PA: IGI Global. doi:10.4018/978-1-4666-2800-7.ch009

Khan, N. A., & Batoo, M. F. (2013). Stone inscriptions of Srinagar: A digital panorama. In T. Ashraf & P. Gulati (Eds.), *Design, development, and management of resources for digital library services* (pp. 58–79). Hershey, PA: IGI Global. doi:10.4018/978-1-4666-2500-6.ch007

Kim, P. (2012). "Stay out of the way! My kid is video blogging through a phone!": A lesson learned from math tutoring social media for children in underserved communities. In *Wireless technologies: Concepts, methodologies, tools and applications* (pp. 1415–1428). Hershey, PA: IGI Global. doi:10.4018/978-1-61350-101-6.ch517

Kisubi, A. T. (2013). A critical perspective on the challenges for blended learning and teaching in Africa's higher education. In E. Jean Francois (Ed.), *Transcultural blended learning and teaching in postsecondary education* (pp. 145–168). Hershey, PA: IGI Global. doi:10.4018/978-1-4666-2014-8.ch009

Koole, M., & Parchoma, G. (2013). The web of identity: A model of digital identity formation in networked learning environments. In S. Warburton & S. Hatzipanagos (Eds.), *Digital identity and social media* (pp. 14–28). Hershey, PA: IGI Global. doi:10.4018/978-1-4666-1915-9.ch002

Kordaki, M., Gorghiu, G., Bîzoi, M., & Glava, A. (2012). Collaboration within multinational learning communities: The case of the virtual community collaborative space for sciences education European project. In A. Juan, T. Daradoumis, M. Roca, S. Grasman, & J. Faulin (Eds.), *Collaborative and distributed e-research: Innovations in technologies, strategies and applications* (pp. 206–226). Hershey, PA: IGI Global. doi:10.4018/978-1-4666-0125-3.ch010

Kovács, J., Bokor, L., Kanizsai, Z., & Imre, S. (2013). Review of advanced mobility solutions for multimedia networking in IPv6. In D. Kanellopoulos (Ed.), *Intelligent multimedia technologies for networking applications: Techniques and tools* (pp. 25–47). Hershey, PA: IGI Global. doi:10.4018/978-1-4666-2833-5.ch002

Kreps, D. (2013). Performing the discourse of sexuality online. In S. Warburton & S. Hatzipanagos (Eds.), *Digital identity and social media* (pp. 118–132). Hershey, PA: IGI Global. doi:10.4018/978-1-4666-1915-9.ch009

Ktoridou, D., Kaufmann, H., & Liassides, C. (2012). Factors affecting WiFi use intention: The context of Cyprus. In *Wireless technologies: Concepts, methodologies, tools and applications* (pp. 1760–1781). Hershey, PA: IGI Global. doi:10.4018/978-1-61350-101-6.ch703

Kumar, N., Nero Alves, L., & Aguiar, R. L. (2013). Employing traffic lights as road side units for road safety information broadcast. In R. Daher & A. Vinel (Eds.), *Roadside networks for vehicular communications: Architectures, applications, and test fields* (pp. 118–135). Hershey, PA: IGI Global. doi:10.4018/978-1-4666-2223-4.ch006

Kvasny, L., & Hales, K. D. (2013). The evolving discourse of the digital divide: The internet, black identity, and the evolving discourse of the digital divide. In *Digital literacy: Concepts, methodologies, tools, and applications* (pp. 1350–1366). Hershey, PA: IGI Global. doi:10.4018/978-1-4666-1852-7.ch071

L'Abate, L. (2013). Of paradigms, theories, and models: A conceptual hierarchical structure for communication science and technoself. In R. Luppicini (Ed.), *Handbook of research on technoself: Identity in a technological society* (pp. 84–104). Hershey, PA: IGI Global. doi:10.4018/978-1-4666-2211-1.ch005

Laghos, A. (2013). Multimedia social networks and e-learning. In D. Kanellopoulos (Ed.), *Intelligent multimedia technologies for networking applications: Techniques and tools* (pp. 365–379). Hershey, PA: IGI Global. doi:10.4018/978-1-4666-2833-5.ch015

Lappas, G. (2012). Social multimedia mining: Trends and opportunities in areas of social and communication studies. In I. Ting, T. Hong, & L. Wang (Eds.), *Social network mining, analysis, and research trends: Techniques and applications* (pp. 1–16). Hershey, PA: IGI Global. doi:10.4018/978-1-61350-513-7.ch001

Lawrence, J. E. (2013). Barriers hindering ecommerce adoption: A case study of Kurdistan region of Iraq. In A. Zolait (Ed.), *Technology diffusion and adoption: Global complexity, global innovation* (pp. 152–165). Hershey, PA: IGI Global. doi:10.4018/978-1-4666-2791-8.ch010

Lawrence, K. F. (2013). Identity and the online media fan community. In S. Warburton & S. Hatzipanagos (Eds.), *Digital identity and social media* (pp. 233–255). Hershey, PA: IGI Global. doi:10.4018/978-1-4666-1915-9.ch017

Lee, M. J., Dalgarno, B., Gregory, S., Carlson, L., & Tynan, B. (2013). How are Australian and New Zealand higher educators using 3D immersive virtual worlds in their teaching? In B. Tynan, J. Willems, & R. James (Eds.), *Outlooks and opportunities in blended and distance learning* (pp. 169–188). Hershey, PA: IGI Global. doi:10.4018/978-1-4666-4205-8.ch013

Lee, S., Alfano, C., & Carpenter, R. G. (2013). Invention in two parts: Multimodal communication and space design in the writing center. In R. Carpenter (Ed.), *Cases on higher education spaces: Innovation, collaboration, and technology* (pp. 41–63). Hershey, PA: IGI Global. doi:10.4018/978-1-4666-2673-7.ch003

Leichsenring, C., Tünnermann, R., & Hermann, T. (2013). Feelabuzz: Direct tactile communication with mobile phones. In J. Lumsden (Ed.), *Developments in technologies for human-centric mobile computing and applications* (pp. 145–154). Hershey, PA: IGI Global. doi:10.4018/978-1-4666-2068-1.ch009

Lemos, A., & Marques, F. P. (2013). A critical analysis of the limitations and effects of the Brazilian national broadband plan. In A. Abdelaal (Ed.), *Social and economic effects of community wireless networks and infrastructures* (pp. 255–274). Hershey, PA: IGI Global. doi:10.4018/978-1-4666-2997-4.ch014

Leung, C. K., Medina, I. J., & Tanbeer, S. K. (2013). Analyzing social networks to mine important friends. In G. Xu & L. Li (Eds.), *Social media mining and social network analysis: Emerging research* (pp. 90–104). Hershey, PA: IGI Global. doi:10.4018/978-1-4666-2806-9.ch006

Li, B. (2012). Toward an infrastructural approach to understanding participation in virtual communities. In H. Li (Ed.), *Virtual community participation and motivation: Cross-disciplinary theories* (pp. 103–123). Hershey, PA: IGI Global. doi:10.4018/978-1-4666-0312-7.ch007

Li, L., Xiao, H., & Xu, G. (2013). Recommending related microblogs. In G. Xu & L. Li (Eds.), *Social media mining and social network analysis: Emerging research* (pp. 202–210). Hershey, PA: IGI Global. doi:10.4018/978-1-4666-2806-9.ch013

Liddell, T. (2013). Historical evolution of adult education in America: The impact of institutions, change, and acculturation. In V. Wang (Ed.), *Handbook of research on technologies for improving the 21st century workforce: Tools for lifelong learning* (pp. 257–271). Hershey, PA: IGI Global. doi:10.4018/978-1-4666-2181-7.ch017

Liljander, V., Gummerus, J., Pihlström, M., & Kiehelä, H. (2013). Mobile services as resources for consumer integration of value in a multi-channel environment. In I. Lee (Ed.), *Strategy, adoption, and competitive advantage of mobile services in the global economy* (pp. 259–282). Hershey, PA: IGI Global. doi:10.4018/978-1-4666-1939-5.ch015

Litaay, T., Prananingrum, D. H., & Krisanto, Y. A. (2013). Indonesian legal perspectives on biotechnology and intellectual property rights. In *Digital rights management: Concepts, methodologies, tools, and applications* (pp. 834–845). Hershey, PA: IGI Global. doi:10.4018/978-1-4666-2136-7.ch039

Little, G. (2013). Collection development for theological education. In S. Holder (Ed.), *Library collection development for professional programs: Trends and best practices* (pp. 112–127). Hershey, PA: IGI Global. doi:10.4018/978-1-4666-1897-8.ch007

Losh, S. C. (2013). American digital divides: Generation, education, gender, and ethnicity in American digital divides. In *Digital literacy: Concepts, methodologies, tools, and applications* (pp. 932–958). Hershey, PA: IGI Global. doi:10.4018/978-1-4666-1852-7.ch048

Lovari, A., & Parisi, L. (2012). Public administrations and citizens 2.0: Exploring digital public communication strategies and civic interaction within Italian municipality pages on Facebook. In F. Comunello (Ed.), *Networked sociability and individualism: Technology for personal and professional relationships* (pp. 238–263). Hershey, PA: IGI Global. doi:10.4018/978-1-61350-338-6.ch012

Maamar, Z., Faci, N., Mostéfaoui, S. K., & Akhter, F. (2013). Towards a framework for weaving social networks into mobile commerce. In D. Chiu (Ed.), *Mobile and web innovations in systems and service-oriented engineering* (pp. 333–347). Hershey, PA: IGI Global. doi:10.4018/978-1-4666-2470-2.ch018

Maia, I. F., & Valente, J. A. (2013). Digital identity built on a cooperative relationship. In S. Warburton & S. Hatzipanagos (Eds.), *Digital identity and social media* (pp. 58–73). Hershey, PA: IGI Global. doi:10.4018/978-1-4666-1915-9.ch005

Maity, M. (2013). Consumer information search and decision-making on m-commerce: The role of product type. In I. Lee (Ed.), *Strategy, adoption, and competitive advantage of mobile services in the global economy* (pp. 73–89). Hershey, PA: IGI Global. doi:10.4018/978-1-4666-1939-5.ch004

Malinen, S., Virjo, T., & Kujala, S. (2012). Supporting local connections with online communities. In A. Lugmayr, H. Franssila, P. Näränen, O. Sotamaa, J. Vanhala, & Z. Yu (Eds.), *Media in the ubiquitous era: Ambient, social and gaming media* (pp. 235–250). Hershey, PA: IGI Global. doi:10.4018/978-1-60960-774-6.ch013

Mantoro, T., Milišic, A., & Ayu, M. (2013). Online authentication using smart card technology in mobile phone infrastructure. In I. Khalil & E. Weippl (Eds.), *Contemporary challenges and solutions for mobile and multimedia technologies* (pp. 127–144). Hershey, PA: IGI Global. doi:10.4018/978-1-4666-2163-3.ch008

Marcato, E., & Scala, E. (2013). Moodle: A platform for a school. In P. Pumilia-Gnarini, E. Favaron, E. Pacetti, J. Bishop, & L. Guerra (Eds.), *Handbook of research on didactic strategies and technologies for education: Incorporating advancements* (pp. 107–116). Hershey, PA: IGI Global. doi:10.4018/978-1-4666-2122-0.ch010

Markaki, O. I., Charalabidis, Y., & Askounis, D. (2013). Measuring interoperability readiness in south eastern Europe and the Mediterranean: The interoperability observatory. In A. Scupola (Ed.), *Mobile opportunities and applications for e-service innovations* (pp. 210–230). Hershey, PA: IGI Global. doi:10.4018/978-1-4666-2654-6.ch012

Martin, J. D., & El-Toukhy, S. (2012). Blogging for sovereignty: An analysis of Palestinian blogs. In T. Dumova & R. Fiordo (Eds.), *Blogging in the global society: Cultural, political and geographical aspects* (pp. 148–160). Hershey, PA: IGI Global. doi:10.4018/978-1-60960-744-9.ch009

Matei, S. A., & Bruno, R. J. (2012). Individualist motivators and community functional constraints in social media: The case of Wikis and Wikipedia. In F. Comunello (Ed.), *Networked sociability and individualism: Technology for personal and professional relationships* (pp. 1–23). Hershey, PA: IGI Global. doi:10.4018/978-1-61350-338-6.ch001

Matsuoka, H. (2013). Acoustic OFDM technology and system. In K. Kondo (Ed.), *Multimedia information hiding technologies and methodologies for controlling data* (pp. 90–103). Hershey, PA: IGI Global. doi:10.4018/978-1-4666-2217-3.ch005

McCarthy, J. (2013). Online networking: Integrating international students into first year university through the strategic use of participatory media. In F. García-Peñalvo (Ed.), *Multiculturalism in technology-based education: Case studies on ICT-supported approaches* (pp. 189–210). Hershey, PA: IGI Global. doi:10.4018/978-1-4666-2101-5.ch012

McDonald, A., & Helmer, S. (2013). A comparative case study of Indonesian and UK organisational culture differences in IS project management. In A. Mesquita (Ed.), *User perception and influencing factors of technology in everyday life* (pp. 46–55). Hershey, PA: IGI Global. doi:10.4018/978-1-4666-1954-8.ch005

McDonough, C. (2013). Mobile broadband: Substituting for fixed broadband or providing value-added. In I. Lee (Ed.), *Mobile services industries, technologies, and applications in the global economy* (pp. 74–86). Hershey, PA: IGI Global. doi:10.4018/978-1-4666-1981-4.ch005

McKeown, A. (2013). Virtual communitas, "digital place-making," and the process of "becoming". In D. Harrison (Ed.), *Digital media and technologies for virtual artistic spaces* (pp. 218–236). Hershey, PA: IGI Global. doi:10.4018/978-1-4666-2961-5.ch016

Medeni, T. D., Medeni, I. T., & Balci, A. (2013). Proposing a knowledge amphora model for transition towards mobile government. In A. Scupola (Ed.), *Mobile opportunities and applications for e-service innovations* (pp. 170–192). Hershey, PA: IGI Global. doi:10.4018/978-1-4666-2654-6.ch010

Melo, A., Bezerra, P., Abelém, A. J., Neto, A., & Cerqueira, E. (2013). PriorityQoE: A tool for improving the QoE in video streaming. In D. Kanellopoulos (Ed.), *Intelligent multimedia technologies for networking applications: Techniques and tools* (pp. 270–290). Hershey, PA: IGI Global. doi:10.4018/978-1-4666-2833-5.ch011

Mendoza-González, R., Rodríguez, F. Á., & Arteaga, J. M. (2013). A usability study of mobile text based social applications: Towards a reliable strategy for design evaluation. In M. Garcia-Ruiz (Ed.), *Cases on usability engineering: Design and development of digital products* (pp. 195–219). Hershey, PA: IGI Global. doi:10.4018/978-1-4666-4046-7.ch009

Metzger, M. J., Wilson, C., Pure, R. A., & Zhao, B. Y. (2012). Invisible interactions: What latent social interaction can tell us about social relationships in social network sites. In F. Comunello (Ed.), *Networked sociability and individualism: Technology for personal and professional relationships* (pp. 79–102). Hershey, PA: IGI Global. doi:10.4018/978-1-61350-338-6.ch005

Millo, G., & Carmeci, G. (2013). Insurance in Italy: A spatial perspective. In G. Borruso, S. Bertazzon, A. Favretto, B. Murgante, & C. Torre (Eds.), *Geographic information analysis for sustainable development and economic planning: New technologies* (pp. 158–178). Hershey, PA: IGI Global. doi:10.4018/978-1-4666-1924-1.ch011

Mingqing, X., Wenjing, X., & Junming, Z. (2012). The future of television. In R. Sharma, M. Tan, & F. Pereira (Eds.), *Understanding the interactive digital media marketplace: Frameworks, platforms, communities and issues* (pp. 406–415). Hershey, PA: IGI Global. doi:10.4018/978-1-61350-147-4.ch032

Miscione, G. (2013). Telemedicine and development: Situating information technologies in the Amazon. In J. Abdelnour-Nocera (Ed.), *Knowledge and technological development effects on organizational and social structures* (pp. 132–145). Hershey, PA: IGI Global. doi:10.4018/978-1-4666-2151-0.ch009

Modegi, T. (2013). Spatial and temporal position information delivery to mobile terminals using audio watermarking techniques. In K. Kondo (Ed.), *Multimedia information hiding technologies and methodologies for controlling data* (pp. 182–207). Hershey, PA: IGI Global. doi:10.4018/978-1-4666-2217-3.ch009

Montes, J. A., Gutiérrez, A. C., Fernández, E. M., & Romeo, A. (2012). Reality mining, location based services and e-business opportunities: The case of city analytics. In *Wireless technologies: Concepts, methodologies, tools and applications* (pp. 1520–1532). Hershey, PA: IGI Global. doi:10.4018/978-1-61350-101-6.ch601

Moreno, A. (2012). The social construction of new cultural models through information and communication technologies. In M. Safar & K. Mahdi (Eds.), *Social networking and community behavior modeling: Qualitative and quantitative measures* (pp. 68–84). Hershey, PA: IGI Global. doi:10.4018/978-1-61350-444-4.ch004

Morris, J. Z., & Thomas, K. D. (2013). Implementing BioSand filters in rural Honduras: A case study of his hands mission international in Copán, Honduras. In H. Muga & K. Thomas (Eds.), *Cases on the diffusion and adoption of sustainable development practices* (pp. 468–496). Hershey, PA: IGI Global. doi:10.4018/978-1-4666-2842-7.ch017

Mura, G. (2012). The MultiPlasticity of new media. In G. Ghinea, F. Andres, & S. Gulliver (Eds.), *Multiple sensorial media advances and applications: New developments in MulSeMedia* (pp. 258–271). Hershey, PA: IGI Global. doi:10.4018/978-1-60960-821-7.ch013

Murray, C. (2012). Imagine mobile learning in your pocket. In *Wireless technologies: Concepts, methodologies, tools and applications* (pp. 2060–2088). Hershey, PA: IGI Global. doi:10.4018/978-1-61350-101-6.ch807

Mutohar, A., & Hughes, J. E. (2013). Toward web 2.0 integration in Indonesian education: Challenges and planning strategies. In N. Azab (Ed.), *Cases on web 2.0 in developing countries: Studies on implementation, application, and use* (pp. 198–221). Hershey, PA: IGI Global. doi:10.4018/978-1-4666-2515-0.ch008

Nandi, B., & Subramaniam, G. (2012). Evolution in broadband technology and future of wireless broadband. In *Wireless technologies: Concepts, methodologies, tools and applications* (pp. 1928–1957). Hershey, PA: IGI Global. doi:10.4018/978-1-61350-101-6.ch801

Naser, A., Jaber, I., Jaber, R., & Saeed, K. (2013). Information systems in UAE education sector: Security, cultural, and ethical issues. In F. Albadri (Ed.), *Information systems applications in the Arab education sector* (pp. 148–162). Hershey, PA: IGI Global. doi:10.4018/978-1-4666-1984-5.ch011

Nemoianu, I., & Pesquet-Popescu, B. (2013). Network coding for multimedia communications. In D. Kanellopoulos (Ed.), *Intelligent multimedia technologies for networking applications: Techniques and tools* (pp. 1–24). Hershey, PA: IGI Global. doi:10.4018/978-1-4666-2833-5.ch001

Nezlek, G., & DeHondt, G. (2013). Gender wage differentials in information systems: 1991 – 2008 a quantitative analysis. In B. Medlin (Ed.), *Integrations of technology utilization and social dynamics in organizations* (pp. 31–47). Hershey, PA: IGI Global. doi:10.4018/978-1-4666-1948-7.ch003

Nishimura, A., & Kondo, K. (2013). Information hiding for audio signals. In K. Kondo (Ed.), *Multimedia information hiding technologies and methodologies for controlling data* (pp. 1–18). Hershey, PA: IGI Global. doi:10.4018/978-1-4666-2217-3.ch001

Norder, J. W., & Carroll, J. W. (2013). Applied geospatial perspectives on the rock art of the lake of the woods region of Ontario, Canada. In D. Albert & G. Dobbs (Eds.), *Emerging methods and multidisciplinary applications in geospatial research* (pp. 77–93). Hershey, PA: IGI Global. doi:10.4018/978-1-4666-1951-7.ch005

O'Brien, M. A., & Rogers, W. A. (2013). Design for aging: Enhancing everyday technology use. In R. Zheng, R. Hill, & M. Gardner (Eds.), *Engaging older adults with modern technology: Internet use and information access needs* (pp. 105–123). Hershey, PA: IGI Global. doi:10.4018/978-1-4666-1966-1.ch006

O'Hanlon, S. (2013). Health information technology and human rights. In J. Lannon & E. Halpin (Eds.), *Human rights and information communication technologies: Trends and consequences of use* (pp. 235–246). Hershey, PA: IGI Global. doi:10.4018/978-1-4666-1918-0.ch014

Odella, F. (2012). Social networks and communities: From traditional society to the virtual sphere. In M. Safar & K. Mahdi (Eds.), *Social networking and community behavior modeling: Qualitative and quantitative measures* (pp. 1–25). Hershey, PA: IGI Global. doi:10.4018/978-1-61350-444-4.ch001

Okazaki, S., Romero, J., & Campo, S. (2012). Capturing market mavens among advergamers: A case of mobile-based social networking site in Japan. In I. Ting, T. Hong, & L. Wang (Eds.), *Social network mining, analysis, and research trends: Techniques and applications* (pp. 291–305). Hershey, PA: IGI Global. doi:10.4018/978-1-61350-513-7.ch017

Omojola, O. (2012). Exploring the impact of Google Igbo in South East Nigeria. In R. Lekoko & L. Semali (Eds.), *Cases on developing countries and ICT integration: Rural community development* (pp. 62–73). Hershey, PA: IGI Global. doi:10.4018/978-1-60960-117-1.ch007

Ovide, E. (2013). Intercultural education with indigenous peoples and the potential of digital technologies to make it happen. In F. García-Peñalvo (Ed.), *Multiculturalism in technology-based education: Case studies on ICT-supported approaches* (pp. 59–78). Hershey, PA: IGI Global. doi:10.4018/978-1-4666-2101-5.ch005

Owusu-Ansah, A. (2013). Exploring Hofstede's cultural dimension using Hollins' structured dialogue to attain a conduit for effective intercultural experiences. In E. Jean Francois (Ed.), *Transcultural blended learning and teaching in postsecondary education* (pp. 52–74). Hershey, PA: IGI Global. doi:10.4018/978-1-4666-2014-8.ch004

Özdemir, E. (2012). Gender and e-marketing: The role of gender differences in online purchasing behaviors. In C. Romm Livermore (Ed.), *Gender and social computing: Interactions, differences and relationships* (pp. 72–86). Hershey, PA: IGI Global. doi:10.4018/978-1-60960-759-3.ch005

Palmer, M. H., & Hanney, J. (2012). Geographic information networks in American Indian governments and communities. In S. Dasgupta (Ed.), *Technical, social, and legal issues in virtual communities: Emerging environments: Emerging environments* (pp. 52–62). Hershey, PA: IGI Global. doi:10.4018/978-1-4666-1553-3.ch004

Pande, R. (2013). Gender gaps and information and communication technology: A case study of India. In *Digital literacy: Concepts, methodologies, tools, and applications* (pp. 1425–1439). Hershey, PA: IGI Global. doi:10.4018/978-1-4666-1852-7.ch075

Park, J., Chung, T., & Hur, W. (2013). The role of consumer innovativeness and trust for adopting internet phone services. In A. Scupola (Ed.), *Mobile opportunities and applications for e-service innovations* (pp. 22–36). Hershey, PA: IGI Global. doi:10.4018/978-1-4666-2654-6.ch002

Parke, A., & Griffiths, M. (2013). Poker gambling virtual communities: The use of computer-mediated communication to develop cognitive poker gambling skills. In R. Zheng (Ed.), *Evolving psychological and educational perspectives on cyber behavior* (pp. 190–204). Hershey, PA: IGI Global. doi:10.4018/978-1-4666-1858-9.ch012

Paschou, M., Sakkopoulos, E., Tsakalidis, A., Tzimas, G., & Viennas, E. (2013). An XML-based customizable model for multimedia applications for museums and exhibitions. In D. Kanellopoulos (Ed.), *Intelligent multimedia technologies for networking applications: Techniques and tools* (pp. 348–363). Hershey, PA: IGI Global. doi:10.4018/978-1-4666-2833-5.ch014

Pauwels, L. (2013). Images, self-images, and idealized identities in the digital networked world: Reconfigurations of family photography in a web-based mode. In S. Warburton & S. Hatzipanagos (Eds.), *Digital identity and social media* (pp. 133–147). Hershey, PA: IGI Global. doi:10.4018/978-1-4666-1915-9.ch010

Peachey, A., & Withnail, G. (2013). A sociocultural perspective on negotiating digital identities in a community of learners. In S. Warburton & S. Hatzipanagos (Eds.), *Digital identity and social media* (pp. 210–224). Hershey, PA: IGI Global. doi:10.4018/978-1-4666-1915-9.ch015

Peixoto, E., Martins, E., Anjo, A. B., & Silva, A. (2012). Geo@NET in the context of the platform of assisted learning from Aveiro University, Portugal. In M. Cruz-Cunha (Ed.), *Handbook of research on serious games as educational, business and research tools* (pp. 648–667). Hershey, PA: IGI Global. doi:10.4018/978-1-4666-0149-9.ch033

Pillay, N. (2013). The use of web 2.0 technologies by students from developed and developing countries: A New Zealand case study. In N. Azab (Ed.), *Cases on web 2.0 in developing countries: Studies on implementation, application, and use* (pp. 411–441). Hershey, PA: IGI Global. doi:10.4018/978-1-4666-2515-0.ch015

Pimenta, M. S., Miletto, E. M., Keller, D., Flores, L. V., & Testa, G. G. (2013). Technological support for online communities focusing on music creation: Adopting collaboration, flexibility, and multiculturality from Brazilian creativity styles. In N. Azab (Ed.), *Cases on web 2.0 in developing countries: Studies on implementation, application, and use* (pp. 283–312). Hershey, PA: IGI Global. doi:10.4018/978-1-4666-2515-0.ch011

Pina, P. (2013). Between scylla and charybdis: The balance between copyright, digital rights management and freedom of expression. In *Digital rights management: Concepts, methodologies, tools, and applications* (pp. 1355–1367). Hershey, PA: IGI Global. doi:10.4018/978-1-4666-2136-7.ch067

Pitsillides, S., Waller, M., & Fairfax, D. (2013). Digital death: What role does digital information play in the way we are (re)membered? In S. Warburton & S. Hatzipanagos (Eds.), *Digital identity and social media* (pp. 75–90). Hershey, PA: IGI Global. doi:10.4018/978-1-4666-1915-9.ch006

Polacek, P., & Huang, C. (2013). QoS scheduling with opportunistic spectrum access for multimedia. In M. Ku & J. Lin (Eds.), *Cognitive radio and interference management: Technology and strategy* (pp. 162–178). Hershey, PA: IGI Global. doi:10.4018/978-1-4666-2005-6.ch009

Potts, L. (2013). Balancing McLuhan with Williams: A sociotechnical view of technological determinism. In J. Abdelnour-Nocera (Ed.), *Knowledge and technological development effects on organizational and social structures* (pp. 109–114). Hershey, PA: IGI Global. doi:10.4018/978-1-4666-2151-0.ch007

Prescott, J., & Bogg, J. (2013). Stereotype, attitudes, and identity: Gendered expectations and behaviors. In *Gendered occupational differences in science, engineering, and technology careers* (pp. 112–135). Hershey, PA: IGI Global. doi:10.4018/978-1-4666-2107-7.ch005

Preussler, A., & Kerres, M. (2013). Managing social reputation in Twitter. In S. Warburton & S. Hatzipanagos (Eds.), *Digital identity and social media* (pp. 91–103). Hershey, PA: IGI Global. doi:10.4018/978-1-4666-1915-9.ch007

Prieger, J. E., & Church, T. V. (2013). Deployment of mobile broadband service in the United States. In I. Lee (Ed.), *Mobile services industries, technologies, and applications in the global economy* (pp. 1–24). Hershey, PA: IGI Global. doi:10.4018/978-1-4666-1981-4.ch001

Puumalainen, K., Frank, L., Sundqvist, S., & Tuppura, A. (2012). The critical mass of wireless communications: Differences between developing and developed economies. In *Wireless technologies: Concepts, methodologies, tools and applications* (pp. 1719–1736). Hershey, PA: IGI Global. doi:10.4018/978-1-61350-101-6.ch701

Rabino, S., Rafiee, D., Onufrey, S., & Moskowitz, H. (2013). Retention and customer share building: Formulating a communication strategy for a sports club. In H. Kaufmann & M. Panni (Eds.), *Customer-centric marketing strategies: Tools for building organizational performance* (pp. 511–529). Hershey, PA: IGI Global. doi:10.4018/978-1-4666-2524-2.ch025

Rahman, H., & Kumar, S. (2012). Mobile computing: An emerging issue in the digitized world. In A. Kumar & H. Rahman (Eds.), *Mobile computing techniques in emerging markets: Systems, applications and services* (pp. 1–22). Hershey, PA: IGI Global. doi:10.4018/978-1-4666-0080-5.ch001

Ratten, V. (2013). Adoption of mobile reading devices in the book industry. In I. Lee (Ed.), *Strategy, adoption, and competitive advantage of mobile services in the global economy* (pp. 203–216). Hershey, PA: IGI Global. doi:10.4018/978-1-4666-1939-5.ch011

Ratten, V. (2013). Mobile banking in the youth market: Implications from an entrepreneurial and learning perspective. In I. Lee (Ed.), *Strategy, adoption, and competitive advantage of mobile services in the global economy* (pp. 112–126). Hershey, PA: IGI Global. doi:10.4018/978-1-4666-1939-5.ch006

Ratten, V. (2013). Social e-enterprise through technological innovations and mobile social networks. In T. Torres-Coronas & M. Vidal-Blasco (Eds.), *Social e-enterprise: Value creation through ICT* (pp. 96–109). Hershey, PA: IGI Global. doi:10.4018/978-1-4666-2667-6.ch006

Rego, P. A., Moreira, P. M., & Reis, L. P. (2012). New forms of interaction in serious games for rehabilitation. In M. Cruz-Cunha (Ed.), *Handbook of research on serious games as educational, business and research tools* (pp. 1188–1211). Hershey, PA: IGI Global. doi:10.4018/978-1-4666-0149-9.ch062

Reinhard, C. D. (2012). Virtual worlds and reception studies: Comparing engagings. In N. Zagalo, L. Morgado, & A. Boa-Ventura (Eds.), *Virtual worlds and metaverse platforms: New communication and identity paradigms* (pp. 117–136). Hershey, PA: IGI Global. doi:10.4018/978-1-60960-854-5.ch008

Rieser, M. (2013). Mobility, liminality, and digital materiality. In D. Harrison (Ed.), *Digital media and technologies for virtual artistic spaces* (pp. 27–45). Hershey, PA: IGI Global. doi:10.4018/978-1-4666-2961-5.ch003

Rodrigues, R. G., Pinheiro, P. G., & Barbosa, J. (2012). Online playability: The social dimension to the virtual world. In M. Cruz-Cunha (Ed.), *Handbook of research on serious games as educational, business and research tools* (pp. 391–421). Hershey, PA: IGI Global. doi:10.4018/978-1-4666-0149-9.ch021

Romm-Livermore, C., Somers, T. M., Setzekorn, K., & King, A. L. (2012). How e-daters behave online: Theory and empirical observations. In C. Romm Livermore (Ed.), *Gender and social computing: Interactions, differences and relationships* (pp. 236–256). Hershey, PA: IGI Global. doi:10.4018/978-1-60960-759-3.ch014

Rosaci, D., & Sarnè, G. M. (2012). An agent-based approach to adapt multimedia web content in ubiquitous environment. In S. Bagchi (Ed.), *Ubiquitous multimedia and mobile agents: Models and implementations* (pp. 60–84). Hershey, PA: IGI Global. doi:10.4018/978-1-61350-107-8.ch003

Rosas, O. V., & Dhen, G. (2012). One self to rule them all: A critical discourse analysis of French-speaking players' identity construction in World of Warcraft. In N. Zagalo, L. Morgado, & A. Boa-Ventura (Eds.), *Virtual worlds and metaverse platforms: New communication and identity paradigms* (pp. 337–366). Hershey, PA: IGI Global. doi:10.4018/978-1-60960-854-5.ch022

Rouibah, K., & Abbas, H. A. (2012). Effect of personal innovativeness, attachment motivation and social norms on the acceptance of camera mobile phones: An empirical study in an Arab country. In W. Hu (Ed.), *Emergent trends in personal, mobile, and handheld computing technologies* (pp. 302–323). Hershey, PA: IGI Global. doi:10.4018/978-1-4666-0921-1.ch018

Ruiz-Mafé, C., Sanz-Blas, S., & Martí-Parreño, J. (2013). Web 2.0 goes mobile: Motivations and barriers of mobile social networks use in Spain. In N. Azab (Ed.), *Cases on web 2.0 in developing countries: Studies on implementation, application, and use* (pp. 109–146). Hershey, PA: IGI Global. doi:10.4018/978-1-4666-2515-0.ch005

Rybas, S. (2012). Community embodied: Validating the subjective performance of an online class. In H. Li (Ed.), *Virtual community participation and motivation: Cross-disciplinary theories* (pp. 124–141). Hershey, PA: IGI Global. doi:10.4018/978-1-4666-0312-7.ch008

Sabelkin, M., & Gagnon, F. (2013). Data transmission oriented on the object, communication media, application, and state of communication systems. In M. Bartolacci & S. Powell (Eds.), *Advancements and innovations in wireless communications and network technologies* (pp. 117–132). Hershey, PA: IGI Global. doi:10.4018/978-1-4666-2154-1.ch009

Sajeva, S. (2013). Towards a conceptual knowledge management system based on systems thinking and sociotechnical thinking. In J. Abdelnour-Nocera (Ed.), *Knowledge and technological development effects on organizational and social structures* (pp. 115–130). Hershey, PA: IGI Global. doi:10.4018/978-1-4666-2151-0.ch008

Salo, M., Olsson, T., Makkonen, M., & Frank, L. (2013). User perspective on the adoption of mobile augmented reality based applications. In I. Lee (Ed.), *Strategy, adoption, and competitive advantage of mobile services in the global economy* (pp. 165–188). Hershey, PA: IGI Global. doi:10.4018/978-1-4666-1939-5.ch009

Samanta, S. K., Woods, J., & Ghanbari, M. (2013). Automatic language translation: An enhancement to the mobile messaging services. In A. Mesquita (Ed.), *User perception and influencing factors of technology in everyday life* (pp. 57–75). Hershey, PA: IGI Global. doi:10.4018/978-1-4666-1954-8.ch006

Santo, A. E., Rijo, R., Monteiro, J., Henriques, I., Matos, A., Rito, C., & Marcelino, L. et al. (2012). Games improving disorders of attention deficit and hyperactivity. In M. Cruz-Cunha (Ed.), *Handbook of research on serious games as educational, business and research tools* (pp. 1160–1174). Hershey, PA: IGI Global. doi:10.4018/978-1-4666-0149-9.ch060

Sarker, S., Campbell, D. E., Ondrus, J., & Valacich, J. S. (2012). Mapping the need for mobile collaboration technologies: A fit perspective. In N. Kock (Ed.), *Advancing collaborative knowledge environments: New trends in e-collaboration* (pp. 211–233). Hershey, PA: IGI Global. doi:10.4018/978-1-61350-459-8.ch013

Sasajima, M., Kitamura, Y., & Mizoguchi, R. (2013). Method for modeling user semantics and its application to service navigation on the web. In G. Xu & L. Li (Eds.), *Social media mining and social network analysis: Emerging research* (pp. 127–139). Hershey, PA: IGI Global. doi:10.4018/978-1-4666-2806-9.ch008

Scheel, C., & Pineda, L. (2013). Building industrial clusters in Latin America: Paddling upstream. In J. Abdelnour-Nocera (Ed.), *Knowledge and technological development effects on organizational and social structures* (pp. 146–166). Hershey, PA: IGI Global. doi:10.4018/978-1-4666-2151-0.ch010

Sell, A., Walden, P., & Carlsson, C. (2013). Segmentation matters: An exploratory study of mobile service users. In D. Chiu (Ed.), *Mobile and web innovations in systems and service-oriented engineering* (pp. 301–317). Hershey, PA: IGI Global. doi:10.4018/978-1-4666-2470-2.ch016

Sermon, P., & Gould, C. (2013). Site-specific performance, narrative, and social presence in multi-user virtual environments and the urban landscape. In D. Harrison (Ed.), *Digital media and technologies for virtual artistic spaces* (pp. 46–58). Hershey, PA: IGI Global. doi:10.4018/978-1-4666-2961-5.ch004

Servaes, J. (2012). The role of information communication technologies within the field of communication for social change. In *Wireless technologies: Concepts, methodologies, tools and applications* (pp. 1117–1135). Hershey, PA: IGI Global. doi:10.4018/978-1-61350-101-6.ch502

Seth, N., & Patnayakuni, R. (2012). Online matrimonial sites and the transformation of arranged marriage in India. In C. Romm Livermore (Ed.), *Gender and social computing: Interactions, differences and relationships* (pp. 272–295). Hershey, PA: IGI Global. doi:10.4018/978-1-60960-759-3.ch016

Shaffer, G. (2013). Lessons learned from grassroots wireless networks in Europe. In A. Abdelaal (Ed.), *Social and economic effects of community wireless networks and infrastructures* (pp. 236–254). Hershey, PA: IGI Global. doi:10.4018/978-1-4666-2997-4.ch013

Shen, J., & Eder, L. B. (2013). An examination of factors associated with user acceptance of social shopping websites. In A. Mesquita (Ed.), *User perception and influencing factors of technology in everyday life* (pp. 28–45). Hershey, PA: IGI Global. doi:10.4018/978-1-4666-1954-8.ch004

Shen, K. N. (2012). Identification vs. self-verification in virtual communities (VC): Theoretical gaps and design implications. In C. El Morr & P. Maret (Eds.), *Virtual community building and the information society: Current and future directions* (pp. 208–236). Hershey, PA: IGI Global. doi:10.4018/978-1-60960-869-9.ch011

Shi, Y., & Liu, Z. (2013). Cultural models and variations. In *Industrial engineering: Concepts, methodologies, tools, and applications* (pp. 1560–1573). Hershey, PA: IGI Global. doi:10.4018/978-1-4666-1945-6.ch083

Shiferaw, A., Sehai, E., Hoekstra, D., & Getachew, A. (2013). Enhanced knowledge management: Knowledge centers for extension communication and agriculture development in Ethiopia. In B. Maumbe & C. Patrikakis (Eds.), *E-agriculture and rural development: Global innovations and future prospects* (pp. 103–116). Hershey, PA: IGI Global. doi:10.4018/978-1-4666-2655-3.ch010

Simão de Vasconcellos, M., & Soares de Araújo, I. (2013). Massively multiplayer online role playing games for health communication in Brazil. In K. Bredl & W. Bösche (Eds.), *Serious games and virtual worlds in education, professional development, and healthcare* (pp. 294–312). Hershey, PA: IGI Global. doi:10.4018/978-1-4666-3673-6.ch018

Simour, L. (2012). Networking identities: Geographies of interaction and computer mediated communication1. In S. Dasgupta (Ed.), *Technical, social, and legal issues in virtual communities: Emerging environments: Emerging environments* (pp. 235–246). Hershey, PA: IGI Global. doi:10.4018/978-1-4666-1553-3.ch016

Singh, G. R. (2013). Cyborg in the village: Culturally embedded resistances to blended teaching and learning. In E. Jean Francois (Ed.), *Transcultural blended learning and teaching in postsecondary education* (pp. 75–90). Hershey, PA: IGI Global. doi:10.4018/978-1-4666-2014-8.ch005

Singh, M., & Iding, M. K. (2013). Does credibility count?: Singaporean students' evaluation of social studies web sites. In R. Zheng (Ed.), *Evolving psychological and educational perspectives on cyber behavior* (pp. 230–245). Hershey, PA: IGI Global. doi:10.4018/978-1-4666-1858-9.ch014

Singh, S. (2013). Information and communication technology and its potential to transform Indian agriculture. In B. Maumbe & C. Patrikakis (Eds.), *E-agriculture and rural development: Global innovations and future prospects* (pp. 140–168). Hershey, PA: IGI Global. doi:10.4018/978-1-4666-2655-3.ch012

Siqueira, S. R., Rocha, E. C., & Nery, M. S. (2012). Brazilian occupational therapy perspective about digital games as an inclusive resource to disabled people in schools. In M. Cruz-Cunha (Ed.), *Handbook of research on serious games as educational, business and research tools* (pp. 730–749). Hershey, PA: IGI Global. doi:10.4018/978-1-4666-0149-9.ch037

Siti-Nabiha, A., & Salleh, D. (2013). Public sector transformation in Malaysia: Improving local governance and accountability. In N. Pomazalová (Ed.), *Public sector transformation processes and internet public procurement: Decision support systems* (pp. 276–290). Hershey, PA: IGI Global. doi:10.4018/978-1-4666-2665-2.ch013

Siwar, C., & Abdulai, A. (2013). Sustainable development and the digital divide among OIC countries: Towards a collaborative digital approach. In *Digital literacy: Concepts, methodologies, tools, and applications* (pp. 242–261). Hershey, PA: IGI Global. doi:10.4018/978-1-4666-1852-7.ch013

Smith, P. A. (2013). Strengthening and enriching audit practice: The socio-technical relevance of "decision leaders". In J. Abdelnour-Nocera (Ed.), *Knowledge and technological development effects on organizational and social structures* (pp. 97–108). Hershey, PA: IGI Global. doi:10.4018/978-1-4666-2151-0.ch006

Smith, P. A., & Cockburn, T. (2013). Generational demographics. In *Dynamic leadership models for global business: Enhancing digitally connected environments* (pp. 230–256). Hershey, PA: IGI Global. doi:10.4018/978-1-4666-2836-6.ch009

Smith, P. A., & Cockburn, T. (2013). Leadership, global business, and digitally connected environments. In *Dynamic leadership models for global business: Enhancing digitally connected environments* (pp. 257–296). Hershey, PA: IGI Global. doi:10.4018/978-1-4666-2836-6.ch010

Sohrabi, B., Gholipour, A., & Amiri, B. (2013). The influence of information technology on organizational behavior: Study of identity challenges in virtual teams. In N. Kock (Ed.), *Interdisciplinary applications of electronic collaboration approaches and technologies* (pp. 79–95). Hershey, PA: IGI Global. doi:10.4018/978-1-4666-2020-9.ch006

Soitu, L., & Paulet-Crainiceanu, L. (2013). Student-faculty communication on Facebook: Prospective learning enhancement and boundaries. In B. Pătruţ, M. Pătruţ, & C. Cmeciu (Eds.), *Social media and the new academic environment: Pedagogical challenges* (pp. 40–67). Hershey, PA: IGI Global. doi:10.4018/978-1-4666-2851-9.ch003

Solvoll, T. (2013). Mobile communication in hospitals: What is the problem? In C. Rückemann (Ed.), *Integrated information and computing systems for natural, spatial, and social sciences* (pp. 287–301). Hershey, PA: IGI Global. doi:10.4018/978-1-4666-2190-9.ch014

Somboonviwat, K. (2013). Topic modeling for web community discovery. In G. Xu & L. Li (Eds.), *Social media mining and social network analysis: Emerging research* (pp. 72–89). Hershey, PA: IGI Global. doi:10.4018/978-1-4666-2806-9.ch005

Speaker, R. B., Levitt, G., & Grubaugh, S. (2013). Professional development in a virtual world. In J. Keengwe & L. Kyei-Blankson (Eds.), *Virtual mentoring for teachers: Online professional development practices* (pp. 122–148). Hershey, PA: IGI Global. doi:10.4018/978-1-4666-1963-0.ch007

Stevenson, G., & Van Belle, J. (2013). Using social media technology to improve collaboration: A case study of micro-blogging adoption in a South African financial services company. In N. Azab (Ed.), *Cases on web 2.0 in developing countries: Studies on implementation, application, and use* (pp. 313–341). Hershey, PA: IGI Global. doi:10.4018/978-1-4666-2515-0.ch012

Strang, K. D. (2013). Balanced assessment of flexible e-learning vs. face-to-face campus delivery courses at an Australian university. In M. Khosrow-Pour (Ed.), *Cases on assessment and evaluation in education* (pp. 304–339). Hershey, PA: IGI Global. doi:10.4018/978-1-4666-2621-8.ch013

Strömberg-Jakka, M. (2013). Social assistance via the internet: The case of Finland in the European context. In J. Lannon & E. Halpin (Eds.), *Human rights and information communication technologies: Trends and consequences of use* (pp. 177–195). Hershey, PA: IGI Global. doi:10.4018/978-1-4666-1918-0.ch011

Sultanow, E., Weber, E., & Cox, S. (2013). A semantic e-collaboration approach to enable awareness in globally distributed organizations. In N. Kock (Ed.), *Interdisciplinary applications of electronic collaboration approaches and technologies* (pp. 1–16). Hershey, PA: IGI Global. doi:10.4018/978-1-4666-2020-9.ch001

Sun, H., Gui, N., & Blondia, C. (2013). A generic adaptation framework for mobile communication. In V. De Florio (Ed.), *Innovations and approaches for resilient and adaptive systems* (pp. 196–207). Hershey, PA: IGI Global. doi:10.4018/978-1-4666-2056-8.ch011

Surgevil, O., & Özbilgin, M. F. (2012). Women in information communication technologies. In C. Romm Livermore (Ed.), *Gender and social computing: Interactions, differences and relationships* (pp. 87–97). Hershey, PA: IGI Global. doi:10.4018/978-1-60960-759-3.ch006

Sylaiou, S., White, M., & Liarokapis, F. (2013). Digital heritage systems: The ARCO evaluation. In M. Garcia-Ruiz (Ed.), *Cases on usability engineering: Design and development of digital products* (pp. 321–354). Hershey, PA: IGI Global. doi:10.4018/978-1-4666-4046-7.ch014

Sylvester, O. A. (2013). Impact of information and communication technology on livestock production: The experience of rural farmers in Nigeria. In B. Maumbe & C. Patrikakis (Eds.), *E-agriculture and rural development: Global innovations and future prospects* (pp. 68–75). Hershey, PA: IGI Global. doi:10.4018/978-1-4666-2655-3.ch007

Taha, K., & Elmasri, R. (2012). Social search and personalization through demographic filtering. In I. Ting, T. Hong, & L. Wang (Eds.), *Social network mining, analysis, and research trends: Techniques and applications* (pp. 183–203). Hershey, PA: IGI Global. doi:10.4018/978-1-61350-513-7.ch012

Tai, Z. (2012). Fame, fantasy, fanfare and fun: The blossoming of the Chinese culture of blogmongering. In T. Dumova & R. Fiordo (Eds.), *Blogging in the global society: Cultural, political and geographical aspects* (pp. 37–54). Hershey, PA: IGI Global. doi:10.4018/978-1-60960-744-9.ch003

Taifi, N., & Gharbi, K. (2013). Technology integration in strategic management: The case of a micro-financing institutions network. In T. Torres-Coronas & M. Vidal-Blasco (Eds.), *Social e-enterprise: Value creation through ICT* (pp. 263–279). Hershey, PA: IGI Global. doi:10.4018/978-1-4666-2667-6.ch015

Talib, S., Clarke, N. L., & Furnell, S. M. (2013). Establishing a personalized information security culture. In I. Khalil & E. Weippl (Eds.), *Contemporary challenges and solutions for mobile and multimedia technologies* (pp. 53–69). Hershey, PA: IGI Global. doi:10.4018/978-1-4666-2163-3.ch004

Tamura, H., Sugasaka, T., & Ueda, K. (2012). Lovely place to buy!: Enhancing grocery shopping experiences with a human-centric approach. In A. Lugmayr, H. Franssila, P. Näränen, O. Sotamaa, J. Vanhala, & Z. Yu (Eds.), *Media in the ubiquitous era: Ambient, social and gaming media* (pp. 53–65). Hershey, PA: IGI Global. doi:10.4018/978-1-60960-774-6.ch003

Tawileh, W., Bukvova, H., & Schoop, E. (2013). Virtual collaborative learning: Opportunities and challenges of web 2.0-based e-learning arrangements for developing countries. In N. Azab (Ed.), *Cases on web 2.0 in developing countries: Studies on implementation, application, and use* (pp. 380–410). Hershey, PA: IGI Global. doi:10.4018/978-1-4666-2515-0.ch014

Teixeira, P. M., Félix, M. J., & Tavares, P. (2012). Playing with design: The universality of design in game development. In M. Cruz-Cunha (Ed.), *Handbook of research on serious games as educational, business and research tools* (pp. 217–231). Hershey, PA: IGI Global. doi:10.4018/978-1-4666-0149-9.ch011

Teusner, P. E. (2012). Networked individualism, constructions of community and religious identity: The case of emerging church bloggers in Australia. In F. Comunello (Ed.), *Networked sociability and individualism: Technology for personal and professional relationships* (pp. 264–288). Hershey, PA: IGI Global. doi:10.4018/978-1-61350-338-6.ch013

Tezcan, M. (2013). Social e-entrepreneurship, employment, and e-learning. In T. Torres-Coronas & M. Vidal-Blasco (Eds.), *Social e-enterprise: Value creation through ICT* (pp. 133–147). Hershey, PA: IGI Global. doi:10.4018/978-1-4666-2667-6.ch008

Thatcher, B. (2012). Approaching intercultural rhetoric and professional communication. In *Intercultural rhetoric and professional communication: Technological advances and organizational behavior* (pp. 1–38). Hershey, PA: IGI Global. doi:10.4018/978-1-61350-450-5.ch001

Thatcher, B. (2012). Borders and etics as units of analysis for intercultural rhetoric and professional communication. In *Intercultural rhetoric and professional communication: Technological advances and organizational behavior* (pp. 39–74). Hershey, PA: IGI Global. doi:10.4018/978-1-61350-450-5.ch002

Thatcher, B. (2012). Core competencies in intercultural teaching and research. In *Intercultural rhetoric and professional communication: Technological advances and organizational behavior* (pp. 318–342). Hershey, PA: IGI Global. doi:10.4018/978-1-61350-450-5.ch011

Thatcher, B. (2012). Distance education and e-learning across cultures. In *Intercultural rhetoric and professional communication: Technological advances and organizational behavior* (pp. 186–215). Hershey, PA: IGI Global. doi:10.4018/978-1-61350-450-5.ch007

Thatcher, B. (2012). Information and communication technologies and intercultural professional communication. In *Intercultural rhetoric and professional communication: Technological advances and organizational behavior* (pp. 97–123). Hershey, PA: IGI Global. doi:10.4018/978-1-61350-450-5.ch004

Thatcher, B. (2012). Intercultural rhetorical dimensions of health literacy and medicine. In *Intercultural rhetoric and professional communication: Technological advances and organizational behavior* (pp. 247–282). Hershey, PA: IGI Global. doi:10.4018/978-1-61350-450-5.ch009

Thatcher, B. (2012). Legal traditions, the universal declaration of human rights, and intercultural professional communication. In *Intercultural rhetoric and professional communication: Technological advances and organizational behavior* (pp. 216–246). Hershey, PA: IGI Global. doi:10.4018/978-1-61350-450-5.ch008

Thatcher, B. (2012). Organizational theory and communication across cultures. In *Intercultural rhetoric and professional communication: Technological advances and organizational behavior* (pp. 159–185). Hershey, PA: IGI Global. doi:10.4018/978-1-61350-450-5.ch006

Thatcher, B. (2012). Teaching intercultural rhetoric and professional communication. In *Intercultural rhetoric and professional communication: Technological advances and organizational behavior* (pp. 343–378). Hershey, PA: IGI Global. doi:10.4018/978-1-61350-450-5.ch012

Thatcher, B. (2012). Website designs as an indicator of globalization. In *Intercultural rhetoric and professional communication: Technological advances and organizational behavior* (pp. 124–158). Hershey, PA: IGI Global. doi:10.4018/978-1-61350-450-5.ch005

Thatcher, B. (2012). Writing instructions and how-to-do manuals across cultures. In *Intercultural rhetoric and professional communication: Technological advances and organizational behavior* (pp. 283–317). Hershey, PA: IGI Global. doi:10.4018/978-1-61350-450-5.ch010

Thirumal, P., & Tartakov, G. M. (2013). India's Dalits search for a democratic opening in the digital divide. In *Digital literacy: Concepts, methodologies, tools, and applications* (pp. 852–871). Hershey, PA: IGI Global. doi:10.4018/978-1-4666-1852-7.ch044

Thomas, G. E. (2013). Facilitating learning with adult students in the transcultural classroom. In E. Jean Francois (Ed.), *Transcultural blended learning and teaching in postsecondary education* (pp. 193–215). Hershey, PA: IGI Global. doi:10.4018/978-1-4666-2014-8.ch011

Tripathi, S. N., & Siddiqui, M. H. (2013). Designing effective mobile advertising with specific reference to developing markets. In I. Lee (Ed.), *Strategy, adoption, and competitive advantage of mobile services in the global economy* (pp. 299–324). Hershey, PA: IGI Global. doi:10.4018/978-1-4666-1939-5.ch017

Truong, Y. (2013). Antecedents of consumer acceptance of mobile television advertising. In A. Mesquita (Ed.), *User perception and influencing factors of technology in everyday life* (pp. 128–141). Hershey, PA: IGI Global. doi:10.4018/978-1-4666-1954-8.ch010

Tsuneizumi, I., Aikebaier, A., Ikeda, M., Enokido, T., & Takizawa, M. (2013). Design and implementation of hybrid time (HT) group communication protocol for homogeneous broadcast groups. In N. Bessis (Ed.), *Development of distributed systems from design to application and maintenance* (pp. 282–293). Hershey, PA: IGI Global. doi:10.4018/978-1-4666-2647-8.ch017

Tzoulia, E. (2013). Legal issues to be considered before setting in force consumer-centric marketing strategies within the European Union. In H. Kaufmann & M. Panni (Eds.), *Customer-centric marketing strategies: Tools for building organizational performance* (pp. 36–56). Hershey, PA: IGI Global. doi:10.4018/978-1-4666-2524-2.ch003

Underwood, J., & Okubayashi, T. (2013). Comparing the characteristics of text-speak used by English and Japanese students. In R. Zheng (Ed.), *Evolving psychological and educational perspectives on cyber behavior* (pp. 258–271). Hershey, PA: IGI Global. doi:10.4018/978-1-4666-1858-9.ch016

Unoki, M., & Miyauchi, R. (2013). Method of digital-audio watermarking based on cochlear delay characteristics. In K. Kondo (Ed.), *Multimedia information hiding technologies and methodologies for controlling data* (pp. 42–70). Hershey, PA: IGI Global. doi:10.4018/978-1-4666-2217-3.ch003

Usman, L. M. (2013). Adult education and sustainable learning outcome of rural widows of central northern Nigeria. In V. Wang (Ed.), *Technological applications in adult and vocational education advancement* (pp. 215–231). Hershey, PA: IGI Global. doi:10.4018/978-1-4666-2062-9.ch017

Usoro, A., & Khan, I. U. (2013). Trust as an aspect of organisational culture: Its effects on knowledge sharing in virtual communities. In R. Colomo-Palacios (Ed.), *Enhancing the modern organization through information technology professionals: Research, studies, and techniques* (pp. 182–199). Hershey, PA: IGI Global. doi:10.4018/978-1-4666-2648-5.ch013

Utz, S. (2012). Social network site use among Dutch students: Effects of time and platform. In F. Comunello (Ed.), *Networked sociability and individualism: Technology for personal and professional relationships* (pp. 103–125). Hershey, PA: IGI Global. doi:10.4018/978-1-61350-338-6.ch006

Vasilescu, R., Epure, M., & Florea, N. (2013). Digital literacy for effective communication in the new academic environment: The educational blogs. In B. Pătruţ, M. Pătruţ, & C. Cmeciu (Eds.), *Social media and the new academic environment: Pedagogical challenges* (pp. 368–390). Hershey, PA: IGI Global. doi:10.4018/978-1-4666-2851-9.ch018

Vladimirschi, V. (2013). An exploratory study of cross-cultural engagement in the community of inquiry: Instructor perspectives and challenges. In Z. Akyol & D. Garrison (Eds.), *Educational communities of inquiry: Theoretical framework, research and practice* (pp. 466–489). Hershey, PA: IGI Global. doi:10.4018/978-1-4666-2110-7.ch023

Vuokko, R. (2012). A practice perspective on transforming mobile work. In *Wireless technologies: Concepts, methodologies, tools and applications* (pp. 1104–1116). Hershey, PA: IGI Global. doi:10.4018/978-1-61350-101-6.ch501

Wall, M., & Kirdnark, T. (2012). The blogosphere in the "land of smiles": Citizen media and political conflict in Thailand. In T. Dumova & R. Fiordo (Eds.), *Blogging in the global society: Cultural, political and geographical aspects* (pp. 19–36). Hershey, PA: IGI Global. doi:10.4018/978-1-60960-744-9.ch002

Warburton, S. (2013). Space for lurking: A pattern for designing online social spaces. In S. Warburton & S. Hatzipanagos (Eds.), *Digital identity and social media* (pp. 149–158). Hershey, PA: IGI Global. doi:10.4018/978-1-4666-1915-9.ch011

Warren, S. J., & Lin, L. (2012). Ethical considerations for learning game, simulation, and virtual world design and development. In H. Yang & S. Yuen (Eds.), *Handbook of research on practices and outcomes in virtual worlds and environments* (pp. 1–18). Hershey, PA: IGI Global. doi:10.4018/978-1-60960-762-3.ch001

Wasihun, T. A., & Maumbe, B. (2013). Information and communication technology uses in agriculture: Agribusiness industry opportunities and future challenges. In B. Maumbe & C. Patrikakis (Eds.), *E-agriculture and rural development: Global innovations and future prospects* (pp. 235–251). Hershey, PA: IGI Global. doi:10.4018/978-1-4666-2655-3.ch017

Webb, L. M., Fields, T. E., Boupha, S., & Stell, M. N. (2012). U.S. political blogs: What aspects of blog design correlate with popularity? In T. Dumova & R. Fiordo (Eds.), *Blogging in the global society: Cultural, political and geographical aspects* (pp. 179–199). Hershey, PA: IGI Global. doi:10.4018/978-1-60960-744-9.ch011

Weeks, M. R. (2012). Toward an understanding of online community participation through narrative network analysis. In H. Li (Ed.), *Virtual community participation and motivation: Cross-disciplinary theories* (pp. 90–102). Hershey, PA: IGI Global. doi:10.4018/978-1-4666-0312-7.ch006

White, J. R. (2013). Language economy in computer-mediated communication: Learner autonomy in a community of practice. In B. Zou, M. Xing, Y. Wang, M. Sun, & C. Xiang (Eds.), *Computer-assisted foreign language teaching and learning: Technological advances* (pp. 75–90). Hershey, PA: IGI Global. doi:10.4018/978-1-4666-2821-2.ch005

Whitworth, B., & Liu, T. (2013). Politeness as a social computing requirement. In J. Bishop (Ed.), *Examining the concepts, issues, and implications of internet trolling* (pp. 88–104). Hershey, PA: IGI Global. doi:10.4018/978-1-4666-2803-8.ch008

Wichowski, D. E., & Kohl, L. E. (2013). Establishing credibility in the information jungle: Blogs, microblogs, and the CRAAP test. In M. Folk & S. Apostel (Eds.), *Online credibility and digital ethos: Evaluating computer-mediated communication* (pp. 229–251). Hershey, PA: IGI Global. doi:10.4018/978-1-4666-2663-8.ch013

Williams, J. (2013). Social cohesion and free home internet in New Zealand. In A. Abdelaal (Ed.), *Social and economic effects of community wireless networks and infrastructures* (pp. 135–159). Hershey, PA: IGI Global. doi:10.4018/978-1-4666-2997-4.ch008

Williams, S., Fleming, S., Lundqvist, K., & Parslow, P. (2013). This is me: Digital identity and reputation on the internet. In S. Warburton & S. Hatzipanagos (Eds.), *Digital identity and social media* (pp. 104–117). Hershey, PA: IGI Global. doi:10.4018/978-1-4666-1915-9.ch008

Winning, R. (2013). Behind the sonic veil: Considering sound as the mediator of illusory life in variable and screen-based media. In D. Harrison (Ed.), Digital media and technologies for virtual artistic spaces (pp. 117-134). Hershey, PA: IGI Global. doi:10.4018/978-1-4666-2961-5.ch009

Wolfe, A. (2012). Network perspective on structures related to communities. In M. Safar & K. Mahdi (Eds.), *Social networking and community behavior modeling: Qualitative and quantitative measures* (pp. 26–50). Hershey, PA: IGI Global. doi:10.4018/978-1-61350-444-4.ch002

Worden, S. (2013). The earth sciences and creative practice: Exploring boundaries between digital and material culture. In D. Harrison (Ed.), *Digital media and technologies for virtual artistic spaces* (pp. 186–204). Hershey, PA: IGI Global. doi:10.4018/978-1-4666-2961-5.ch014

Xing, M., Zou, B., & Wang, D. (2013). A wiki platform for language and intercultural communication. In B. Zou, M. Xing, Y. Wang, M. Sun, & C. Xiang (Eds.), *Computer-assisted foreign language teaching and learning: Technological advances* (pp. 1–15). Hershey, PA: IGI Global. doi:10.4018/978-1-4666-2821-2.ch001

Xu, G., Gu, Y., & Yi, X. (2013). On group extraction and fusion for tag-based social recommendation. In G. Xu & L. Li (Eds.), *Social media mining and social network analysis: Emerging research* (pp. 211–223). Hershey, PA: IGI Global. doi:10.4018/978-1-4666-2806-9.ch014

Yakura, E. K., Soe, L., & Guthrie, R. (2012). Women in IT careers: Investigating support for women in the information technology workforce. In C. Romm Livermore (Ed.), *Gender and social computing: Interactions, differences and relationships* (pp. 35–49). Hershey, PA: IGI Global. doi:10.4018/978-1-60960-759-3.ch003

Yang, Y., Rahim, A., & Karmakar, N. C. (2013). 5.8 GHz portable wireless monitoring system for sleep apnea diagnosis in wireless body sensor network (WBSN) using active RFID and MIMO technology. In N. Karmakar (Ed.), *Advanced RFID systems, security, and applications* (pp. 264–303). Hershey, PA: IGI Global. doi:10.4018/978-1-4666-2080-3.ch012

Yu, Z., Liang, Y., Yang, Y., & Guo, B. (2013). Supporting social interaction in campus-scale environments by embracing mobile social networking. In G. Xu & L. Li (Eds.), *Social media mining and social network analysis: Emerging research* (pp. 182–201). Hershey, PA: IGI Global. doi:10.4018/978-1-4666-2806-9.ch012

Zaman, M., Simmers, C. A., & Anandarajan, M. (2013). Using an ethical framework to examine linkages between "going green" in research practices and information and communication technologies. In B. Medlin (Ed.), *Integrations of technology utilization and social dynamics in organizations* (pp. 243–262). Hershey, PA: IGI Global. doi:10.4018/978-1-4666-1948-7.ch015

Zarmpou, T., Saprikis, V., & Vlachopoulou, M. (2013). Examining behavioral intention toward mobile services: An empirical investigation in Greece. In A. Scupola (Ed.), *Mobile opportunities and applications for e-service innovations* (pp. 37–56). Hershey, PA: IGI Global. doi:10.4018/978-1-4666-2654-6.ch003

Zavala Pérez, J. M. (2012). Registry culture and networked sociability: Building individual identity through information records. In F. Comunello (Ed.), *Networked sociability and individualism: Technology for personal and professional relationships* (pp. 41–62). Hershey, PA: IGI Global. doi:10.4018/978-1-61350-338-6.ch003

Zemliansky, P., & Goroshko, O. (2013). Social media and other web 2.0 technologies as communication channels in a cross-cultural, web-based professional communication project. In B. Pătruţ, M. Pătruţ, & C. Cmeciu (Eds.), *Social media and the new academic environment: Pedagogical challenges* (pp. 256–272). Hershey, PA: IGI Global. doi:10.4018/978-1-4666-2851-9.ch013

Zervas, P., & Alexandraki, C. (2013). The realisation of online music services through intelligent computing. In D. Kanellopoulos (Ed.), *Intelligent multimedia technologies for networking applications: Techniques and tools* (pp. 291–317). Hershey, PA: IGI Global. doi:10.4018/978-1-4666-2833-5.ch012

Zhang, J., & Mao, E. (2013). The effects of consumption values on the use of location-based services on smartphones. In I. Lee (Ed.), *Strategy, adoption, and competitive advantage of mobile services in the global economy* (pp. 1–18). Hershey, PA: IGI Global. doi:10.4018/978-1-4666-1939-5.ch001

Zhang, S., Köbler, F., Tremaine, M., & Milewski, A. (2012). Instant messaging in global software teams. In N. Kock (Ed.), *Advancing collaborative knowledge environments: New trends in e-collaboration* (pp. 158–179). Hershey, PA: IGI Global. doi:10.4018/978-1-61350-459-8.ch010

Zhang, T., Wang, C., Luo, Z., Han, S., & Dong, M. (2013). RFID enabled vehicular network for ubiquitous travel query. In D. Chiu (Ed.), *Mobile and web innovations in systems and service-oriented engineering* (pp. 348–363). Hershey, PA: IGI Global. doi:10.4018/978-1-4666-2470-2.ch019

Zhang, W. (2012). Virtual communities as subaltern public spheres: A theoretical development and an application to the Chinese internet. In H. Li (Ed.), *Virtual community participation and motivation: Cross-disciplinary theories* (pp. 143–159). Hershey, PA: IGI Global. doi:10.4018/978-1-4666-0312-7.ch009

Zhang, X., Wang, L., Li, Y., & Liang, W. (2013). Global community extraction in social network analysis. In G. Xu & L. Li (Eds.), *Social media mining and social network analysis: Emerging research* (pp. 156–171). Hershey, PA: IGI Global. doi:10.4018/978-1-4666-2806-9.ch010

Zhang, X., Wang, L., Li, Y., & Liang, W. (2013). Local community extraction in social network analysis. In G. Xu & L. Li (Eds.), *Social media mining and social network analysis: Emerging research* (pp. 172–181). Hershey, PA: IGI Global. doi:10.4018/978-1-4666-2806-9.ch011

Zulu, S. F. (2013). Emerging information and communication technology policy framework for Africa. In B. Maumbe & J. Okello (Eds.), *Technology, sustainability, and rural development in Africa* (pp. 236–256). Hershey, PA: IGI Global. doi:10.4018/978-1-4666-3607-1.ch016

Compilation of References

A murderous attack on Western freedoms. (2015, January 8). *Daily Mail*, p.14.

Abdallah, M. (2013). A Community of practice facilitated by Facebook for integrating new online EFL writing forms into Assiut university college of education. *Journal of New Valley Faculty of Education, 12*(1). ERIC Document No. ED545728.

About Roden Crater. (n.d.). Retrieved from http://rodencrater.com/about/

Ahn, J. (2010). *The influence of social networking sites on high school students' social and academic development* (Unpublished doctoral dissertation). University of Southern California, Los Angeles, CA.

Alavi, M., Marakas, G. M., & Yoo, Y. (2002). A comparative study of distributed learning environments on learning outcomes. *Information Systems Research, 13*(4), 404–415. doi:10.1287/isre.13.4.404.72

Al-Jarf, R. (2007). Online instruction and creative writing by Saudi EFL freshman students. *The Asian EFL Journal Professional Teaching Articles, 22.*

Al-Jarf, R. (2012). *Learning English on Facebook.* Paper presented at the 11th Asia CALL Conference, Ho Chi Minh City Open University, Vietnam.

Al-Jarf, R. (2014). Integrating ethnic culture Facebook pages in EFL instruction. In M. V. Makarych (Ed.), *Proceedings of the International Scientific Conference on Ethnology: Genesis of Hereditary Customs* (41-43). Minsk, Belarus: Belarusian National Technical University (BNTU).

Allwein, G., & Barwise, J. (Eds.). (1996). *Logical reasoning with diagrams.* Oxford, UK: Oxford University Press.

Alon, I., & Herath, R. K. (2014). Teaching international business via social media projects. *Journal of Teaching in International Business, 25*(1), 44–59. doi:10.1080/08975930.2013.847814

Al-Sayed, R. K., Abdel-Haq, E. M., El-Deeb, M. A., & Ali, M. A. (2016). *Fostering the memoir writing skills as a creative non-fiction genre using a WebQuest model.* ERIC Document No. ED565329.

American Association for the Advancement of Science. (2001). Special Issue: The Human Genome. *Science, 291*(5507). Retrieved from http://science.sciencemag.org/content/291/5507

Ames, W. S. (1966). The development of a classification scheme of contextual aids. *Reading Research Quarterly, 2*(1), 57–82. doi:10.2307/747039

An irreverent French institution (2015, January 8). *Financial Times*, p. 3.

Anae, N. (2014). Creative writing as freedom, education as exploration: Creative writing as literary and visual arts pedagogy in the first-year teacher-education experience. *Australian Journal of Teacher Education, 39*(8). doi:10.14221/ajte.2014v39n8.8

Andersen, P. B. (1992). Computer semiotics. *Scandinavian Journal of Information Systems, 4*(1), 1–30.

Andriessen, J. (2006). Arguing to learn. In R. K. Sawyer (Ed.), *The Cambridge Handbook of the Learning Sciences*. New York: Cambridge University Press.

Anstey, J., Seyed, A. P., Bay-Cheng, S., Pape, D., Shapiro, S. C., Bona, J., & Hibit, S. (2009). The agent takes the stage. *International Journal of Arts and Technology, 2*(4), 277–296. doi:10.1504/IJART.2009.029236

Aristotle. (1982). Poetics: Vol. 23. *Loeb Classical Library*. Cambridge, MA: Harvard University Press.

Aristotle. (350 BCE). *On interpretation, 1.16a 4-9* (E. M. Edgehill, E.M., Trans.). Retrieved from http://classics.mit.edu/Aristotle/interpretation.html

Arshavskaya, E. (2015). Creative writing assignments in a second language course: A way to engage less motivated students. *InSight: A Journal of Scholarly Teaching, 10*, 68-78.

Artzt, A., & Newman, C. M. (1997). *How to use cooperative learning in the mathematics class* (2nd ed.). Reston: National Council of Teachers of Mathematics.

Assange, J. (2012). *Cypherpunks: Freedom and the future of the Internet*. New York: OR Books.

Atkin, A. (2013). Peirce's theory of signs. *The Stanford Encyclopedia of Philosophy*. Retrieved July 15, 2017 at 11:01 AM EDT from https://plato.stanford.edu/cgi-bin/encyclopedia/archinfo.cgi?entry=peirce-semiotics

Atwood, G., & Stolorow, R. (1979/1993). *Faces in a cloud: Intersubjectivity in personality theory*. Northvale, NJ: Jason Aronson.

Awada, G. (2016). Effect of WhatsApp on critique writing proficiency and perceptions toward learning. *Cogent Education, 3*(1).

Baldwin, J. M. (1902). *Dictionary of philosophy and psychology*. New York: Macmillan.

Balogun, O. M. (2012). Cultural and cosmopolitan: Idealized femininity and embodied nationalism in Nigerian beauty pageants. *Gender & Society, 26*(3), 357–381. doi:10.1177/0891243212438958

Banet-Weiser, S. (1999). *The most beautiful girl in the world: Beauty pageants and national identity*. Berkeley, CA: University of California Press.

Banet-Weiser, S., & Portwood-Stacer, L. (2006). 'I just want to be me again!' Beauty pageants, reality television and post-feminism. *Feminist Theory, 7*(2), 255–272. doi:10.1177/1464700106064423

Barbieri, M. (Ed.). (2007). *Introduction to biosemiotics. The new biological synthesis*. Heidelberg, Germany: Springer. doi:10.1007/1-4020-4814-9

Baresh, B., Hsu, S., & Reese, S. D. (2009). The power of framing: new challenges for researching the structure of meaning in news. In S. Allan (Ed.), *The Routledge Companion to News and Journalism* (pp. 637–647).

Barker-Plummer, D., & Bailin, S. C. (1997). The role of diagrams in mathematical proofs. *Machine Graphics and Vision, 8*, 25–58.

Barker-Plummer, D., & Bailin, S. C. (2001). On the practical semantics of mathematical diagrams. In M. Anderson (Ed.), *Reasoning with diagrammatic representations*. New York: Springer.

Barnes, N. B. (1994). Face of the nation: Race, nationalisms and identities in Jamaican beauty pageants. *The Massachusetts Review*, *35*(3/4), 471–492.

Baroody, A. J. (2000). Does mathematics instruction for three-to-five-year-olds really make sense? *Young Children*, *55*(4), 61–67.

Barrs, M., & Horrocks, S. (2014). *Educational blogs and their effects on pupils' writing*. CFBT Education Trust. ERIC Document No. ED546797.

Barsalou, L. W. (1999). Perceptual symbol systems. *Behavioral and Brain Sciences*, *22*, 577–660. PMID:11301525

Barthes, R. (1957). *Mythologies. Translated, Lavers, A.* London: Paladin Press.

Bartscher, M., Lawler, K., Ramirez, A. & Schinault, K. (2001). *Improving student's writing ability through journals and creative writing exercises*. ERIC Document No. ED455525.

Baumgarten, A. G. (1750). *Aesthetica*. Frankfurt-an-der-Oder, Germany: Kley.

Bayat, S. (2016). The effectiveness of the creative writing instruction program based on speaking activities (CWIPSA). *International Electronic Journal of Elementary Education*, *8*(4), 617–628.

Beatty, R., & Moss, J. (2006b). Multiple vs. numeric approaches to developing functional understanding through patterns: Affordances and limitations for grade 4 students. In S. Alatorre, J. L. Cortina, M. Sáiz, & A. Méndes (Eds.), *Proceedings of the 28th Annual Meeting of the North American Chapter of the International Group for the Psychology of Mathematics Education*. Mérida, México: Universidad Pedagógica Nacional.

Beatty, R., & Moss, J. (2009). *I think for every problem you can solve it...for every pattern there is a rule: Tracking young students' developing abilities to generalize in the context of patterning problems*. Paper presented at the Annual Meeting of the American Educational Research Association, San Diego, CA.

Beatty, R., & Moss, J. (2006a). Connecting grade 4 students from diverse urban classrooms: Virtual collaboration to solve generalizing problems. *Proceedings of the Seventeenth ICMI Study: Technology Revisited*.

Benjamin, W. (2010). The work of art in the age of its technological reproducibility (M. W. Jennings, Trans.). *Grey Room, 39*, 11-37. (Original work published 1935)

Bentley, J. & Bourret, R. (1991). *Using emerging technology to improve instruction in college transfer*. ERIC Document No. ED405011.

Bergin, T. G., & Fisch, M. (1984). *The New Science of Giambattista Vico*. Ithaca, NY: Cornell University Press.

Berkower, L. (1970). The military influence among Freud's dynamic psychiatry. *The American Journal of Psychiatry*, *127*(2), 85–92. doi:10.1176/ajp.127.2.167 PMID:4919635

Berners-Lee, T. (1998). *Semantic web roadmap*. Retrieved on March 31, 2011 from http://www.w3.org/DesignIssues/Semantic.html

Biesenbach-Lucas, S., & Weasenforth, D. (2001). E-Mail and word processing in the ESL classroom: How the medium affects the message. *Language Learning & Technology*, *5*(1), 135–165.

Bigi, S., & Greco Morasso, S. (2012). Keywords, frames and the reconstruction of material starting points in argumentation. *Journal of Pragmatics*, *44*(10), 1135–1149. doi:10.1016/j.pragma.2012.04.011

Bikowski, D., & Vithanage, R. (2016). Effects of web-based collaborative writing on individual l2 writing development. *Language Learning & Technology*, *20*(1), 79–99.

Bintz, W. P. (2017). Writing etheree poems across the curriculum. *The Reading Teacher*, *70*(5), 605–609. doi:10.1002/trtr.1544

Bion, W. (1983). *Elements of psycho-analysis*. Northvale, NJ: Jason Aronson.

Black, I. (2015, January 8). Paris terror attack. Reaction: leaders condemn attack. *The Guardian*, p. 5.

Blonsky, M. (Ed.). (1985). On signs. Baltimore, MD: John Hopkins University Press.

Boaler, J. (1998). Open and closed mathematics: Student experiences and understandings. *Journal for Research in Mathematics Education*, *29*(1), 41–62. doi:10.2307/749717

Boaler, J. (2002). Paying the price for "sugar and spice:" Shifting the analytical lens in equity research. *Mathematical Thinking and Learning*, *4*(2/3), 127–144. doi:10.1207/S15327833MTL04023_3

Bognar, B., Bognár, B., & Kuma, K. (2009). *Material immaterial: The new work of Kengo Kuma*. New York: Princeton Architectural Press.

Boonen, A., Kolkman, M., & Kroesbergen, E. (2011). The relation between teachers' math talk and the acquisition of number sense within kindergarten classrooms. *Journal of School Psychology*, *49*(1), 281–299. doi:10.1016/j.jsp.2011.03.002 PMID:21640245

Bouissac, P. (1977). *Circus and culture: A semiotic approach (Advances in Semiotics)*. Bloomington, IN: Indiana University Press.

Bouissac, P. (2010). *Saussure. A guide for the perplexed*. New York: Continuum International Publishing.

Bredekamp, S. (2004). Standards for preschool and kindergarten mathematics education. In D. H. Clements & J. Sarama (Eds.), *Engaging young children in mathematics: Standards for early childhood mathematics education* (pp. 77–87). Mahwah, NJ: Lawrence Erlbaum Associates.

Bremner, S., Peirson-Smith, A., Jones, R., & Bhatia, V. (2014). Task design and interaction in collaborative writing: The students' story. *Business and Professional Communication Quarterly*, *77*(2), 150–168. doi:10.1177/2329490613514598

Brenner, J. B., & Cunningham, J. G. (1992). Gender differences in eating attitudes, body concept, and self-esteem among models. *Sex Roles*, *27*(7), 413–437. doi:10.1007/BF00289949

Brentano, F. (1874). *Von empirische Standpunkt. In Psychologie* (Vols. 1–2). Leipzig: Von Duncker & Humblot.

Briggs, J., & Peat, D. (2000). *Seven life lessons of chaos: Spiritual wisdom from the science of change*. New York, NY: Harper Perennial.

Brighton, H. (2002). Compositional syntax from cultural transmission. *Artificial Life*, *8*(1), 25–54. doi:10.1162/106454602753694756 PMID:12020420

Broadbent, S., & Gallotti, M. (2015). *Collective intelligence: How does it emerge?* Nesta School of Advanced Study, University of London.

Bromberg, P. (1998). *Standing in the spaces*. Hillsdale, NJ: Analytic Press.

Brown, A. L. (1992). Design experiments: Theoretical and methodological challenges in creating complex interventions in classroom settings. *Journal of the Learning Sciences*, *2*(2), 141–178. doi:10.1207/s15327809jls0202_2

Brown, S. R. (2002). Peirce, Searle, and the Chinese room argument. *Cybernetics & Human Knowing*, *9*(1), 23–38.

Bruce, C.D (2007). *Student interaction in the math class: Stealing ideas or building understanding*. OISE Presentation.

Bruno, M. (2002). *Creative writing: The warm-up*. ERIC Document No. ED464335.

Bucci. (2011). The interplay of subsymbolic and symbolic processes in psychoanalytic treatment: It takes two to tango—But who knows the steps, who's the leader? The choreography of the psychoanalytic interchange. *Psychoanalytic Dialogues*, *21*, 45-54.

Burge, T. (1986). Individualism and psychology. *The Philosophical Review*, *95*(1), 3–45. doi:10.2307/2185131

Butt, B., & White, K. (2005). Mail fraud. *Corner Gas*. First broadcast in Canada, October 24, 2005.

Byerly, A. (2007). A prodigious map beneath his feet: Virtual travel and the panoramic perspective. *Nineteenth-Century Contexts*, *29*(2-3), 151–168. doi:10.1080/08905490701584643

Campbell, J. (1949/1973). *The hero with a thousand faces*. Princeton, NJ: Bollingen Series, Princeton University.

Cariani, P. (2001). Symbols and dynamics in the brain. *Bio Systems*, *60*(1-3), 59–83. doi:10.1016/S0303-2647(01)00108-3 PMID:11325504

Carroll, L. (2015). *Through the looking glass*. Retrieved July 12, 2017 at 12:23 PM EDT from http://pdfreebooks.org/0-carroll.htm

Carvajal, D., & Daley, S. (2015, January 8). Proud to offend. Paper carries torch of political provocation. *The New York Times*.

Casella, V. (1989). Poetry and word processing inspire good writing. *Instructor*, *98*(9), 28.

Cassirer, E. (1921). *Zur Einstein'sche Relativitätstheorie. In Substance and function and Einstein's theory of relativity*. Chicago: Open Court Publishing Co.

Cassirer, E. (1955). *Philosophy of symbolic forms* (R. Manheim, Trans.). New Haven, CT: Yale University Press.

Chalmers, D. J. (1996). *The conscious mind: In search of a fundamental theory*. New York: Oxford University Press.

Chandrasekaran, B., Glasgow, J., & Narayanan, N. H. (Eds.). (1995). *Diagrammatic reasoning: Cognitive and computational perspectives*. Cambridge, MA: MIT Press.

Chen, B., & Hong, H. Y. (2016). Schools as knowledge-building organizations: Thirty years of design research. *Educational Psychologist*, *51*(2), 266–288. doi:10.1080/00461520.2016.1175306

Cheney, D. L., & Seyfarth, R. M. (1992). *How monkeys see the world*. Chicago: Chicago University Press.

Cheung, W., Tse, S., & Tsang, H. (2003). Teaching creative writing skills to primary school children in Hong Kong: Discordance between the views and practices of language teachers. *The Journal of Creative Behavior*, *37*(2), 77–98. doi:10.1002/j.2162-6057.2003.tb00827.x

Chu, K., Grundy, Q., & Bero, L. (2017). "Spin" in published biomedical literature. A methodological system review. *PLoS Biology*, *16*(9). Retrieved from http://journals.plos.org/plosbiology/article?id=10.1371/journal.pbio.2002173

Church, A. (1936). An unsolvable problem of elementary number theory. *American Journal of Mathematics*, *58*(2), 345–363. doi:10.2307/2371045

Cleland, C. E. (1993). Is the Church-Turing thesis true? *Minds and Machines, 3*(3), 283-312.

Coffa, J. A. (1991). *The semantic tradition from Kant to Carnap: To the Vienna Station*. Cambridge, UK: Cambridge University Press. doi:10.1017/CBO9781139172240

Cohen, D. (2007). The worrying world of eating disorder wannabes. *The British Medical Journal*. doi: https://doi-org. myaccess.library.utoronto.ca/10.1136/bmj.39328.510880.59

Confrey, J., & Kazak, S. (2006). A thirty-year reflection on constructivism in mathematics education in PME. In A. Gutiérrez & P. Boero (Eds.), *Handbook of research on the psychology of mathematics education: past, present and future* (pp. 305–345). Rotterdam: Sense Publications.

Conway, J. H. (1970). See Gardner, M. (October 1970) Mathematical games—The fantastic combinations of John Conway's new solitaire game "life". *Scientific American, 223*, 120-123. Retrieved from http://rosettacode.org/wiki/Conway%27s_Game_of_Life

Cooke, L., & Adams, V. (1998). "Math talk" in the classroom. *Middle School Journal, 29*(5), 35–40. doi:10.1080/00940771.1998.11495918

Cortina, M., & Liotti, G. (2007). New approaches to understanding unconscious processes: Implicit and explicit memory systems. *International Forum of Psychoanalysis, 16*(4), 204–212. doi:10.1080/08037060701676326

Crane, T. (1990). The language of thought: No syntax without semantics. *Mind & Language, 5*(3), 187–212. doi:10.1111/j.1468-0017.1990.tb00159.x

Creto, J. (2004). Cold plums and the old men in the water: Let children read and write great poetry. *The Reading Teacher, 58*(3), 266–271. doi:10.1598/RT.58.3.4

Crossley, J. N., & Henry, A. S. (1990). Thus spake al-Khwārizmī: A translation of the text of Cambridge University Library Ms. Ii.vi.5. *Historia Mathematica, 17*(2), 103–131. doi:10.1016/0315-0860(90)90048-I

Cummins, R. (1996). *Representations, targets, and attitudes*. Cambridge, MA: MIT Press.

Dam Christensen, H. (2010). The repressive logic of a profession: On the use of reproductions in art history. *Konsthistorik tidskrift/Journal of Art History, 79*(4), 200-215.

Danesi, M. (1993). *Vico, metaphor and the origin of language*. Bloomington, IN: Indiana University Press.

Danesi, M. (1994). *Cool: The signs and meanings of Adolescence*. Toronto: University of Toronto Press. doi:10.3138/9781442673472

Danesi, M. (2013). *Discovery in mathematics: An interdisciplinary approach*. Munich: Lincom Europa.

Danesi, M. (2016). *Language, society and new media: Sociolinguistics today*. New York: Routledge.

Danesi, M., & Perron, P. (1999). *Analyzing cultures: An introduction and handbook*. Bloomington, IN: Indiana University Press.

Davidson, N. (1990). *Cooperative learning in mathematics: A handbook for teachers*. Menlo Park, CA: Addison-Wesley.

de Saussure, F. (1959). *Course in general linguistics* (W. Baskin, Trans.). Philosophical Library. Retrieved on July 23, 2017 at 12:59 PM EDT from https://archive.org/stream/courseingeneral00saus/courseingeneral00saus_djvu.txt

De Vreese, C. H. (2005). News framing: theory and typology. *Information Design Journal, Document Design, 13*(1), 51-62.

Deacon, T. (1997). *The symbolic species*. New York: W. W. Norton & Co.

Del Nero, J. R. (2017). Fun while showing, not telling: Crafting vivid detail in writing. *The Reading Teacher, 71*(1), 83–87. doi:10.1002/trtr.1575

Deledalle, G. (1979). *Théorie et pratique du signe: Introduction à la sémiotique de Charles S. Peirce. Paris: Payot.*

Dennett, D. (2009). Darwin's 'Strange Inversion of Reasoning.' *Proceedings of the National Academy of Sciences, 106*(Suppl 1), 10061-10065.

Deutsch, D. (2012). *The Beginning of Infinity*. New York: Penguin Books.

Dewart, L. (2016). *Hume's Challenge and the Renewal of Modern Philosophy*. BookBaby Publishers.

Dewulf, A., Gray, B., Putnam, L., Lewicki, R., Aarts, N., Bouwen, R., & Van Woerkum, C. (2009). Disentangling approaches to framing in conflict and negotiation research: A meta-paradigmatic perspective. *Human Relations, 62*(2), 155–193. doi:10.1177/0018726708100356

Dietrich, E. (1990). Computationalism. *Social Epistemology, 4*(2), 135–154. doi:10.1080/02691729008578566

Dingguo, S. (2008). *The Wisdom of Chinese Characters (English Edition)*. Beiging Language and Culture University.

Dipert, R. R. (1984). Peirce, Frege, the logic of telations, and Church's Theorem. *History and Philosophy of Logic, 5*(1), 49–66. doi:10.1080/01445348408837062

Dizon, G. (2016). A comparative study of Facebook vs. paper-and-pencil writing to improve L2 writing skills. *Computer Assisted Language Learning, 29*(8), 1249–1258. doi:10.1080/09588221.2016.1266369

Dollins, C. A. (2016). Crafting creative nonfiction: From close reading to close writing. *The Reading Teacher, 70*(1), 49–58. doi:10.1002/trtr.1465

Donald, M. (2001). *A mind so rare*. New York: W.W. Norton & Co.

Donovan, M. (2015). *14 types of creative writing*. Retrieved August 15, 2017 from https://www.writingforward.com/creative-writing/types-of-creative-writing

Dresner, E. (2010). Measurement-theoretic representation and computation-theoretic Realization. *The Journal of Philosophy, 107*(6), 275–292. doi:10.5840/jphil2010107622

Dreyfus, H. L. (1965). *Alchemy and artificial intelligence*. Report P-3244. RAND Corp. Accessed 2 November 2011 from http://tinyurl.com/4hczr59

Dreyfus, H. L. (1972). *What computers still can't do: A critique of artificial reason*. New York: Harper & Row.

Dulin, K. L. (1970). Using context clues in word recognition and comprehension. *The Reading Teacher, 23*(5), 440–445, 469.

Duncan, G., Dowsett, C., Claessens, A., Magnuson, K., Huston, A., Klebanov, P., & Japel, C. et al. (2007). School readiness and later achievement. *Developmental Psychology, 43*(6), 428–1446. doi:10.1037/0012-1649.43.6.1428 PMID:18020822

Dzekoe, R. (2017). Computer-based multimodal composing activities, self-revision, and L2 acquisition through writing. *Language Learning & Technology, 21*(2), 73–95.

Eberhart, M. (2012). Sensing, time, and the aural imagination in Titian's 'Venus with Organist and Dog.'. *Artibus et Historiae, 33*(65), 79–95. Retrieved from http://www.jstor.org/stable/23509712

Eco, U. (1975). Preface. In S. Chapman, U. Eco, & J.-M. Klinkenberg (Eds.), *A Semiotic Landscape/Panorama Sémiotique*. The Hague, The Netherlands: Mouton.

Eco, U. (1976). *A theory of wemiotics (Advances in Semiotics)*. Bloomington, IN: Indiana University Press. doi:10.1007/978-1-349-15849-2

Eco, U. (1979). *A theory of semiotics*. Bloomington, IN: Indiana University Press.

Eco, U. (1979). *The role of the reader: Explorations in the semiotics of texts*. Bloomington, IN: Indiana University Press.

Eco, U. (1988). On truth: A fiction. In U. Eco, M. Santambrogio, & P. Violi (Eds.), *Meaning and mental representations* (pp. 41–59). Bloomington, IN: Indiana University Press.

Eco, U. (2017). *The philosophy of Umberto Eco* (S. Beardsworth & E. A. Randall, Eds.). Chicago: Open Court Publishing.

Edelman, M. (1993). Contestable categories and public opinion. *Political Communication, 10*(3), 231–242. doi:10.1080/10584609.1993.9962981

Edelman, S. (2008a). On the nature of minds, or: Truth and consequences. *Journal of Experimental & Theoretical Artificial Intelligence, 20*(3), 181–196. doi:10.1080/09528130802319086

Edelman, S. (2008b). *Computing the Mind*. New York: Oxford University Press.

Edelman, S. (2011). Regarding reality: Some consequences of two incapacities. *Frontiers in Theoretical and Philosophical Psychology, 2*, 1–8. doi:10.3389/fpsyg.2011.00044

Eglash, R. (1999). *African fractals: Modern computing and indigenous design*. Rutgers University Press.

Ehrlich, K. (1995). *Automatic vocabulary expansion through narrative context*. Technical Report 95-09. Buffalo, NY: SUNY Buffalo Department of Computer Science.

Elgort, I. (2017). Blog posts and traditional assignments by first- and second-language writers. *Language Learning & Technology, 21*(2), 52–72.

El-Hani, C., Queiroz, J., & Stjernfelt, F. (2010). Firefly femmes fatales: A case study in the semiotics of deception. *Journal of Biosemiotics, 3*(1), 33–55. doi:10.1007/s12304-009-9048-2

Ellenberger, H. (1981). *The discovery of the unconscious*. New York, NY: Basic Books.

Ellis, M. (2008). Spectacles within doors: Panoramas of London in the 1790s. *Romanticism, 14*(2), 133–148. doi:10.3366/E1354991X0800024X

Emmeche, C., Kull, K., & Stjernfelt, F. (2002). *Reading Hoffmeyer, rethinking biology. Tartu Semiotics Library 3*. Tartu: Tartu University Press.

Entman, R. M. (1993). Framing: Toward clarification of a fractured paradigm. *Journal of Communication, 43*(4), 51–58. doi:10.1111/j.1460-2466.1993.tb01304.x

Entman, R. M. (2004). *Projections of power: Framing news, public opinion, and US foreign policy*. Chicago: The University of Chicago Press.

Eriksen, J.-M., & Stjernfelt, F. (2012). *The democratic contradictions of multiculturalism*. New York: Telos Press.

Ernest, P. (1991). *The philosophy of mathematics education*. London: Falmer Press.

Ernest, P. (1994). Social constructivism and the psychology of mathematics education. In P. Ernest (Ed.), *Constructing mathematical knowledge: Epistemology and mathematics education* (pp. 68–77). London: Falmer Press.

Essex, C. (1996). *Teaching creative writing in the elementary school*. ERIC Document No. ED391182.

Falle, J. (2004). Let's Talk Maths: A model for teaching to reveal student understandings. *Australian Senior Mathematics Journal, 18*(2), 17–27.

Fallows, J. (2015). That weirdo announcer-voice accent: Where it came from and why It went away. *The Atlantic*. Retrieved July 23, 2017 at 2:58 EDT from https://www.theatlantic.com/national/archive/2015/06/that-weirdo-announcer-voice-accent-where-it-came-from-and-why-it-went-away/395141/

Favareau, D. (2009). *Essential readings in biosemiotics: Anthology and commentary (Biosemiotics, 3)*. Dordrecht, The Netherlands: Springer. doi:10.1007/978-1-4020-9650-1

Feder, L. (1974). Adoption trauma: Oedipus myth/clinical reality. In G. Pollock & J. Ross (Eds.), *The Oedipus papers*. Madison, CT: International Universities Press.

Ferguson, C. (2010). *Online social networking goes to college: Two case studies of higher education institutions that implemented college-created social networking sites for recruiting undergraduate students*. ERIC Document No. ED516904.

Fetzer, J. H. (2008, July 11). *Computing vs. cognition: Three dimensional differences*. Unpublished.

Fetzer, J. H. (2010). *Limits to simulations of thought and action*. Talk given at the 2010 North American Conference on Computing and Philosophy (NA-CAP), Carnegie-Mellon University.

Fetzer, J. H. (1990). *Artificial intelligence: Its scope and limits*. Dordrecht, The Netherlands: Kluwer Academic Publishers. doi:10.1007/978-94-009-1900-6

Fetzer, J. H. (1994). Mental algorithms: Are minds computational systems? *Pragmatics & Cognition*, *2*(1), 1–29. doi:10.1075/pc.2.1.01fet

Fetzer, J. H. (1998). People are not computers, (most) thought processes are not computational procedures. *Journal of Experimental & Theoretical Artificial Intelligence*, *10*(4), 371–391. doi:10.1080/095281398146653

Fetzer, J. H. (2001). *Computers and cognition: Why minds are not machines*. Dordrecht, The Netherlands: Kluwer Academic Publishers. doi:10.1007/978-94-010-0973-7

Fetzer, J. H. (2011). Minds and machines: Limits to simulations of thought and action. *International Journal of Signs and Semiotic Systems*, *1*(1), 39–48. doi:10.4018/ijsss.2011010103

Feuer, A. (2011). Developing foreign language skills, competence and identity through a collaborative creative writing project. *Language, Culture and Curriculum*, *24*(2), 125–139. doi:10.1080/07908318.2011.582873

Fillmore, C. J. (2003). *Form and meaning in language*. Leland, CA: CSLI Publications.

Fillmore, C. J. (2006). Frame semantics. In D. Geeraerts (Ed.), *Cognitive Linguistics: Basic readings* (pp. 373–400). Berlin: De Gruyter. (Original work published 1982) doi:10.1515/9783110199901.373

Fillmore, C. J., Kay, P., & O'Connor, M. C. (1988). Regularity and Idiomaticity in Grammatical Constructions: The Case of Let Alone. *Language*, *64*(3), 501–538. doi:10.2307/414531

Finlayson, M. (2014). Addressing math anxiety in the classroom. *Improving Schools*, *17*(1), 99–115. doi:10.1177/1365480214521457

Fisher, P. (1991). *Making and effacing art: Modern American art in a culture of museums*. Cambridge, MA: Harvard University Press.

Flavell, J. H. (1963). *The developmental psychology of Jean Piaget*. New York, NY: Van Nostrand; doi:10.1037/11449-000

Fodor, J. A. (1975). *The language of thought*. New York: Crowell.

Fodor, J. A. (1980). Methodological solipsism considered as a research strategy in cognitive psychology. *Behavioral and Brain Sciences*, *3*(01), 63–109. doi:10.1017/S0140525X00001771

Fodor, J. A. (2000). *The mind doesn't work that way: The scope and limits of computational psychology*. Cambridge, MA: MIT Press.

Fodor, J. A. (2008). *LOT 2: The language of thought revisited*. Oxford, UK: Clarendon. doi:10.1093/acprof:oso/9780199548774.001.0001

Forbus, K. D. (2010). AI and Cognitive Science: The Past and Next 30 Years. *Topics in Cognitive Science*, 2(3), 345–356. doi:10.1111/j.1756-8765.2010.01083.x PMID:25163864

Ford, K., & Hayes, P. (1998). On computational wings. Scientific American Presents, 9(4), 78-83.

Fosshage. (2011). How do we "know" what we "know?" and change what we "know?" *Psychoanalytic Dialogues, 21*, 55-74.

Fowler, R. (1996). *Language in the news: Discourse and ideology in the press*. London: Routledge.

Frank, R. H., & Cook, P. J. (2010). *The winner-take-all society: Why the few at the top get so much more than the rest of us*. New York: Random House.

Franzén, T. (2005). *Gödel's theorem: An incomplete guide to its use and abuse*. Wellesley, MA: A.K. Peters. doi:10.1201/b10700

Fraser, D. (2006). The creative potential of metaphorical writing in the literacy classroom. *English Teaching*, 5(2), 93–108.

Freeman, W. (1999). Consciousness, intentionality, and causality. *Journal of Consciousness Studies*, 6, 143–172.

Freud, S. (1900). *The interpretation of dreams* (J. Strachey, Trans.). New York, NY: Basic Books.

Gallese, V. (2001). The "Shared Manifold" hypothesis. From mirror neurons to empathy. *Journal of Consciousness Studies*, 8, 5–7, 33.

Gallese, V. (2009). Mirror neurons, embodied simulation, and the neural basis of social identification. *Psychoanalytic Dialogues*, 19(5), 519–536. doi:10.1080/10481880903231910

Gammon, G. (1989). You won't lay an egg with the bald-headed chicken. B. C. *The Journal of Special Education*, 13(2), 183–187.

Gandy, R. (1988). The confluence of ideas in 1936. In R. Herken (Ed.), The universal Turing Machine: A half-century survey (2nd ed.; pp. 51-102). Vienna: Springer-Verlag.

Gardner, M. (1999). *The annotated Alice*. New York, NY: Norton.

Gelman, R., & Gallistel, C. R. (1986). *The child's understanding of number*. Cambridge, MA: Harvard University Press.

Gimlin, D. (2007). What is 'body work'? A review of the literature. *Sociology Compass*, 1(1), 353–370. doi:10.1111/j.1751-9020.2007.00015.x

Ginot, E. (2007). Intersubjectivity and neuroscience: Understanding enactments and their therapeutic significance within emerging paradigms. *Psychoanalytic Psychology*, 24(2), 317–332. doi:10.1037/0736-9735.24.2.317

Ginsberg, A. (1956) *Howl*. Retrieved July 24, 2017 at 10:27 EDT from https://www.poetryfoundation.org/poems/49303/howl

Ginsburg, H. P., Inoue, N., & Seo, K. H. (1999). Young children doing mathematics: Observations of everyday activities. In J. V. Copley (Ed.), *Mathematics in the early years* (pp. 88–99). Reston, VA: NCTM.

Gödel, K. (1931). Über formal unentscheidbare Sätze der Principia Mathematica und verwandter Systeme, Teil I. *Monatshefte für Mathematik und Physik, 38*(38), 173–189. doi:10.1007/BF01700692

Goffman, E. (1974). *Frame analysis: An essay on the organization of experience.* London: Harper and Row.

Goguen, J. (1999). An introduction to algebraic semiotics, with application to user interface design. In C. L. Nehaniv (Ed.), Lecture Notes in Computer Science: Vol. 1562. *Computation for metaphors, analogy, andaAgents* (pp. 242–291). Berlin: Springer. doi:10.1007/3-540-48834-0_15

Goldfain, A. (2006). Embodied enumeration: Appealing to activities for mathematical explanation. In M. Beetz, K. Rajan, M. Thielscher, & R. Bogdan Rusu (Eds.), *Cognitive robotics: Papers from the AAAI Workshop (CogRob2006)* (pp. 69-76). Technical Report WS-06-03. Menlo Park, CA: AAAI Press.

Goldfain, A. (2008). *A computational theory of early mathematical cognition* (PhD dissertation). Buffalo, NY: SUNY Buffalo Department of Computer Science & Engineering. Retrieved from http://www.cse.buffalo.edu/sneps/Bibliography/GoldfainDissFinal.pdf

Gonçalves, F., Campos, P., & Garg, A. (2015). Understanding UI design for creative writing: A pilot evaluation. In *Adjunct Proceedings of the INTERACT 2015 Conference* (179-186). University of Bamberg Press.

Gopnik, A. (2009). *The philosophical baby: What children's minds tell us about truth, love, and the meaning of life.* New York: Farrar, Straus and Giroux.

Gorilla Foundation. (2017). Retrieved July 12, 2017 at 12:33 PM EDT from http://www.koko.org/sign-language

Goswami, A. (1995). *The self-aware universe.* Los Angeles, CA: Tarcher.

Gould, S. J., & Eldredge, N. (1977). Punctuated equilibria: The tempo and mode of evolution reconsidered. *Paleobiology, 3*(02), 115–151. doi:10.1017/S0094837300005224

Greco Morasso, S. (2012). Contextual frames and their argumentative implications: A case study in media argumentation. *Discourse Studies, 14*(2), 197–216. doi:10.1177/1461445611433636

Guba, E. G., & Lincoln, Y. S. (2005). Paradigmatic controversies, contradictions and emerging confluences. In N. K. Denzin & Y. S. Lincoln (Eds.), *The SAGE handbook of qualitative research* (2nd ed.; pp. 191–215). Thousand Oaks, CA: Sage.

Gudwin, R., & Queiroz, J. (2005). Towards an introduction to computational semiotics. *Proceedings of the 2005 IEEE International Conference on Integration of Knowledge Intensive Multi-Agent Systems*, 393-398. doi:10.1109/KIMAS.2005.1427113

Gundle, S. (1997). *Bellissima: Feminine beauty and the idea of Italy.* New Haven, CT: Yale University Press.

Guzzetti, B., & Gamboa, M. (2005). Online journaling: The informal writings of two adolescent girls. *Research in the Teaching of English, 40*(2), 168–206.

Habib, S. (1983). Emulation. In A. Ralston & E. D. Reilly Jr., (Eds.), *Encyclopedia of Computer Science and Engineering* (2nd ed.; pp. 602–603). New York: Van Nostrand Reinhold.

Häfliger, M. (2015, January 8). Satire-Freiheit relativiert; Leuthard erntet Empörung. *Neue Zürcher Zeitung,* 10.

Haines, S. (2015). Picturing words: Using photographs and fiction to enliven writing for ELL students. *Schools: Studies in Education, 12*(1), 9–32. doi:10.1086/680692

Hammer, E. (1995). Reasoning with sentences and diagrams. *Notre Dame Journal of Formal Logic, 35,* 73–87.

Hammer, E., & Shin, S. (1996). Euler and the role of visualization in logic. In J. Seligman & D. Westerståhl (Eds.), *Logic, language and computation* (Vol. 1). Stanford, CA: CSLI Publications.

Hammer, E., & Shin, S. (1998). Euler's visual logic. *History and Philosophy of Logic, 19*(1), 1–29. doi:10.1080/01445349808837293

Harnad, S. (1990). The symbol grounding problem. *Physica D. Nonlinear Phenomena, 42*(1-3), 335–346. doi:10.1016/0167-2789(90)90087-6

Harris, P. (2000). *The work of the imagination*. Malden, MA: Blackwell.

Haskell, F., & Penny, N. (1981). *Taste and the antique: The lure of classical sculpture 1500-1900*. London, UK: Yale University Press.

Hayman, D. (1999). *The life of Jung*. New York, NY: W.W. Norton.

Heims, S. (1991). *The cybernetics group*. Cambridge, MA: The MIT Press.

Henry, L. (2003). *Creative writing through wordless picture books*. ERIC Document No. ED477997.

Hesmondhalgh, D. (2013). *Why music matters*. John Wiley & Sons.

Hilpinen, R. (1992). On Peirce's philosophical Logic: Propositions and their objects. *Transactions of the Charles S. Peirce Society, 28*(3), 467.

Hintikka, J. (1997). *Lingua universalis vs. calculus ratiocinator: An ultimate presupposition of twentieth-century philosophy*. Dordrecht, The Netherlands: Kluwer. doi:10.1007/978-94-015-8601-6

Hobbes, T. (1651). *Leviathan*. Indianapolis, IN: Bobbs-Merrill Library of Liberal Arts.

Hochschild, A. (1983). *The managed heart*. Berkeley, CA: University of California Press.

Hodges, B. (1999). Electronic books: Presentation software makes writing more fun. *Learning and Leading with Technology, 27*(1), 18–21.

Hoffmeyer, J. (Ed.). (2008). *A legacy for living systems. Gregory Bateson as precursor to biosemiotics*. Berlin: Springer.

Hoffmeyer, J., & Stjernfelt, F. (2015). The great chain of semiosis: Investigating the steps in the evolution of biosemiotic competence. *Biosemiotics, 9*(1), 7–29. doi:10.1007/s12304-015-9247-y

Hofstadter, D. (2007). *I am a strange loop*. New York: Basic Books.

Hong, E., Peng, Y., & O'Neil, H. Jr. (2014). Activities and accomplishments in various domains: Relationships with creative personality and creative motivation in adolescence. *Roeper Review, 36*(2), 92–103. doi:10.1080/02783193.2014.884199

Horst, S. (2009). The computational theory of mind. In E. N. Zalta (Ed.), Stanford Encyclopedia of Philosophy. Accessed 6 May 2011 from http://plato.stanford.edu/archives/win2009/entries/computational-mind/

Howarth, P. (2007). Creative writing and Schiller's aesthetic education. *Journal of Aesthetic Education, 41*(3), 41–58. doi:10.1353/jae.2007.0025

Huemer, M. (2011). Sense-data. In E. N. Zalta (Ed.), Stanford Encyclopedia of Philosophy. Accessed 6 May 2011 from http://plato.stanford.edu/archives/spr2011/entries/sense-data/

Hufferd-Ackles, K., Fuson, K. C., & Sherin, M. G. (2004). Describing levels and components of a math-talk learning community. *Journal for Research in Mathematics Education, 35*(2), 81–116. doi:10.2307/30034933

Hume, D. (1739). A treatise of human nature (L. A. Selby-Bigge, Ed.). London: Oxford University Press.

Hume, D. (1777). An enquiry concerning human understanding (L. A. Selby-Bigge, Ed.). London: Oxford University Press. doi:10.1093/oseo/instance.00046350

Hung, H., & Yuen, S. (2010). Educational use of social networking technology in higher education. *Teaching in Higher Education*, *15*(6), 703–714. doi:10.1080/13562517.2010.507307

Hurford, J. (2007). *The origin of meaning*. Oxford, UK: Oxford University Press.

Hurme, T., & Jarvela, S. (2005). Students' activity in computer-supported collaborative problem solving in mathematics. *International Journal of Computers for Mathematical Learning*, *10*(1), 49–73. doi:10.1007/s10758-005-4579-3

Hussey, A. (2015, January 8). French humor, turned into tragedy. *The New York Times*, p. 23.

Hutchins, E. (1996). *Cognition in the wild*. Cambridge, MA: MIT Press.

Hutton, K., Chen, B., & Moss, J. (2013). *A knowledge building discourse analysis of proportional reasoning in grade 1*. Paper presented at the 2013 Knowledge Building Summer Institute, Puebla, Mexico.

Hyland, K. (1993). ESL computer writers: What can we do to help? *System*, *2*(1), 21–30. doi:10.1016/0346-251X(93)90004-Z

Hylton, W. S. (2013, June 13). *How James Turrell knocked the art world off its feet*. Retrieved from http://www.nytimes.com/2013/06/16/magazine/how-james-turrell knocked-the-art-world-off-its-feet.html

Imm, K., & Stylianou, A. (2012). Talking mathematically: An analysis of discourse. *The Journal of Mathematical Behavior*, *31*(1), 130–148. doi:10.1016/j.jmathb.2011.10.001

Incardona, L. (2012). *Semiotica e web semantico: Basi teoriche e metodologiche per la semiotica computazionale* [Semiotics and semantic web: Theoretical and methodological foundations for computational semiotics] (PhD dissertation). University of Bologna.

Investopedia. (2017). *Staycation*. Retrieved on July 24, 2017 at 4:52 PM from http://www.investopedia.com/terms/s/staycation.asp

Jackendoff, R. (2002). *Foundations of language: Brain, meaning, grammar, evolution*. Oxford, UK: Oxford University Press. doi:10.1093/acprof:oso/9780198270126.001.0001

Jacobsen, M., Lock, J., & Friesen, S. (2013). *Strategies for engagement: Knowledge building and intellectual engagement in participatory learning environments*. Toronto: Education Canada.

James Turrell's Roden Crater. (2013). Los Angeles County Museum of Art. Retrieved from https://vimeo.com/67926427

James, D. (2008). A short take on evaluation and creative writing. *Community College Enterprise*, *14*(1), 79–82.

Jaworski, P. (2013). In defense of fakes and artistic treason: Why visually-indistinguishable duplicates of paintings are just as good as the originals. *The Journal of Value Inquiry*, *47*(4), 391–405. doi:10.1007/s10790-013-9383-z

Jayaron, J., & Abidin, M. J. (2016). A pedagogical perspective on promoting English as a foreign language writing through online forum discussions. *English Language Teaching*, *9*(2), 84–101. doi:10.5539/elt.v9n2p84

Jaynes, J. (1976). *The origin of consciousness in the breakdown of the bicameral mind*. Boston, MA: Houghton Mifflin.

Jelić, A. (2015). Designing "pre-reflective" architecture. *Ambiances, Experimental Simulation*. Retrieved from http://ambiances.revues.org/628

Johnson-Laird, P. N. (1983). *Mental models: Towards a cognitive science of language, inference, and consciousness.* Cambridge, MA: Harvard University Press.

Johnson-Laird, P. N. (1988). *The computer and the mind: An introduction to cognitive science.* Cambridge, MA: Harvard University Press.

Johnson, M. (1987). *The body in the mind: The bodily basis of meaning, imagination and reason.* Chicago: University of Chicago Press.

Jones, D. (2010). *Animal liberation.* Paper presented at the 2010 North American Conference on Computing and Philosophy (NA-CAP), Carnegie-Mellon University.

Jonesa, A., & Issroffb, K. (2005). Learning technologies: Affective and social issues in computer-supported collaborative learning. *Computers & Education, 44*(4), 395–408. doi:10.1016/j.compedu.2004.04.004

Julesz, B. (1971). *Foundations of Cyclopean perception.* Chicago: University of Chicago Press.

Jung, C. (1956). Symbols of transformation. In *Collected works.* London, UK: Routledge & Kegan Paul.

Jung, C. (1961). *Memories, dreams, reflections.* New York, NY: Random House.

Jung, C. (2009). *The red book.* New York, NY: Norton.

Kauffman, S. (2011). *There are more uses for a screwdriver than you can calculate.* Retrieved from http://www.wbur.org/npr/135113346/there-are-more-uses-for-a-screwdriver-than-you-can-calculate

Kauffman, L. K. (2001). The mathematics of Charles Sanders Peirce. *Cybernetics & Human Knowing, 8,* 79–110.

Kaufman, J., Gentile, C., & Baer, J. (2005). Do gifted student writers and creative writing experts rate creativity the same way? *Gifted Child Quarterly, 49*(3), 260–265. doi:10.1177/001698620504900307

Kay, P., & Kempton, W. (1984). *What is the Sapir Whorf Hypothesis?* American Anthropological Association. Retrieved July 23, 2017 at 12:42 PM from http://www1.icsi.berkeley.edu/~kay/Kay&Kempton.1984.pdf

Keiner, J. (1996). *Real audiences-worldwide: A case study of the impact of WWW publication on a child writer's development.* ERIC Document No. ED427664.

Keneally, T. (1982). *Schindler's List.* Hodder and Stoughton.

Kerr, J. (1995). *A most dangerous method.* New York, NY: Vintage Books/Random House.

Ketner, K. L. (1988). Peirce and Turing: Comparisons and conjectures. *Semiotica, 68*(1/2), 33–61.

Khalaf, R. S. (2014). Lebanese youth narratives: A bleak post-war landscape. *Compare: A Journal of Comparative Education, 44*(1), 97–116. doi:10.1080/03057925.2013.859899

Kilpatrick, J., Swafford, J., & Findell, B. (2001). *Adding it up: Helping children learn mathematics.* Washington, DC: National Academies Press.

Kirby, S. (2000). Syntax without natural selection: How compositionality emerges from vocabulary in a population of learners. In C. Knight (Ed.), *The evolutionary emergence of language: Social function and the origins of linguistic form* (pp. 303–323). Cambridge, UK: Cambridge University Press. doi:10.1017/CBO9780511606441.019

Kiryushchenko, V. (2012). The visual and the virtual in theory, life and scientific practice: The case of Peirce's quincuncial map projection. In M. Bockarova, M. Danesi, & R. Núñez (Eds.), *Semiotic and cognitive science essays on the nature of mathematics.* Munich: Lincom Europa.

Kleene, S. C. (1952). *Introduction to metamathematics*. Princeton, NJ: D. Van Nostrand.

Kleene, S. C. (1967). *Mathematical logic*. New York: Wiley.

Klein, M. (1932). *The psycho-analysis of children*. London, UK: Hogarth.

Klibanoff, R. S., Levine, S., Huttenlocher, J., Hedges, L. V., & Vasilyeva, M. (2006). Preschool children's mathematical knowledge: The effect of teacher "Math Talk." *Developmental Psychology, 42*(1), 59–69. doi:10.1037/0012-1649.42.1.59 PMID:16420118

Knizek, I. (1993). Walter Benjamin and the mechanical reproducibility of art works revisited. *British Journal of Aesthetics, 33*(4), 357–366. doi:10.1093/bjaesthetics/33.4.357

Knowledge Building Gallery. (in press). Knowledge building in math: How do we help students to think like mathematicians? In *Leading Student Achievement: Networks for Learning* (pp. 51-79). Academic Press.

Knuth, D. E. (1972). Ancient Babylonian algorithms. *Communications of the ACM, 15*(7), 671–677. doi:10.1145/361454.361514

Knuth, D. E. (1973). Basic concepts: Algorithms. In *The Art of Computer Programming* (2nd ed.; pp. xiv-9). Reading, MA: Addison-Wesley.

Koblin, A. (2008-2015). *Aaron Koblin Website: Information*. Retrieved on April 1, 2017 from http://www.aaronkoblin.com/info.html

Kohut, H. (1971). *The analysis of the self*. New York, NY: International Universities Press.

Kohut, H. (1977). *The restoration of the self*. New York, NY: International Universities Press.

Kopp, B., & Mandl, H. (2011). Fostering argument justification using collaboration scripts and content schemes. *Learning and Instruction, 21*(5), 636–649. doi:10.1016/j.learninstruc.2011.02.001

Kornhaber, D. (2009). Regarding the Eidophusikon: Spectacle, scenography, and culture in eighteenth century England. *TAJ - Theatre Arts Journal: Studies in Scenography and Performance, 1*(1). Retrieved from http://myaccess.library.utoronto.ca/login?url=http://search.proquest.com/docview/1086486005?accountid=14771

Kothari, A. (2010). The framing of the Darfur Conflict in *The New York Times*: 2003-2006. *Journalism Studies, 11*(2), 209-224.

Kristof, N. (2015, January 8). Satire, Terrorism and Islam. *The New York Times*, p. 23.

Kubler, G. (2008). *The shape of time: Remarks on the history of things*. New Haven, CT: Yale University Press. (Original work published 1962)

Kugel, P. (2002). Computing machines can't be intelligent (…and Turing said so). *Minds and Machines, 12*(4), 563–579. doi:10.1023/A:1021150928258

Kuhn, T. (1962). *The structure of scientific revolutions*. Chicago, IL: University of Chicago Press.

Kull, K., Deacon, T., Emmeche, C., Hoffmeyer, F., & Stjernfelt, F. (2009). Theses on biosemiotics: Prolegomena to a theoretical biology. *Biological Theory, 4*(2), 167–173. doi:10.1162/biot.2009.4.2.167

Kull, K., & Velmezova, E. (2014). What is the main challenge for contemporary semiotics? *Sign Systems Studies, 42*(4), 530–548. doi:10.12697/SSS.2014.42.4.06

Kulpa, Z. (2004). On diagrammatic representation of mathematical knowledge. In A. Sperti, G. Bancerek, & A. Trybulec (Eds.), *Mathematical knowledge management*. New York: Springer. doi:10.1007/978-3-540-27818-4_14

Kunstmuseum Wolfsburg. (2009, November 25). *James Turrell: The Wolfsburg Project*. Retrieved from https://www.youtube.com/watch?v=QWekIcZaKns

Lakatos, I. (1970). Criticism and the methodology of scientific research programmes. *Proceedings of the Aristotelian Society, 69*(1968–1969), 149-186.

Lakoff, G. (2001). *Metaphors of terror*. Retrieved July 12, 2016, from http://www.press.uchicago.edu/sites/daysafter/911lakoff.html

Lakoff, G. (1987). *Women, fire and dangerous things: What categories reveal about the mind*. Chicago: University of Chicago Press. doi:10.7208/chicago/9780226471013.001.0001

Lakoff, G. (2012). Explaining embodied cognition results. *Topics in Cognitive Science, 4*(4), 773–785. doi:10.1111/j.1756-8765.2012.01222.x PMID:22961950

Lakoff, G., & Johnson, M. (1980). *Metaphors we live by*. Chicago, IL: University of Chicago Press.

Lakoff, G., & Johnson, M. (1999). *Philosophy in flesh: The embodied mind and Its challenge to western thought*. New York: Basic.

Lakoff, G., & Johnson, M. (1999). *Philosophy in the flesh: The embodied mind and its challenge to Western thought*. New York, NY: Basic Books.

Lakoff, G., & Núñez, R. (2000). *Where mathematics comes from: How the embodied mind brings mathematics into being*. New York: Basic Books.

Landauro, I., Bisserbe, N., & Gauthier-Villars, D. (2015, January 8). Terror strikes heart of Paris. Gunmen attack French magazine, killing at least 12. *The Wall Street Journal*.

Lane, R. D., Ryan, L., Nadel, L., & Greenberg, L. (2015). Memory reconsolidation, emotional arousal, and the process of change in psychotherapy: New insights from brain science. *Behavioral and Brain Sciences*, 38. PMID:24827452

Langer, S. K. (1948). *Philosophy in a new key*. New York: Mentor Books.

Lanius, R., Vermetten, E., & Pain, C. (2010). *The impact of early life trauma on health and disease: The hidden epidemic*. Cambridge, UK: Cambridge University Press; doi:10.1017/CBO9780511777042

Latham, A. J. (1995). Packaging woman: The concurrent rise of beauty pageants, public bathing, and other performances of female "nudity.". *Journal of Popular Culture, 29*(3), 149–167. doi:10.1111/j.0022-3840.1995.00149.x

Lazarus, J., & Roulet, G. (2013). *Blended math-talk community: Extending the boundaries of classroom collaboration*. OAME/AOEM Gazette.

Lecheler, S., & de Vreese, C. H. (2012). What a difference a day makes? The effects of repetitive and competitive news framing over time. *Communication Research, 40*(2), 147–175. doi:10.1177/0093650212470688

Lee, K., & Ranta, L. (2014). Facebook: Facilitating social access and language acquisition for international students? *TESL Canada Journal, 31*(2), 22–50. doi:10.18806/tesl.v31i2.1175

Lehrer, R., & Chazan, D. (1998). *Designing learning environments for developing understanding of geometry and space*. Mahwah, NJ: Lawrence Erlbaum Associates.

Levesque, H. (2012). *Thinking as Computation: A First Course*. Cambridge, MA: MIT Press. doi:10.7551/mitpress/9780262016995.001.0001

Lévi-Strauss, C. (1977). *Structural anthropology* (C. Jacobson & B. G. Schoepf, Trans.). Harmondsworth, UK: Penguin.

Lobina, D. J. (2010). Recursion and the competence/performance distinction in AGL tasks. *Language and Cognitive Processes*. Accessed 4 November 2011 from http://tinyurl.com/Lobina2010

Lobina, D. J., & García-Albea, J. E. (2009). Recursion and cognitive science: Data structures and mechanisms. In N. Taatgen & H. van Rijn (Eds.), *Proceedings of the 31st Annual Conference of the Cognitive Science Society* (pp. 1347-1352). Academic Press.

Locke, J. (1690). *An essay concerning human understanding*. Chapt. XXI, Book 4.

Lohr, S. (2010, September 24). Frederick Jelinek, who gave machines the key to human speech, dies at 77. *New York Times*, p. B10.

Lorusso, A. M. (2015). *Cultural semiotics*. London: Palgrave Macmillan. doi:10.1057/9781137546999

Lotman, J. M. (1990). *Universe of the mind: A semiotic theory of culture* (A. Shukman, Trans.). Bloomington, IN: Indiana University Press.

Lucas, J. R. (1961). Minds, machines and Gödel. *Philosophy (London, England)*, *36*(137), 112–127. doi:10.1017/S0031819100057983

Lunberry, C. (2009). Soliloquies of silence: James Turrell's theatre of installation. *Mosaic*, *42*(1), 33–50. Retrieved from http://myaccess.library.utoronto.ca/login?url=http://search.proquest.com/docview/205371831?accountid=14771

Lyons-Ruth, K., Bruschweiler-Stern, N., Harrison, A. M., Morgan, A. C., Nahum, J. P., Sander, L., & Tronick, E. Z. et al. (1998). Implicit relational knowing: Its role in development and psychoanalytic treatment. *Infant Mental Health Journal*, *19*(3), 282–289. doi:10.1002/(SICI)1097-0355(199823)19:3<282::AID-IMHJ3>3.0.CO;2-O

Makuuchi, M., Bahlmann, J., Anwander, A., & Friederici, A. D. (2009). Segregating the core computational faculty of human language from working memory. *Proceedings of the National Academy of Sciences*, *106*(20), 8362-8367. doi:10.1073/pnas.0810928106

Malesky, L., & Peters, C. (2012). Defining appropriate professional behavior for faculty and university students on social networking websites. *Higher Education: The International Journal of Higher Education and Educational Planning*, *63*(1), 135–151. doi:10.1007/s10734-011-9451-x

Mancia, M. (2006). Implicit memory and early unrepressed unconscious: Their role in the therapeutic process (How the neurosciences can contribute to psychoanalysis). *The International Journal of Psycho-Analysis*, *87*, 83–103. PMID:16635862

Mann, G. A. (2010). A machine that daydreams. *IFIP Advances in Information and Communication Technology*, *333*, 21–35. doi:10.1007/978-3-642-15214-6_3

Margulis, L. (1995). Gaia is a tough bitch. In J. Brockman (Ed.), *The third culture: Beyond the scientific revolution*. New York: Simon and Schuster. Retrieved from https://www.edge.org/conversation/lynn_margulis-chapter-7-gaia-is-a-tough-bitchhttp

Markov, A. A. (1954). Theory of algorithms. Tr. Mat. Inst. Steklov, 42.

Marks-Tarlow, T. (2003). The certainty of uncertainty. *Psychological Perspectives*, *45*(1), 118–130. doi:10.1080/00332920308403045

Marks-Tarlow, T. (2008a). *Psyche's veil: Psychotherapy, fractals and complexity*. London, UK: Routledge.

Marks-Tarlow, T. (2008b). Alan Turing meets the sphinx: Some new and old riddles. *Chaos & Complexity Letters*, *3*(1), 83–95.

Marks-Tarlow, T. (2008c). Riddle of the sphinx: A paradox of self-reference revealed and reveiled. In *Reflecting interfaces: The complex coevolution of information technology ecosystems*. Hershey, PA: Idea Group; doi:10.4018/978-1-59904-627-3.ch002

Marks-Tarlow, T. (2012). *Clinical intuition in psychotherapy: The neurobiology of embodied response*. New York, NY: Norton.

Marks-Tarlow, T. (2013). *Awakening clinical intuition*. New York, NY: Norton.

Marks-Tarlow, T. (2015). The nonlinear dynamics of clinical intuition. *Chaos & Complexity Letters*, *8*(2-3), 1–24.

Marks-Tarlow, T. (2017). I am an avatar of myself: Fantasy, trauma and self deception. *American Journal of Play*, *9*(2), 169–201.

Marr, D. (1982). *Vision: A computational investigation into the human representation and processing of visual information*. New York: W.H. Freeman.

Martinovic, D. (2004). *Communicating mathematics online: The case of online help* (PhD thesis). University of Toronto.

Marty, R. (1990). *L' Algèbre des signes: Essai de sémiotique scientifique d'après Charles Sanders Peirce (Foundations of Semiotics, 24)*. Amsterdam: J. Benjamins Publishing Co. doi:10.1075/fos.24

Masson, J. (1985). *The assault on truth: Freud's suppression of the seduction theory*. New York: Penguin Press.

Matts, T., & Tynan, A. (2012). Earthwork and eco-clinic: Notes on James Turrell's 'Roden Crater' Project. *Design Ecologies*, *2*(2), 188–211. doi:10.1386/des.2.2.188_1

McCorduck, P. (1990). *Aaron's code: Meta-Art, artificial intelligence, and the work of Harold Cohen*. New York: W.H. Freeman.

McCulloch, W. S., & Pitts, W. H. (1943). A logical calculus of the ideas immanent in nervous activity. *The Bulletin of Mathematical Biophysics*, *5*(4), 115–133. doi:10.1007/BF02478259

McGee, K. (2011). Review of K. Tanaka-Ishii's *Semiotics of programming*. *Artificial Intelligence*, *175*, 930–931. doi:10.1016/j.artint.2010.11.023

McGilchrist, I. (2009). The Master and his emissary: The divided brain and the making of the Western world. New Haven, CT: Yale University Press.

McNutt, J. K. (1990). Plaster casts after antique sculpture: Their role in the elevation of public taste and in American art instruction. *Studies in Art Education*, *31*(3), 158–167. doi:10.2307/1320763

McWhorter, J. (2003). *Doing our own thing: The degradation of language and music and why we should like care*. New York: Gotham Books.

Mercer, N., & Sams, C. (2006). Teaching children how to use language to solve maths problems. *Language and Education*, *20*(6), 507–528. doi:10.2167/le678.0

Merriam Webster Dictionary. (2017). Retrieved various dates and times from http://www.merriam-webster.com/dictionary/

Michell, S. (1988). *Relational concepts in psychoanalysis: An integration*. Cambridge, MA: Harvard University Press.

Mill, J. S. (1843). *A system of logic.* Accessed 4 November 2011 from http://tinyurl.com/MillSystemLogic

Miller, G. A., Galanter, E., & Pribram, K. H. (1960). *Plans and the structure of behavior.* New York: Henry Holt. doi:10.1037/10039-000

Minsky, M. (1975). A framework for representing knowledge. In P. H. Winston (Ed.), *The psychology of computer vision* (pp. 211–277). New York: McGraw-Hill.

Mitchell, C. J., De Houwer, J., & Lovibond, P. F. (2009). The propositional nature of human associative learning. *Behavioral and Brain Sciences*, *32*(2), 183–198. doi:10.1017/S0140525X09000855 PMID:19386174

Modell, A. (2003). *Imagination and the meaningful brain.* Cambridge, MA: MIT Press.

Mohapatra, S., & Mohanty, S. (2017). Assessing the overall value of an online writing community. *Education and Information Technologies*, *22*(3), 985–1003. doi:10.1007/s10639-016-9468-y

Mongré, P. (1897). Sant' Ilario. Thoughts from Zarathustra's landscape. Leipzig: C.G. Nauman.

Monte, C., & Sollod, R. (2003). *Beneath the mask: An introduction to theories of personality.* New York: John Wiley & Sons.

Morgan, C. (2014). Mathematical language. In S. Lerman (Ed.), *Encyclopedia of Mathematics Education* (pp. 388–391). New York: Springer.

Morgan, W. (2006). Poetry makes nothing happen: Creative writing and the English classroom. *English Teaching*, *5*(2), 17–33.

Morris, C. (1938). *Foundations of the theory of signs.* Chicago: University of Chicago Press.

Morris, C. W. (1938). *Foundation of the theory of signs* (1st ed.). Chicago: University of Chicago Press.

Morris, R., & Ward, G. (Eds.). (2005). *The cognitive psychology of planning.* London: Psychology Press.

Mosco, V., & McKercher, C. (2009). *The Laboring of communication: Will knowledge workers of the world unite?* Lanham, MD: Rowman & Littlefield.

Moss, J., & Beatty, R. (2006). Knowledge building in mathematics: Supporting collaborative learning in pattern problems. *International Journal of Computer-Supported Collaborative Learning*, *1*(4), 441–465. doi:10.1007/s11412-006-9003-z

Moss, J., & Beatty, R. (2010). Knowledge building and mathematics: Shifting the responsibility for knowledge advancement and engagement. *Canadian Journal of Learning and Technology*, *36*(1). doi:10.21432/T24G6B

Moss, J., Beatty, R., McNab, S. L., & Eisenband, J. (2006). The potential of geometric sequences to foster young students ability to generalize in mathematics. Paper Presented at the *Annual Meeting of the American Educational Research Association*, San Francisco, CA.

Mountz, R. (2011) *Let's talk math: Investigating the use of mathematical discourse, or "math talk," in a lower-achieving elementary class* (Dissertation). Penn State University.

Mueller, E. T. (1990). *Daydreaming in humans and machines: A computer model of the stream of thought.* Norwood, NJ: Ablex.

Mumford, L. (1967). *The myth of the machine: Technics and human development.* New York: Harcourt, Brace, Jovanovich.

Nadin, M. (1986). Pragmatics in the semiotic framework. In H. Stachowiak (Ed.), Pragmatik, II The rise of pragmatic thought in the 19th and 20th centuries (pp. 148–170). Academic Press.

Nadin, M. (2007). Semiotic machine. *Public Journal of Semiotics, 1*(1), 85-114. Accessed 5 May 2011 from http://www.nadin.ws/archives/760

Nadin, M. (2010). Anticipation and dynamics: Rosen's anticipation in the perspective of time. *International Journal of General Systems, 39*(1), 3-33.

Nadin, M. (2010). Remembering Peter Bøgh Andersen: Is the computer a semiotic machine? A discussion never finished. *Semiotix: New Series*, (1). Retrieved from http://www.semioticon.com/semiotix/2010/03/

Nadin, M. (1977). Sign and fuzzy automata. *Semiosis, 1*, 5.

Nadin, M. (1980). The logic of vagueness and the category of synechism. *The Monist, 63*(3), 351–363. doi:10.5840/monist198063326

Nadin, M. (1983). The logic of vagueness and the category of synechism. In E. Freeman (Ed.), *The relevance of Charles Peirce. La Salle* (pp. 154–166). The Monist Library of Philosophy.

Nadin, M. (1991). *Mind—anticipation and chaos.* Stuttgart, Germany: Belser.

Nadin, M. (1997). *The civilization of illiteracy.* Dresden, Germany: Dresden University Press.

Nadin, M. (2011). Information and semiotic processes. The semiotics of computation. *Cybernetics & Human Knowing, 18*(1-2), 153–175.

Nadin, M. (2012). Reassessing the foundations of semiotics: Preliminaries. *International Journal of Signs and Semiotic Systems, 2*(1), 43–75. doi:10.4018/ijsss.2012010101

Nadin, M. (2014). Semiotics is fundamental science. In M. Jennex (Ed.), *Knowledge discovery, transfer, and management in the information age* (pp. 76–125). Hershey, PA: IGI Global. doi:10.4018/978-1-4666-4711-4.ch005

Nadin, M. (2016). *Anticipation and the brain, Anticipation and medicine.* Cham, Switzerland: Springer International Publishers.

Nadin, M. (2017). Semiotic engineering – An opportunity or an opportunity missed. In *Conversations around semiotic engineering.* Berlin: *Springer.*

Nason, R. A., Woodruff, E., & Lesh, R. (2002). Fostering authentic, sustained and progressive mathematical knowledge-building activity in CSCL communities. In B. Barton, C. Irwin, M. Pfannkuch, & M. O. J. Thomas (Eds.), *Mathematics education in the South Pacific: Proceedings of the Annual Conference of the Mathematics Education Research Group of Australasia* (pp. 504-511). Sydney, Australia: MERGA.

Nason, R. A., Brett, C., & Woodruff, E. (1996). Creating and maintaining knowledge-building communities of practice during mathematical investigations. In P. Clarkson (Ed.), *Technology in mathematics education* (pp. 20–29). Melbourne: Mathematics Education Research Group of Australia.

Nason, R. A., & Woodruff, E. (2004). Online collaborative learning in mathematics: Some necessary innovations. In T. Roberts (Ed.), *Online learning: Practical and theoretical considerations* (pp. 103–131). Hershey, PA: Idea Group Inc. doi:10.4018/978-1-59140-174-2.ch005

Nason, R., & Woodruff, E. (2002). New ways of learning mathematics: Are we ready for it? *Proceedings of the International Conference on Computers in Education.* doi:10.1109/CIE.2002.1186334

Nathan, M. J., & Knuth, E. J. (2003). A study of whole classroom mathematical discourse and teacher change. *Cognition and Instruction, 21*(2), 175–207. doi:10.1207/S1532690XCI2102_03

National Council of Teachers of Mathematics. (2000). *Principles and standards for school mathematics*. Reston, VA: National Council of Teachers of Mathematics.

Neumann, E. (1954/1993). *The origins and history of consciousness*. Princeton, NJ: Princeton.

Newell, A., Shaw, J. C., & Simon, H. A. (1958). Elements of a theory of human problem solving. *Psychological Review*, *65*(3), 151–166. doi:10.1037/h0048495

Newell, A., & Simon, H. A. (1976). Computer science as empirical inquiry: Symbols and search. *Communications of the ACM*, *19*(3), 113–126. doi:10.1145/360018.360022

Nicolelis, M. A. L., & Shuler, M. (2001). Thalamocortical and corticocortical interactions in the somatosensory system. In M. A. L. Nicolelis (Ed.), Progress in brain research: Vol. 130. *Advances in neural population coding* (pp. 89–110). Amsterdam: Elsevier Science. doi:10.1016/S0079-6123(01)30008-0

Nietzsche, F. (1975). *Kritische Gesamtausgabe Briefwechsel* (G. Colli & M. Montinari, Eds.). Berlin: Walter de Gruyter.

Nietzsche, F. (1999). *The birth of tragedy and other writings (Cambridge texts in the history of philosophy)*. Cambridge, UK: Cambridge University Press. (Original work published 1871)

Norman, D. A. (1993). *Things that make us smart*. Reading: Addison-Wesley Publishing.

Nöth, W. (2003). Semiotic machines. *S.E.E.D. Journal (Semiotics, Evolution, Energy, and Development)*, *3*(3), 81-99. Retrieved from http://www.library.utoronto.ca/see/SEED/Vol3-3/Winfried.pdf

Nöth, W. (1985). *Handbuch der Semiotik. Stuttgart: Metzler. Published in English (1995), Handbook of semiotics (Advances in Semiotics)*. Bloomington, IN: University of Indiana Press.

Nuns, C. A. A., Nunes, M. M. R., & Davis, C. (2003). Assessing the inaccessible: Metacognition and attitudes. *Assessment in Education, 10*(3), 375-388.

Nye, D. (1981). *The Semiotics of Biography: The Lives and Deeds of Thomas Edison*. A Monograph of The Toronto Semiotics Circle.

O'Day, S. (2006). *Setting the stage for creative writing: Plot scaffolds for beginning and intermediate writers*. ERIC Document No. ED493378.

Ontario Education Quality and Accountability Office. (2016). Toronto: Ontario Government.

Ontario Ministry of Education and Training. (2005). *The Ontario curriculum, grades 1-8: Mathematics, revised*. Retrieved from http://www.edu.gov.on.ca/eng/curriculum/elementary/math18curr.pdf

Ontario Ministry of Education and Training. (2009). *Growing success document*. Retrieved from www.edu.gov.on.ca/eng/policyfunding/growsuccess.pdf

Ornstein, R. (Ed.). (1973). The nature of human consciousness. San Francisco: W.H. Freeman.

Orsucci, F., & Sala, N. (2008). *Reflexing interfaces: The complex coevolution of information technology ecosystems*. Hershey, PA: IGI Global; doi:10.4018/978-1-59904-627-3

Orsucci, F., & Sala, N. (2012). *Complexity science, living systems, and reflexing interfaces: New models and perspectives*. Hershey, PA: IGI Global; doi:10.4018/978-1-4666-2077-3

Osta, I. (2014). Mathematics curriculum evaluation. In S. Lerman (Ed.), *Encyclopedia of Mathematics Education* (pp. 417–423). New York: Springer.

Owen, T. (1995). Poems that change the world: Canada's wired writers. *English Journal, 84*(6), 48–52. doi:10.2307/820891

Oxford English Dictionary. (2017). *A-Level: Introduction to OED online.* Retrieved August 1, 2017 at 1:21 PM EDT from http://public.oed.com/resources/for-students-and-teachers/a-level/

Oxford English Dictionary. (2017). Retrieved on July 18, 2017 at 11:51 AM EDT from https://en.oxforddictionaries.com/definition/staycation

Paci, F. (2015, January 8). Islam diviso: le autorità condannano ma nel Web dilaga l'odio per Parigi. *La Stampa*, p. 5.

Pan, Z., & Kosicki, G. M. (1993). Framing analysis: An approach to news discourse. *Political Communication, 10*(1), 55–75. doi:10.1080/10584609.1993.9962963

Parisien, C., & Thagard, P. (2008). Robosemantics: How Stanley represents the world. *Minds and Machines, 18*(2), 169–178. doi:10.1007/s11023-008-9098-2

Pattee, H. H. (1969). How does a molecule become a message? *Developmental Biology*, (Supplement 3), 1–16.

Pattee, H. H. (1982). Cell psychology: An evolutionary approach to the symbol-matter problem. *Cognition and Brain Theory, 5*(4), 325–341.

Paul, H. (1880). Prinzipien der Sprachgeschichte. Halle: Max Niemeyer.

Peck, M. (2012). *Prosthetics of the future: Driven by thoughts, powered by bodily fluids.* Retrieved from http://spectrum.ieee.org/tech-talk/biomedical/devices/prosthetics-of-the-future-driven-by-thoughts-powered-by-bodily-fluids

Peirce, C. S. (1908). Semiotic and significs: The correspondence between Charles S. Peirce and Victoria Lady Welby. Bloomington, IN: Indiana University Press.

Peirce, C. S. (1931-1958). Collected papers of Charles Sanders Peirce (vols. 1-8). Cambridge, MA: Harvard University Press.

Peirce, C. S. (1934-58). *Collected papers, I-VIII.* Cambridge, MA: Belknap Press of the Harvard University Press.

Peirce, C. S. (1998). The essential Peirce: Selected philosophical writings (Vol. 2). Bloomington, IN: Indiana University Press.

Peirce, C. S. (1887). Logical machines. *The American Journal of Psychology, 1*, 165–170.

Peirce, C. S. (1931-1958). In C. Hartshorne, P. Weiss, & A. W. Burks (Eds.), *Collected papers of Charles Sanders Peirce* (Vols. 1-8). Cambridge, MA: Harvard University Press.

Peirce, C. S. (1953). *Letters to Lady Welby* (I. C. Lieb, Ed.). New Haven, CT: Whitlock.

Peirce, C. S. (1992). Some consequences of four incapacities. In N. Houser & C. Kloesel (Eds.), *The essential Peirce: Selected philosophical writings* (Vol. 1, pp. 186–199). Bloomington, IN: Indiana University Press.

Peirce, C. S. (1992). *The essential Peirce, I.* Bloomington, IN: Indiana University Press.

Peirce, C. S. (1998). *The essential Peirce, II.* Bloomington, IN: Indiana University Press.

Penrose, R. (1989). *The Emperor's new mind: Concerning computers, minds and the laws of physics.* New York: Oxford University Press.

Pepperberg, I. (2002). *The Alex studies: Cognitive and communicative abilities of Grey Parrots.* Cambridge, MA: Harvard University Press.

Perlovsky, L. I. (2007). Symbols: Integrated cognition and language. In R. Gudwin & J. Queiroz (Eds.), *Semiotics and intelligent systems development* (pp. 121–151). Hershey, PA: Idea Group. doi:10.4018/978-1-59904-063-9.ch005

Peyton, J., & Rigg, P. (1999). *Poetry in the adult ESL classroom.* ERIC Document No. ED439626.

Phillips, A. L. (2011). The algorists. *American Scientist, 99*(2), 126. doi:10.1511/2011.89.126

Piaget, J. (1955). *The child's construction of reality.* London: Routledge and Kegan Paul.

Picard, R. (2010). *Affective computing.* Accessed 6 August 2010 from http://affect.media.mit.edu/

Picard, R. W. (1997). *Affective computing.* Cambridge, MA: MIT Press. doi:10.1037/e526112012-054

Piccinini, G. (2005). *Symbols, strings, and spikes: The empirical refutation of computationalism.* Retrieved from http://tinyurl.com/Piccinini2005

Piccinini, G. (2007). Computational explanation and mechanistic explanation of mind. In M. Marraffa, M. De Caro, & F. Ferretti (Eds.), *Cartographies of the mind: Philosophy and psychology in intersection* (pp. 23–36). Dordrecht, The Netherlands: Springer. doi:10.1007/1-4020-5444-0_2

Piccinini, G. (2008). Computation without representation. *Philosophical Studies, 137*(2), 205–241. doi:10.1007/s11098-005-5385-4

Piccinini, G. (2010). The mind as neural software? Understanding functionalism, computationalism, and computational functionalism. *Philosophy and Phenomenological Research, 81*(2), 269–311. doi:10.1111/j.1933-1592.2010.00356.x

Piccinini, G., & Scarantino, A. (2011). Information processing, computation, and cognition. *Journal of Biological Physics, 37*(1), 1–38. doi:10.1007/s10867-010-9195-3 PMID:22210958

Pifarré, M., Marti, L., & Guijosa, A. (2014). *Collaborative creativity processes in a wiki: A study in secondary education.* Paper presented at the Conference on Cognition and Exploratory Learning in Digital Age (CELDA), Porto, Portugal. Retrieved August 15, 2017 from http://www.scopus.com/record/display.uri?eid=2-s2.0-84925220858&origin=inward&txGid=28E2C4265FE0D85B8C5C3946BE703BAB.WlW7NKKC52nnQNxjqAQrA %3a2

Pimm, D. (1987). *Speaking mathematically: Communications in mathematics classrooms.* London: Routledge & Kegan Paul.

Pimm, D. (1988). Mathematical metaphor. *For the Learning of Mathematics, 8*(1), 30–34.

Pinegger, A., Hiebel, H., Wriessnegger, S. C., & Müller-Putz, G. R. (2017). Composing only by thought: Novel application of the P300 brain-computer interface. *PLoS, 12*(9). Retrieved from https://www.ncbi.nlm.nih.gov/pubmed/28877175

Plato. (360 BCE). *Cratylus* (B. Jowett, Trans.). Retrieved on April 10, 2011 from, http://classics.mit.edu/Plato/cratylus.html

Pollock, G., & Ross, J. (1988). *The Oedipus papers.* Madison, CT: International Universities Press.

Pollock, J. L. (2008). What am I? Virtual machines and the mind/body problem. *Philosophy and Phenomenological Research, 76*(2), 237–309. doi:10.1111/j.1933-1592.2007.00133.x

Popper, K. R. (1934). *Logik der Forschung. Vienna: Mohr Siebeck.*

Posner, R. (1992). Origins and development of contemporary syntactics. *Languages of Design, 1*, 37–50.

Posner, R. (2004). Basic tasks of cultural semiotics. In G. Witham & J. Wallmannsberger (Eds.), *Signs of power – Power of signs. Essays in honor of Jeff Bernard* (pp. 56–89). Vienna: INST.

Putnam, H. (1961). *Brains and behavior.* Presented at the American Association for the Advancement of Science, Section L (History and Philosophy of Science).

Putnam, H. (1975). The meaning of 'meaning.' In K. Gunderson (Ed.), Language, mind, and knowledge (pp. 131-193). Minneapolis, MN: University of Minnesota Press.

Putnam, H. (1960). Minds and machines. In S. Hook (Ed.), *Dimensions of mind: A symposium* (pp. 148–179). New York: New York University Press.

Putnam, H. (1988). *Representation and reality.* Cambridge, MA: MIT Press.

Putnam, R. (1987). Structuring and adjusting content for students: A study of live and simulated tutoring of addition. *American Educational Research Journal, 24*(1), 13–28. doi:10.3102/00028312024001013

Pylyshyn, Z. W. (1980). Computation and cognition: Issues in the foundations of cognitive science. *Behavioral and Brain Sciences, 3*(01), 111–169. doi:10.1017/S0140525X00002053

Pylyshyn, Z. W. (1985). *Computation and cognition: Toward a foundation for cognitive science* (2nd ed.). Cambridge, MA: MIT Press.

Racco, R. G. (2010). Creative writing: An instructional strategy to improve literacy. Attitudes of the intermediate English student. *Journal of Classroom Research in Literacy, 3,* 3–9.

Radford, L. (2010). Algebraic thinking from a cultural semiotic perspective. *Research in Mathematics Education, 12*(1), 1–19. doi:10.1080/14794800903569741

Ramachandran, V., & Rogers-Ramachandran, D. (2009). Two eyes, two views. *Scientific American Mind, 20*(5), 22–24. doi:10.1038/scientificamericanmind0909-22

Rambe, P. (2011). Exploring the impacts of social networking sites on academic relations in the university. *Journal of Information Technology Education, 10,* 271–293.

Rapaport, W. J. (1985/1986). Non-existent objects and epistemological ontology. Grazer Philosophische Studien, 25/26, 61-95.

Rapaport, W. J. (1988). Syntactic semantics: Foundations of computational natural-language understanding. In J. H. Fetzer (Ed.), Aspects of artificial intelligence (pp. 81-131). Dordrecht, The Netherlands: Kluwer Academic Publishers. Retrieved from http://tinyurl.com/SynSemErrata1

Rapaport, W. J. (1990). *Computer processes and virtual persons: Comments on Cole's 'Artificial Intelligence and Personal Identity.'* Technical Report 90-13. Buffalo, NY: SUNY Buffalo Department of Computer Science. Retrieved from http://tinyurl.com/Rapaport1990

Rapaport, W. J. (1995). Understanding understanding: Syntactic semantics and computational cognition. In J. E. Tomberlin (Ed.), AI, connectionism, and philosophical psychology (pp. 49-88). Atascadero, CA: Ridgeview.

Rapaport, W. J. (2000). How to pass a Turing Test: Syntactic semantics, natural-language understanding, and first-person cognition. Journal of Logic, Language, and Information, 9(4), 467-490.

Rapaport, W. J. (2005a). In defense of contextual vocabulary acquisition: How to do things with words in context. In A. Dey, B. Kokinov, D. Leake, & R. Turner (Eds.), *Modeling and using context: 5th International and Interdisciplinary Conference, CONTEXT 05, Paris, France, July 2005, Proceedings* (pp. 396-409). Berlin: Springer-Verlag.

Rapaport, W. J. (2015). On the relation of computing to the world. In T. M. Powers (Ed.), *Philosophy and computing: Essays in epistemology, philosophy of mind, logic, and ethics.* Springer. Retrieved from http://www.cse.buffalo. edu/~rapaport/Papers/covey.pdf

Rapaport, W. J. (2017a). Semantics as syntax. *American Philosophical Association Newsletter on Philosophy and Computers.* Retrieved from http://www.cse.buffalo.edu/~rapaport/Papers/synsemapa.pdf

Rapaport, W. J. (2017b). *Philosophy of computer science.* Retrieved from https://www.cse.buffalo.edu//~rapaport/Papers/phics.pdf

Rapaport, W. J., & Kibby, M. W. (2010). *Contextual vocabulary acquisition: From algorithm to curriculum.* Retrieved from http://tinyurl.com/RapaportKibby2010

Rapaport, W. J. (1979). (1978. Meinongian theories and a Russellian paradox. *Noûs 12*, 153-180. *Noûs (Detroit, Mich.),* *13*, 125. doi:10.2307/2214805

Rapaport, W. J. (1979). An adverbial Meinongian theory. *Analysis, 39*(2), 75–81. doi:10.1093/analys/39.2.75

Rapaport, W. J. (1981). How to make the world fit our language: An essay in Meinongian semantics. *Grazer Philosophische Studien, 14*, 1–21. doi:10.5840/gps1981141

Rapaport, W. J. (1986). Searle's experiments with thought. *Philosophy of Science, 53*(2), 271–279. doi:10.1086/289312

Rapaport, W. J. (1998). How minds can be computational systems. *Journal of Experimental & Theoretical Artificial Intelligence, 10*(4), 403–419. doi:10.1080/095281398146671

Rapaport, W. J. (1999). Implementation is semantic interpretation. *The Monist, 82*(1), 109–130. doi:10.5840/monist19998212

Rapaport, W. J. (2002). Holism, conceptual-role semantics, and syntactic semantics. *Minds and Machines, 12*(1), 3–59. doi:10.1023/A:1013765011735

Rapaport, W. J. (2003a). What did you mean by that? Misunderstanding, negotiation, and syntactic semantics. *Minds and Machines, 13*(3), 397–427. doi:10.1023/A:1024145126190

Rapaport, W. J. (2003b). What is the 'context' for contextual vocabulary acquisition? In P. P. Slezak (Ed.), *Proceedings of the 4th International Conference on Cognitive Science/7th Australasian Society for Cognitive Science Conference (ICCS/ASCS-2003; Sydney, Australia)* (Vol. 2, pp. 547-552). Sydney: University of New South Wales.

Rapaport, W. J. (2005b). Implementation is semantic interpretation: Further thoughts. *Journal of Experimental & Theoretical Artificial Intelligence, 17*(4), 385–417. doi:10.1080/09528130500283998

Rapaport, W. J. (2006). How Helen Keller used syntactic semantics to escape from a Chinese room. *Minds and Machines, 16*(4), 381–436. doi:10.1007/s11023-007-9054-6

Rapaport, W. J. (2011). Yes, she was! Reply to Ford's 'Helen Keller was never in a Chinese room.'. *Minds and Machines, 21*(1), 3–17. doi:10.1007/s11023-010-9213-z

Rapaport, W. J., & Ehrlich, K. (2000). A computational theory of vocabulary acquisition. In Ł. M. Iwańska & S. C. Shapiro (Eds.), *Natural language processing and knowledge representation: Language for knowledge and knowledge for language* (pp. 347–375). Menlo Park, CA: AAAI Press/MIT Press. Retrieved from http://tinyurl.com/RapaportEhrlichErrata

Rapaport, W. J., & Kibby, M. W. (2007). Contextual vocabulary acquisition as computational philosophy and as philosophical computation. *Journal of Experimental & Theoretical Artificial Intelligence, 19*(1), 1–17. doi:10.1080/09528130601116162

Ray, K. (2010). *Web 3.0.* Accessed 1 March 2011 from http://kateray.net/film

Reichelt, J., Wickert, U., Pointner, N., & Vehlewald, H. (2015, January 8). Blutiger Terroranschlag in Frankreich seit Jahrzehnten. *Bild.*

Reinhardt, J., & Zander, V. (2011). Social networking in an intensive English program classroom: A language socialization perspective. *CALICO Journal, 28*(2), 326–344. doi:10.11139/cj.28.2.326-344

Rescorla, M. (2017). The computational theory of mind. In *The Stanford Encyclopedia of Philosophy.* Retrieved from https://plato.stanford.edu/archives/spr2017/entries/computational-mind/

Resnick, L. B. (1988). Treating mathematics as an ill-structured discipline. In R. I. Charles & E. A. Silver (Eds.), *The teaching and assessing of mathematical problem solving* (pp. 32–60). Hillsdale, NJ: Lawrence Erlbaum Associates, Inc.

Ricoeur, P. (1970). *Freud and philosophy.* Cambridge, MA: Yale University Press.

Rieger, B. B. (1997). Computational semiotics and fuzzy linguistics. On meaning constitution and soft categories. In A. Meystel (Ed.), *A learning perspective. International Conference on Intelligent Systems and Semiotics.* Washington, DC: NIST Special Publication.

Rieger, B. B. (2003). Semiotic cognitive information processing: Learning to understand discourse. A systemic model of meaning constitution. In R. Kühn et al. (Eds.), *Adaptivity and learning. An interdisciplinary debate* (pp. 347–403). Berlin: Springer. doi:10.1007/978-3-662-05594-6_24

Rigotti, E. (1993). La sequenza testuale. Definizione e procedimenti di analisi con esemplificazione in lingue diverse. *L'analisi linguistica e letteraria, 1*(2), 43-148.

Rigotti, E., & Cigada, S. (2013). La comunicazione verbale (2nd ed.). Santarcangelo di Romagna: Maggioli.

Rigotti, E., & Rocci, A. (2001). Sens – non-sens – contresens. *Studies in Communication Sciences, 2*, 45–80.

Ripley's Believe it or Not Museum [Billboard]. (2017). Toronto Transit Commission Spadina Station Line 1 Subway, South Bound Platform on July 27 2017 at 9:38 AM EDT.

Ritchhart, R. (2015). *Creating Cultures of Thinking: The 8 forces We Must Master to Truly Transform our Schools.* Hoboken, NJ: Wiley.

Roberts, D. D. (2009). *The Existential Graphs of Charles S. Peirce.* The Hague, The Netherlands: Mouton. doi:10.1515/9783110226225

Robey, E. (2014). John Vanderlyn's view of Versailles: Spectacle, landscape, and the visual demands of panorama painting. *Early Popular Visual Culture, 12*(1), 1–21. doi:10.1080/17460654.2013.876922

Rocci, A. (2005). Are manipulative texts 'coherent'? Manipulation, presuppositions and (in)congruity. In L. Saussure & P. Schulz (Eds.), *Manipulation and ideologies in the twentieth century: Discourse, language, mind* (pp. 85–112). Amsterdam: Benjamins. doi:10.1075/dapsac.17.06roc

Rogers, M. (1998). Spectacular bodies: Folklorization and the politics of identity in Ecuadorian beauty pageants. *Journal of Latin American and Caribbean Anthropology, 3*(2), 54–85. doi:10.1525/jlat.1998.3.2.54

Rosaspina, E. (2015, January 8). Liberi e irriverenti: la redazione che non c'è più. *Corriere della Sera*, p. 8.

Rosenblat, D. (2000). *Re: Etymology of wannabe.* Retrieved July 19, 2017 at 4:29 EDT from http://listserv.linguistlist.org/pipermail/linganth/2000-February/000147.html

Rosen, R. (1985). Organisms as causality systems which are not machines: An essay on the nature of complexity. In R. Rosen (Ed.), *Rosen: Theoretical biology and complexity* (pp. 165–203). Orlando, FL: Academic Press. doi:10.1016/B978-0-12-597280-2.50008-8

Rosen, R. (1987). On the scope of syntactics in mathematics and science: The machine metaphor. In J. L. Casti & A. Karlqvist (Eds.), *Real brains, artificial minds* (pp. 1–23). New York: Elsevier Science.

Rosser, B. (1939). An informal exposition of proofs of Gödel's theorem. *The Journal of Symbolic Logic*, *4*(2), 53–60. doi:10.2307/2269059

Roth, P. F. (1983). Simulation. In A. Ralston & E. D. Reilly Jr., (Eds.), *Encyclopedia of Computer Science and Engineering* (2nd ed.; pp. 1327–1341). New York: Van Nostrand Reinhold.

Rothschild, B. (2000). *The body remembers: The psychophysiology of trauma and trauma treatment*. New York, NY: W. W. Norton.

Royal Academy of Arts Collections. (2015). *After Francesco Laurana: Bust of a woman, possibly Ippolita Maria Sforza, c. 1473, 19th-century plaster cast*. Retrieved from http://www.racollection.org.uk/ixbin/indexplus?record=ART13883

Ruan, Z. (2014). Metacognitive awareness of EFL student writers in a Chinese ELT context. *Language Awareness*, *23*(1-2), 76–90. doi:10.1080/09658416.2013.863901

Ryan, M. (2014). Writers as performers: Developing reflexive and creative writing identities. *English Teaching*, *13*(3), 130–148.

Sadowski, P. (2010). *Towards systems semiotics: Some remarks and (hopefully useful) definitions*. Retrieved from http://www.semioticon.com/semiotix/2010/03/towards-systems-semiotics-some-remarks-and-hopefully-useful-definitions/

Sahbaz, N., & Duran, G. (2011). The efficiency of cluster method in improving the creative writing skill of 6th grade students of primary school. *Educational Research Review*, *6*(11), 702–709.

Salcito, A. (2012). *Exploring creative writing through technology*. Retrieved August 15, 2017 from http://dailyedventures.com/index.php/2012/01/30/1369/

Santore, J. F., & Shapiro, S. C. (2003). Crystal Cassie: Use of a 3-D gaming environment for a cognitive agent. In R. Sun (Ed.), *Papers of the IJCAI 2003 Workshop on Cognitive Modeling of Agents and Multi-Agent Interactions (IJCAII; Acapulco, Mexico, August 9, 2003)* (pp. 84-91). Academic Press.

Sarar Kuzu, T. (2016). The impact of a semiotic analysis theory-based writing activity on students' writing skills. *Eurasian Journal of Educational Research*, *63*, 37–54.

Scardamalia, M., & Bereiter, C. (1983). Child as coinvestigator: Helping children gain insight into their own mental processes. In S. Paris, G. Olson, G., & H. Stevenson (Eds.), Learning and motivation in the classroom (pp. 61-82). Hillsdale, NJ: Lawrence Erlbaum Associates.

Scardamalia, M., & Bereiter, C. (2003). Knowledge building. Encyclopedia of Education, 2, 1370-1373.

Scardamalia, M., Bransford, J., Kozma, B., & Quellmalz, E. (2012). New Assessments and Environments for Knowledge Building. In P. Griffin, P., B. McGaw, & E. Care (Eds.), Assessment and teaching of 21st century skills (pp. 231-300). New York: Springer. doi:10.1007/978-94-007-2324-5_5

Scardamalia, M., & Bereiter, C. (1991). Higher levels of agency for children in knowledge-building: A challenge for the design of new knowledge media. *Journal of the Learning Sciences*, *1*(1), 37–68. doi:10.1207/s15327809jls0101_3

Scardamalia, M., Bereiter, C., McLean, R. S., Swallow, J., & Woodruff, E. (1989). Computer supported intentional learning environments. *Journal of Educational Computing Research, 5*(1), 51–68. doi:10.2190/CYXD-6XG4-UFN5-YFB0

Schagrin, M. L., Rapaport, W. J., & Dipert, R. R. (1985). *Logic: A computer approach.* New York: McGraw-Hill.

Scheufele, D. A. (1999). Framing as a theory of media effects. *Journal of Communication, 49*(1), 103–122. doi:10.1111/j.1460-2466.1999.tb02784.x

Schneider, S. (2009). The paradox of fiction. *Internet Encyclopedia of Philosophy.* Accessed 6 May 2011 from http://www.iep.utm.edu/fict-par/

Schore, A. (2011). The right brain implicit self lies at the core of psychoanalytic psychotherapy. *Psychoanalytic Dialogues, 21,* 75–100. doi:0481885.2011.54532910.1080/1

Schore, A. (2001). Minds in the making: Attachment, the self-organizing brain, and developmentally-oriented psychoanalytic psychotherapy. *British Journal of Psychotherapy, 17*(3), 299–328. doi:10.1111/j.1752-0118.2001.tb00593.x

Schore, A. (2010). The right-brain implicit self: A central mechanism of the psychotherapy change process. In J. Petrucelli (Ed.), *Knowing, not-knowing and sort of knowing: Psychoanalysis and the experience of uncertainty* (pp. 177–202). London, UK: Karnac.

Schore, A. (2012). *The science of the art of psychotherapy.* New York, NY: Norton.

Schunk, D. H. (2012). *Learning theories: An educational perspective.* New York, NY: Macmillan Publishing Co, Inc.

Scott, V. (1990). Task-oriented creative writing with system-D. *CALICO Journal, 7*(3), 58–67.

Searle, J. R. (1980). Minds, brains, and programs. *Behavioral and Brain Sciences, 3*(03), 417–457. doi:10.1017/S0140525X00005756

Searle, J. R. (1990). Is the brain a digital computer? *Proceedings and Addresses of the American Philosophical Association, 64*(3), 21–37. doi:10.2307/3130074

Segal, G. (1989). Seeing what is not there. *The Philosophical Review, 98*(2), 189–214. doi:10.2307/2185282

Serra, M. (2015, January 8). La brigata dei libertini. *La Stampa.*

Sessions, L., Kang, M. O., & Womack, S. (2016). The neglected "R": Improving writing instruction through iPad apps. *TechTrends, 60*(3), 218–225. doi:10.1007/s11528-016-0041-8

Shafie, L. A., Yaacob, A., & Singh, P. K. (2016). Facebook activities and the investment of L2 learners. *English Language Teaching, 9*(8), 53–61. doi:10.5539/elt.v9n8p53

Shanley, W. (Ed.). (2011). *Alice and the quantum cat.* Pari, Italy: Pari Publishing.

Shannon, C. E., & Weaver, W. (1949). *The mathematical theory of communication.* Urbana, IL: University of Illinois Press.

Shapiro, S. C. (1998). Embodied Cassie. In *Cognitive Robotics: Papers from the 1998 AAAI Fall Symposium* (pp. 136-143). Technical Report FS-98-02. Menlo Park, CA: AAAI Press.

Shapiro, S. C. (2001). *Computer science: The study of procedures.* Accessed 5 May 2011 from http://www.cse.buffalo.edu/~shapiro/Papers/whatiscs.pdf

Shapiro, S. C., & Rapaport, W. J. (1991). Models and minds: Knowledge representation for natural-language competence. In R. Cummins & J. Pollock (Eds.), Philosophy and AI: Essays at the interface (pp. 215-259). Cambridge, MA: MIT Press.

Shapiro, S. C., Amir, E., Grosskreutz, H., Randell, D., & Soutchanski, M. (2001). *Commonsense and embodied agents. A panel discussion.* At Common Sense 2001: The 5th International Symposium on Logical Formalizations of Commonsense Reasoning, Courant Institute of Mathematical Sciences, New York University. Accessed 20 October 2011 from http://www.cse.buffalo.edu/~shapiro/Papers/commonsense-panel.pdf

Shapiro, S. C. (1992). Artificial Intelligence. In S. C. Shapiro (Ed.), *Encyclopedia of Artificial Intelligence* (2nd ed.; pp. 54–57). New York: John Wiley & Sons.

Shapiro, S. C. (1993). Belief spaces as sets of propositions. *Journal of Experimental & Theoretical Artificial Intelligence*, *5*(2-3), 225–235. doi:10.1080/09528139308953771

Shapiro, S. C. (2006). Natural-language-competent robots. *IEEE Intelligent Systems*, *21*(4), 76–77.

Shapiro, S. C., & Bona, J. P. (2010). The GLAIR cognitive architecture. *International Journal of Machine Consciousness*, *2*(02), 307–332. doi:10.1142/S1793843010000515

Shapiro, S. C., & Ismail, H. O. (2003). Anchoring in a grounded layered architecture with integrated reasoning. *Robotics and Autonomous Systems*, *43*(2-3), 97–108. doi:10.1016/S0921-8890(02)00352-4

Shapiro, S. C., & Rapaport, W. J. (1987). SNePS Considered as a fully intensional propositional semantic network. In N. Cercone & G. McCalla (Eds.), *The knowledge frontier: Essays in the representation of knowledge* (pp. 262–315). New York: Springer-Verlag. doi:10.1007/978-1-4612-4792-0_11

Shapiro, S. C., Rapaport, W. J., Kandefer, M., Johnson, F. L., & Goldfain, A. (2007). Metacognition in SNePS. *AI Magazine*, *28*(1), 17–31.

Shin, S. (1994). *The logical status of diagrams.* Cambridge, UK: Cambridge University Press.

Shmueli, D. F. (2008). Framing in geographical analysis of environmental conflicts: Theory, methodology and three case studies. *Geoforum*, *39*(6), 2048–2061. doi:10.1016/j.geoforum.2008.08.006

Short, T. (2007). *Peirce's theory of signs.* Cambridge, UK: Cambridge University Press. doi:10.1017/CBO9780511498350

Sieckenius de Souza, C. (2005). *The semiotic engineering of human-computer interaction.* Cambridge, MA: MIT Press.

Siegel, D. (2001). Memory: An overview, with emphasis on developmental, interpersonal, and neurobiological aspects. *Journal of the Academy of Child & Adolescent Psychiatry*, *40*(9), 997–1011. doi:10.1097/00004583-200109000-00008 PMID:11556645

Silverman, K. (1984). The subject of semiotics. Oxford University Press.

Simon, H. A. (1956). Rational choice and the structure of the environment. *Psychological Review*, *63*(2), 129–138. doi:10.1037/h0042769 PMID:13310708

Simon, H. A. (1967). Motivational and emotional controls of cognition. *Psychological Review*, *74*(1), 29–39. doi:10.1037/h0024127 PMID:5341441

Sinclair, M. P. (2005). Peer interactions in a computer lab: Reflections on results of a case study involving web-based dynamic geometry sketches. *The Journal of Mathematical Behavior*, *24*(1), 89–107. doi:10.1016/j.jmathb.2004.12.003

Sinclair, N., & Bruce, C. (2015). New opportunities in geometry education at the primary school. *ZDM Mathematics Education*, *47*(3), 319–329. doi:10.1007/s11858-015-0693-4

Sloman, A. (2004). *What are emotion theories about?* Workshop on Architectures for Modeling Emotion, AAAI Spring Symposium, Stanford University.

Sloman, A. (2009). *The cognition and affect project*. Accessed 6 May 2011 from http://www.cs.bham.ac.uk/research/projects/cogaff/cogaff.html

Sloman, A., & Croucher, M. (1981). Why robots will have emotions. *Proceedings of the International Joint Conference on Artificial Intelligence*. Retrieved from http://www.cs.bham.ac.uk/research/projects/cogaff/81-95.html#36

Smalley, R. E. (2001, September). Of chemistry, love and nanobots (nanofallacies). *Scientific American, 285*, 76–77. doi:10.1038/scientificamerican0901-76 PMID:11524973

Smith, B. C. (1985). Limits of correctness in computers. ACM SIGCAS Computers and Society, 14-15(1-4), 18-26.

Smith, B. C. (2001). True grid. In D. Montello (Ed.), *Spatial Information Theory: Foundations of Geographic Information Science, Proceedings of COSIT 2001, Morro Bay, CA, September 2001* (pp. 14-27). Berlin: Springer.

Smith, B. (2005). Against fantology. In M. Reicher & J. Marek (Eds.), *Experience and analysis*. Vienna: ÖBV & HPT.

Smith, B. C. (1982). Linguistic and computational semantics. In *Proceedings of the 20th Annual Meeting of the Association for Computational Linguistics* (pp. 9-15). Morristown, NJ: Association for Computational Linguistics. doi:10.3115/981251.981254

Smith, B. C. (2010). Introduction. In B. C. Smith (Ed.), *Age of Significance*. Cambridge, MA: MIT Press. Retrieved from http://www.ageofsignificance.org/aos/en/aos-v1c0.html

Soare, R. I. (2009). Turing oracle machines, online computing, and three displacements in computability theory. *Annals of Pure and Applied Logic, 160*(3), 368–399. doi:10.1016/j.apal.2009.01.008

Solms, M. (2004). Freud returns. *Scientific American*, 83–89. PMID:15127665

Solms, M., & Turnbull, O. (2002). *The brain and the inner world: An introduction to the neuroscience of subjective experience*. London, UK: Karnac Books.

Solomon, J. F. (1988). *The signs of our time. Semiotics: The hidden messages of environments, objects, and cultural images*. New York: St Martin's Press.

Sowa, J. (2017). *The virtual reality of the mind*. Retrieved from http://222.jfsowa.com/talks/vrmind.pdf

Srihari, R. K., & Rapaport, W. J. (1990). Combining linguistic and pictorial information: Using captions to interpret newspaper photographs. In D. Kumar (Ed.), Current Trends in SNePS—Semantic Network Processing System (pp. 85-96). Berlin: Springer-Verlag.

Srihari, R. K., & Rapaport, W. J. (1989). Extracting visual information from text: Using captions to label human faces in newspaper photographs. In *Proceedings of the 11th Annual Conference of the Cognitive Science Society* (pp. 364-371). Hillsdale, NJ: Lawrence Erlbaum Associates.

Stahl, G. (2005). Group cognition in computer-assisted collaborative learning. *Journal of Computer Assisted Learning, 21*(2), 79–90. doi:10.1111/j.1365-2729.2005.00115.x

Steffe, L. P., & Kieren, T. (1994). Radical constructivism and mathematics education. *Journal for Research in Mathematics Education, 25*(6), 711–733. doi:10.2307/749582

Steinbuch, Y. (2017). *The New York Times*. Retrieved on July 24, 2017 at 10:12 EDT from http://nypost.com/2017/04/24/this-is-whats-left-after-the-mother-of-all-bombs-hit-afghanistan/

Stephan, P. (1996). Auf dem Weg zu Computational Semiotics. In D. Dotzler (Ed.), *Computer als Faszination* (p. 209). Frankfurt, Germany: CAF Verlag.

Stern, D. (1983). Unformulated experience: From familiar chaos to creative disorder. *Contemporary Psychoanalysis, 19*, 71–99. doi:10.1080/00107530.1983.10746593

Sternberg, R. (1990). *Metaphors of mind: Conceptions of the nature of intelligence*. Cambridge, UK: Cambridge University Press.

Stigler, J., & Hiebert, J. (1999). *The teaching gap: Best ideas from the world's teachers for improving education in the classroom*. New York: The Free Press.

Stillar, S. (2013). Raising critical consciousness via creative writing in the EFL classroom. *TESOL Journal, 4*(1), 164–174. doi:10.1002/tesj.67

Stjernfelt, F. (2007). *Diagrammatology. An investigation on the borderlines of phenomenology, ontology, and semiotics*. Dordrecht, The Netherlands: Springer.

Stjernfelt, F. (2007). *Diagrammatology: An investigation on the borderlines of phenomenology, ontology, and semiotics*. New York: Springer. doi:10.1007/978-1-4020-5652-9

Stjernfelt, F. (2012). The evolution of semiotic self-control. In T. Deacon, T. Schilhab, & F. Stjernfelt (Eds.), *The symbolic species evolved* (pp. 39–63). Dordrecht, The Netherlands: Springer. doi:10.1007/978-94-007-2336-8_3

Stjernfelt, F. (2014). *Natural propositions: The actuality of Peirce's doctrine of Dicisigns*. Boston: Docent Press.

Stjernfelt, F. (2014a). Dicisigns and cognition: The logical interpretation of the ventral-dorsal split in animal perception. *Cognitive Semiotics, 7*(1), 61–82.

Stjernfelt, F. (2015). Iconicity of logic and the roots of the "iconicity" concept. In K. Masako, W. J. Hiraga, K. Z. Herlofsky, & A. Kimi (Eds.), *Iconicity: East meets West* (pp. 35–56). Amsterdam: John Benjamins. doi:10.1075/ill.14.02stj

Stjernfelt, F. (2015a). Dicisigns. *Synthese, 192*(4), 1019–1054. doi:10.1007/s11229-014-0406-5

Stjernfelt, F. (2016). Blocking evil infinites: A note on a note on a Peircean strategy. *Sign Systems Studies, 42*(4), 518–522.

Stjernfelt, F. (2016a). Dicisigns and habits: Implicit propositions and habit-taking in Peirce's pragmatism. In D. West & M. Anderson (Eds.), *Consensus on Peirce's concept of Habit: Before and beyond consciousness* (pp. 241–264). New York: Springer. doi:10.1007/978-3-319-45920-2_14

Stoddart, A., Chan, J. Y., & Liu, G. (2016). Enhancing successful outcomes of wiki-based collaborative writing: A state-of-the-art. *Review of Facilitation Frameworks. Interactive Learning Environments, 24*(1), 142–157. doi:10.1080/10494820.2013.825810

Stolorow, R., Brandchaft, B., & Atwood, G. (1987). *Psychoanalytic treatment: An intersubjective approach*. Hillsdale, NJ: The Analytic Press.

Strawson, G. (2010). *Mental reality* (2nd ed.). Cambridge, MA: MIT Press.

Struhsaker, T. (1967). Auditory communication among vervet monkeys (*Cercopithecus aethiops*). In S. A. Altmann (Ed.), *Social communication among primates*. Chicago: University of Chicago Press.

Tarski, A. (1944). The semantic conception of truth and the foundations of semantics. *Philosophy and Phenomenological Research, 4*(3), 341–376. doi:10.2307/2102968

Taylor, R., & Wiles, A. (1995). Ring-theoretic properties of certain Hecke algebras. *Annals of Mathematics, 141*(3), 553–572. doi:10.2307/2118560

Teen Eating Disorders. (2017). *What is wannarexia.* Retrieved on August 8, 2017 at 11:59 AM EDT from http://www. teeneatingdisorders.us/content/what-is-wannarexia.html

Teflpedia. (2017). Retrieved on July 23, 2017 at 1:36 PM from http://teflpedia.com/IPA_phoneme_/%CA%83/

Tenenbaum, A. M., & Augenstein, M. J. (1981). *Data structures using Pascal.* Englewood Cliffs, NJ: Prentice-Hall.

Tewksbury, D., & Scheufele, D. A. (2009). News framing theory and research. In J. Bryant & M. B. Oliver (Eds.), *Media effects: Advances in theory and research* (3rd ed.; pp. 17–33). London: Routledge.

Thagard, P. (2006). *Hot thought: Mechanisms and applications of emotional cognition.* Cambridge, MA: MIT Press.

The final word. (2015, January 8). *The Washington Post.*

Thomson, A. (2015, January 8). France grapples with state of shock. Terror in Paris. *The Financial Times,* p. 8.

Tin, T. (2011). Language creativity and co-emergence of form and meaning in creative writing tasks. *Applied Linguistics, 32*(2), 215–235. doi:10.1093/applin/amq050

Tomasello, M. (2008). *Origins of human communication.* Cambridge, MA: MIT Press.

Toussaint, G. (1993). A new look at Euclid's second proposition. *The Mathematical Intelligencer, 15*(3), 12–23. doi:10.1007/BF03024252

Towers, J. (1999). *In what ways do teachers' interventions interact with and occasion the growth of students' mathematical understanding* (Unpublished doctoral dissertation). University of British Columbia.

Trench, L. (2010). *The Victoria and Albert Museum.* London, UK: V&A Publishing.

Trubetzkoy, N. S. (1939). Grundzüge der Phonologie. Traveaux du Cercle Linguistique du Prague, 7.

Trusted, M. (Ed.). (2007). *The making of sculpture: The materials and techniques of European sculpture.* London, UK: V&A Publishing.

Turing, A. M. (1936). On computable numbers, with an application to the *Entscheidungsproblem. Proceedings of the London Mathematical Society, Ser. 2, 42,* 230–265.

Turing, A. M. (1950). Computing machinery and intelligence. *Mind, 59*(236), 433–460. doi:10.1093/mind/LIX.236.433

United States Holocaust Museum. (2017). Retrieved August 1, 2017 at 12:58 PM EDT from https://www.ushmm.org/ wlc/en/article.php?ModuleId=10007962

Untitled photograph of the original bust of *Portrait of a Woman.* (n.d.). Retrieved from http://www.croatia.org/crown/ articles/10219/

Urbancic, A. (1994). Urban semiotics: Where is downtown? *Signifying Behaviour, 1*(1), 23–34.

V&A Collections. (n.d.). *Bust of a woman* (artifact record). Retrieved from http://collections.vam.ac.uk/item/O39862/

Van Boven, L., & Ashworth, L. (2007). Looking forward, looking back: Anticipation is more evocative than retrospection. *Journal of Experimental Psychology. General, 136*(2), 289–300. doi:10.1037/0096-3445.136.2.289 PMID:17500652

Van den Akker, J., Graveemeijer, K., McKenney, S., & Nieveen, N. (2006). *Educational design research.* London: Routledge.

van Dijk, T. A. (1985). Structure of news in the press. In T. A. van Dijk (Ed.), *Discourse and communication. New approaches to the analysis of mass media discourse and communication* (pp. 69–93). Berlin: De Gruyter. doi:10.1515/9783110852141.69

van Dijk, T. A. (1988). *News as discourse*. New York: Routledge.

van Eemeren, F. H. (2010). *Strategic maneuvering in argumentative discourse: extending the pragma-dialectical theory of argumentation*. Amsterdam: Benjamins. doi:10.1075/aic.2

Van Gelder, T. (1995). What might cognition be, if not computation? *The Journal of Philosophy*, *92*(7), 345–381. doi:10.2307/2941061

Vass, E. (2002). Friendship and collaborative creative writing in the primary classroom. *Journal of Computer Assisted Learning*, *18*(1), 102–110. doi:10.1046/j.0266-4909.2001.00216.x

Vass, E. (2007). Exploring processes of collaborative creativity--The role of emotions in children's joint creative writing. *Thinking Skills and Creativity*, *2*(2), 107–117. doi:10.1016/j.tsc.2007.06.001

Vass, E., Littleton, K., Miell, D., & Jones, A. (2008). The discourse of collaborative creative writing: Peer collaboration as a context for mutual inspiration. *Thinking Skills and Creativity*, *3*(3), 192–202. doi:10.1016/j.tsc.2008.09.001

Venn, J. (1880). On the employment of geometrical diagrams for the sensible representation of logical propositions. *Proceedings of the Cambridge Philosophical Society*, *4*, 47–59.

Venn, J. (1881). *Symbolic logic*. London: Macmillan. doi:10.1037/14127-000

Vera, A. H., & Simon, H. A. (1993). Situated action: A symbolic interpretation. *Cognitive Science*, *17*(1), 7–48. doi:10.1207/s15516709cog1701_2

von Foerster, H. (1981). *Observing systems*. Seaside, CA: Intersystems Publications.

Von Frisch, K. (1993). *The dance language and orientation of bees*. Cambridge, MA: Harvard University Press. doi:10.4159/harvard.9780674418776

von Uxeküll, J. (1934). Streifzüge durch die Umwelt von Tieren und Menschen. Berlin: Julius Springer Verlag. doi:10.1007/978-3-642-98976-6

Vygotsky, L. S. (1978). *Mind in society: The development of higher psychological processes*. Cambridge, MA: Harvard University Press.

Wagner, T., & Dintersmith, T. (2015). *Most likely to succeed: Preparing our kids for the innovation era*. New York: Scribner's.

Wang, S., & Vásquez, C. (2014). The effect of target language use in social media on intermediate-level Chinese language learners' writing performance. *CALICO Journal*, *31*(1), 78–102. doi:10.11139/cj.31.1.78-102

Wegner, P. (1997). Why interaction is more powerful than algorithms. *Communications of the ACM*, *40*(5), 80–91. doi:10.1145/253769.253801

Weizenbaum, J. (1976). *Computer power and human reason: From judgment to calculation*. New York: W. H. Freeman.

Whiteley, W. (1999). The decline and rise of Geometry in 20th century North America. In J. Grant (Ed.), *Proceedings of CMESG 1999 Annual Meeting* (pp. 7–30). Academic Press.

Wichadee, S. (2013). Peer feedback on Facebook: The use of social networking websites to develop writing ability of undergraduate students. *Turkish Online Journal of Distance Education*, *14*(4), 260–270.

widdowquinn (Photographer). (2010). *Portrait of a woman* [Photograph]. Retrieved from https://flic.kr/p/8kfZP6

Widdowson, H. G. (2004). *Text, Context, Pretext*. Malden, MA: Blackwell. doi:10.1002/9780470758427

Wierzbicka, A. (1997). *Understanding cultures through their keywords. English, Russian, Polish, German, and Japanese.* New York: Oxford University Press.

Wikipedia. (2017). Retrieved on August 1, 2017 at 1:50 PM EDT from https://en.wikipedia.org/wiki/Online_encyclopedia

Wikitionary. (2017). Retrieved on August 1, 2017 at 3:17 PM EDT from https://en.wiktionary.org/wiki/lizardy

Wiles, A. (1995). Modular elliptic curves and Fermat's Last Theorem. *Annals of Mathematics, 141*(3), 443–551. doi:10.2307/2118559

Williams, B. (1998). The end of explanation? *The New York Review of Books, 45*(18), 40-44.

Wilson, E. O. (1984). *Biophilia: The human bond with other species.* Cambridge, MA: Harvard University Press.

Wilson, P. (2011). Creative writing and critical response in the university literature class. *Innovations in Education and Teaching International, 48*(4), 439–446. doi:10.1080/14703297.2011.617091

Windelband, W. (1894). Geschicthte und Naturwissenschaft, Präludien. Aufsätze und Reden zur Philosophie und ihrer Geschichte. Tubingen: J.C.B. Mohr.

Winograd, T., & Flores, F. (1987). *Understanding computers and cognition: A new foundation for design.* Reading, MA: Addison-Wesley.

Wittgenstein, L. (1953). *Philosophical investigations* (G. E. M. Anscombe, Trans.). London: Basil Blackwell.

Wittgenstein, L. (2003). *Diktat fur Schlick (Moritz Schlick, Dec. 1932). In The voices of Wittgenstein: The Vienna Circle: Ludwig Wittgenstein and Friedrich Waisman.* London: Routledge.

Witty, P., & Labrant, L. (1946). *Teaching the people's language.* New York: Hinds, Hayden & Eldredge. Retrieved August 15, 2017 from http://archive.org/stream/teachingpeoplesl00witt#page/n3/mode/2up

Wolf, F. (1995). *The dreaming universe: A mind-expanding journey into the realm where psyche and physics meet.* New York, NY: Touchstone.

Wood, D., Bruner, J. S., & Ross, G. (1976). The role of tutoring in problem solving. *Journal of Child Psychology and Psychiatry, and Allied Disciplines, 17*(2), 89–100. doi:10.1111/j.1469-7610.1976.tb00381.x PMID:932126

Wood, D., & Wood, H. (1996). Vygotsky, tutoring and learning. *Oxford Review of Education, 22*(1), 5–16. doi:10.1080/0305498960220101

Woods, W. A. (2010). The right tools: Reflections on computation and language. *Computational Linguistics, 36*(4), 601–630. doi:10.1162/coli_a_00018

Wright, I., Sloman, A., & Beaudoin, L. (1996). Towards a design-based analysis of emotional episodes. *Philosophy, Psychiatry, & Psychology, 3*(2), 101–126. doi:10.1353/ppp.1996.0022

Wu, H. H. (2011). *Understanding numbers in Elementary school mathematics.* Providence, RI: American Mathematical Society. doi:10.1090/mbk/079

Wu, J. T. C. (1997). "Loveliest daughter of our ancient Cathay!:" Representations of ethnic and gender identity in the Miss Chinatown USA beauty pageant. *Journal of Social History, 31*(1), 5–31. doi:10.1353/jsh/31.1.5

Yackel, E., Gravemeijer, K., & Sfard, A. (2011). *A journey in mathematics education research: Insights from the work of Paul Cobb.* Dordrecht: Springer.

Your Dictionary. (2015). Retrieved August 15, 2017 from http://reference.yourdictionary.com/word-definitions/definition-of-creativewriting.html

Yuhua, X. (2002). Shu: Naxi nature goddess archetype. *Gender, Technology and Development, 6*(3), 409–426. doi:10.1177/097185240200600305

Yunus, M., & Salehi, H. (2012). Integrating social networking tools into ESL writing classroom: Strengths and weaknesses. *English Language Teaching, 5*(8), 42–48. doi:10.5539/elt.v5n8p42

Zack, V., & Graves, B. (2001). Making mathematical meaning through dialogue: Once you think of it, the z minus three seems pretty weird. *Educational Studies in Mathematics, 46*(1/3), 229–271. doi:10.1023/A:1014045408753

Zeller, C. (2012). Language of immediacy: Authenticity as a premise in Benjamin's *The Work of Art in the Age of Its Technological Reproducibility. Monatshefte, 104*(1), 70–85. doi:10.1353/mon.2012.0015

Zhang, J., Scardamalia, M., Reeve, R., & Messina, R. (2009). Designs for collective cognitive responsibility in knowledge building communities. *Journal of the Learning Sciences, 18*(1), 7–44. doi:10.1080/10508400802581676

Zobrist, A. L. (2000). Computer games: Traditional. In A. Ralston, E. D. Reilly, & D. Hemmendinger (Eds.), Encyclopedia of Computer Science (4th ed.; pp. 364-368). New York: Grove's Dictionaries.

Zumthor, P. (2006). *Thinking architecture*. Basel: Birkhäuser.

About the Contributors

Marcel Danesi is Full Professor of Anthropology and Semiotics at the University of Toronto. He has published extensively in both linguistics and semitoics. His most recent book, *The Semiotics of Emoji: The Rise of Visual Language in the Age of the Internet*, has been nominated for several prizes. He is currently editor of *International Journal of Semiotics and Visual Rhetoric*.

* * *

Reima Al-Jarf has taught ESL, ESP and translation for 26 years. She is author of 240 books, e-books, book chapters, journal and encyclopedia articles, and book reviews. She has given 330 conference presentations in 69 countries. She reviews articles for numerous journals worldwide including some ISI journals, in addition to reviewing grant proposals, faculty promotion works and Ph.D. theses. A few years ago, she won 3 Excellence in Teaching Awards and the Best Faculty Website Award at King Saud University. Her areas of interest are: Foreign language teaching and learning, technology integration in education, lexicography, Arabization and translation studies.

Victoria Bigliardi is a student at the University of Toronto, currently completing her graduate studies in knowledge management and archival practice. She holds an undergraduate degree in Linguistics, Semiotics, and Material Culture from Victoria College in the University of Toronto. Her work has previously appeared in *The Criminal Humanities: An Introduction* (2016), *Acta Victoriana*, and the *International Journal of Semiotics and Visual Rhetoric*.

Mariana Bockarova received her PhD from the University of Toronto. She also holds a Master's degree from Harvard University from which she graduated with highest distinction in Psychology. She has researched mostly psychological phenomena including PTSD in criminal attorneys, student test taking anxiety, stress in the workplace, and math anxiety in teachers. She has also worked as a senior consultant at a boutique consultancy with Fortune 500 clients including Nestlé, General Mills, New Balance, Bayer Pharmaceuticals, Teva Pharmaceuticals, Janssen Inc., and Scotiabank.

Caterina Clivio has a Master of Science from Columbia University where her research focused on the psychoneuroimmunology of Generalized Anxiety Disorder. She is an assistant coordinator for the Fields Cognitive Science Network and has an interest in the implications of cognitive science in mathematics and interdisciplinary fields.

Stacy Costa is currently a Doctoral Student in a collaborative program in the department of Curriculum, Studies and Teacher development at the Ontario Institute for Studies in Education (OISE), and the Collaborative Program in Engineering Education (ENgEd). She is currently a lab member at the Institute for Knowledge Innovation and Technology (IKIT). In this role she conducts research regarding advancing knowledge work in education, innovation for the Knowledge Age, and developing related technology.

Sara Greco is Assistant Professor of argumentation at USI, Università della Svizzera Italiana (Lugano, Switzerland). Her research is focused on the analysis of contextualized argumentative discourse; her approach to argumentation integrates methods from discourse analysis and linguistic semantics. In particular, she focuses on argumentative dialogue as a means to prevent and solve conflict through dispute mediation (see the monograph *Argumentation* in *dispute mediation: A reasonable way to handle conflict* John Benjamins 2011) and other formal and informal practices of dispute resolution.

Terry Marks-Tarlow, Ph.D., is a clinical psychologist in private practice in Santa Monica, California, and an independent scholar. She is part of Italian Università Niccolò Cusano London, and a research associate at the Institute for Fractal Research at Kassel, Germany. Dr. Marks-Tarlow does workshops on interpersonal neurobiology and nonlinear science nationally and internationally and has authored numerous articles and books, including *Psyche's Veil, Clinical Intuition in Psychotherapy, Awakening Clinical Intuition*, and most recently is first editor of *Play and Creativity in Psychotherapy*.

Mihai Nadin is Director of the Institute for Research in Anticipatory Systems and Ashbel Smith University Professor at the University of Texas at Dallas. An expert in Peircean semiotics, he was the first to investigate semiotics related to the visual, to artificial intelligence, to digital technology—in particular human-computer interaction--and has influenced many others working in these fields. He is currently working on A New Foundation for Semiotics.

Aaron Rambhajan is at Victoria College in the University of Toronto, currently completing his studies in Psychology and Art History. His work has previously appeared in the *International Journal of Semiotics and Visual Rhetoric*.

William J. Rapaport is CSE Eminent Professor Emeritus in the Department of Computer Science and Engineering, an affiliated faculty member emeritus in the Departments of Philosophy and of Linguistics, and a member emeritus of the Center for Cognitive Science, all at State University of New York at Buffalo. Prior to that, he was Associate Professor of Philosophy at SUNY Fredonia. His research interests are in cognitive science, artificial intelligence, computational linguistics, knowledge representation and reasoning, contextual vocabulary acquisition, philosophy of mind, philosophy of language, critical thinking, and cognitive development. His research has been supported by the National Science Foundation and the National Endowment for the Humanities. He is a recipient of the SUNY Chancellor's Award for Excellence in Teaching, the International Association for Computing and Philosophy's Covey Award for "senior scholars with a substantial record of innovative research in the field of computing and philosophy broadly conceived", and the American Philosophical Association's Barwise Prize "for significant and sustained contributions to areas relevant to philosophy and computing by an APA member".

Frederik Stjernfelt is Professor of semiotics, intellectual history and philosophy of science, University of Aalborg, Copenhagen. He is also critic at the Copenhagen weekly Weekendavisen. Recent publications include: *The Democratic Contradictions of Multiculturalism* (2012, with Jens-Martin Eriksen), *Natural Propositions: The Actuality of Peirce's Doctrine of Dicisigns* (2014); *Mapping Frontier Research in the Humanities* (2016); *MEN - ytringsfrihedens historie i Danmark* ("BUT - the History of Free Speech in Denmark", with Jacob Mchangama, 2016).

Miriam Tribastone is a Research Master's student in Communication Science at the University of Amsterdam. Her domains of interest are political communication and journalism. In particular, her research focuses on news framing and media effects using both qualitative and quantitative methodological approaches.

Joel West studies at Victoria College at the University of Toronto. His interests include Forensic Semiotics, Sociolinguistics, especially the hyper indexicality of neologisms, and Religion as a sign making phenomenon. His last published work was: "The Dialogues of Bernie Madoff's Ponzi Fraud: An Exploration of the Discourses of Greed, Cliques, Peer Pressure, and Error," in the *International Journal of Semiotics and Visual Rhetoric*. Mr. West also has an upcoming publication, "The Ontology of Yentl: Umberto Eco, Semiosis, Mimesis, Closets, Limits and How to Read Yentl the Yeshiva Boy," to be published in *Semiotica*.

Index

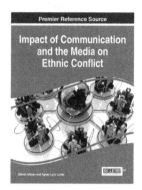

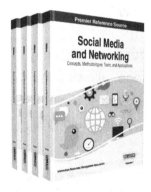

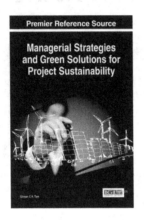

Printed in the United States
By Bookmasters